D1544458

Elementary Concepts
of Mathematics

Under the editorship of

Carl B. Allendoerfer

Burton W. Jones

Professor of Mathematics
University of Colorado

The Macmillan Company
Collier-Macmillan Limited, London

Elementary Concepts of Mathematics

Third Edition

Department of Mathematics
Texas Christian University
Fort Worth, Texas 76129

First Printing

Earlier editions copyright 1940 and 1947 and © 1963 by Burton W. Jones.

Library of Congress catalog card number: 70–80306

THE MACMILLAN COMPANY
COLLIER-MACMILLAN CANADA, LTD.,
TORONTO, ONTARIO

Printed in the United States of America

Preface
to the Third Edition

The basis of this revision has been the comments of various persons who have used the text in its previous forms, as well as the author's experience in teaching from it. The number of exercises has been increased by about 30 per cent and the answers to most of the odd-numbered problems have been supplied "in response to popular demand."

In Chapter 1 there is a new section on universal and complementary sets. In Chapter 2 there are two new sections having to do with switching circuits.

In Chapters 4 much of the material on groups has been rewritten, to conform more closely with the usual form of the definition of a group and to improve clarity. In Chapter 5 the sections on real and complex numbers have been revised somewhat.

In the section on compound interest in Chapter 6, the concept of continuous interest is introduced, and in Chapter 7 the section on curve fitting has been completely rewritten.

The section on combinations in Chapter 8 has been largely rewritten. The teacher may wonder why the formula for permutations is not given. The author's reason is that in this introduction to combinatorics it is very important that the student take each problem on its own merits rather than try to see what known pattern it fits. Thus the fewer formulas one has, the fewer patterns one has to refer to. In any case, the basic difficulty is with counting; we try to give help in this respect.

Besides the above, there have been many little changes introduced in the interest of clarity.

The author is chiefly indebted to three readers: Carl B. Allendoerfer of the University of Washington; C. Ralph Verno of West Chester State College; and Mrs. Jean Ferris, one of his colleagues; the latter two have taught from the text. The author's gratitude to The Macmillan Company is attested to by his continued association with them in this endeavor.

B. W. J.

The University of Colorado

Preface
to the Second Edition

The revision has been extensive. The first chapter, on sets, is entirely new, and many of the ideas formulated there are used throughout the book. The chapter on logic has been revised a number of times to put it into its present form. More properties of groups are introduced and proved, and the definition of a field is given and used. Inequalities and their graphs are included, and there is a section on mathematical induction. The chapter on "mirror geometry" is rounded out by new sections on Euclidean transformations and orientation. In Chapter X there is a new section on rotations and lorotations which relates the material of the chapter to relativity theory. In the last chapter Euler's formula is used to prove the existence of no more than five regular polyhedrons. The bibliography has been enlarged and brought up to date.

Throughout the book many of the results have been sharpened. The author has changed terminology in a few places to conform to that which seems to be evolving in the new curricula. In fact, the revision of the book as a whole has been made with an eye toward new developments in mathematics.

An attempt has been made to make the book more closely knit than before, and interrelationships are stressed. This means that for the revision, more than in the original, care must be taken especially in the first part of the book if parts are to be omitted. For instance, although in the first edition the chapter on logic could be postponed, here some of the ideas of logic recur throughout the book. The same is true of the chapter on sets. However, for the teacher's convenience there follows a revised table of topics which may be omitted without affecting the treatment of those which follow except perhaps in minor references and certain exercises:

Topic	Chapters	Sections
Causality	II	8
Numbers to various bases, Nim, tests for divisibility	III	5, 6, 8, 9, 10
Compound interest, progressions, and annuities	VI	5, 6, 8
Antifreeze formula	VI	11
Puzzle problems and Diophantine equations	VI	12, 13
Finding a formula which fits or almost fits	VII	8
Mirror geometry	IX	Entire chapter
Lorentz geometry	X	Entire chapter
Topology	XI	Entire chapter

There are other topics which can be omitted if certain supplementary material is supplied by the teacher. This is true, for instance, of the chapter on mathematical systems.

It was a remark of R. J. Walker which led to the section on Euclidean transformations; of all the process of revision, the author most enjoyed writing this section. He wishes to acknowledge especially the encouragement and detailed editorial assistance of Carl B. Allendoerfer, who worked closely with him throughout the process of revision. He is also grateful to The Macmillan Company in general, and A. H. McLeod in particular, for their assistance far beyond what an author has a right to expect and for their painstaking care and skill in the many details of publication.

B. W. J.

Boulder, Colorado

Preface
to the First Edition

In 1939 it became evident to members of the mathematics department of Cornell University that there was need for a course designed for students who have had a minimum of training, who do not expect to take other courses in the subject, but who want a firmer grounding in what useful mathematics they have had and wish such additional training as they, as nonmathematicians, may find interesting and/or useful in later life. Toward this end several members of the department cooperated: R. P. Agnew wrote a chapter on rational and real numbers, F. A. Ficken one on graphs, Mark Kac one on probability, Wallace Givens chapters on "mirror" and Lorentz geometries, R. L. Walker one on topology, and this author chapters on integers and algebra. With the advice and criticism of W. B. Carver and W. A. Hurwitz, the book was edited and prepared for lithoprinting by Wallace Givens and the author and appeared in that form in 1940 under the authorship of "the department of mathematics." From its beginning the course has had a clientele from all over the university and seems to have been growing in favor with students and faculty. In the light of experience the text has been thoroughly revised by the author and numerous additions made, including a chapter on logic in the preparation of which he wishes to acknowledge the assistance of his father, Arthur J. Jones.

While this text is designed for college students, almost no previous training is assumed beyond an acquaintance with integers and fractions and, for Chapters IX, X, XI, a little plane geometry.

First, much space and effort have been given to an attempt to cultivate an *understanding* of the material. To this end stress is laid on the distributive law for numbers, since it and the use of parentheses is at the root of much difficulty in algebra. Letters are used almost from the beginning to stand for numbers so that the student may become used to the idea. At times the subject is admittedly made harder by efforts to make it understood. For instance, in the treatment of annuities the student would more quickly solve the problems at hand if formulas were developed for annuities certain and various types of problems classified, but he is more likely to be led to an understanding of the subject if he thinks through each problem on its own merits. Second, an attempt is made to straighten out in the student's mind certain mathematical concepts of his everyday life which are usually only dimly understood: compound interest, the graph, averages, probability and games of chance, cause versus coincidence. Third, there is an effort to cultivate *appreciation* of mathematics—not awe of mathematics which he can never understand, but a joy in it and a respect for its methods. From this point of view, algebra is chiefly a tool for finding formulas—a method of expressing certain things clearly and succinctly so that, after the process is gone through with, the rest is mere substitution. It is hoped that the student will feel that a proof is a way of showing something rather than an exercise imposed by the teacher. Numerous references to other books are given in an effort to encourage the student to study some topic further and/or to acquire a little acquaintance with mathematical literature. In fact, in teaching the course the author has often required a term paper on some one of the suggested topics. Three of the references given (the books of Bell and of Hardy and the article by Weyl) are especially useful in showing how mathematicians think about themselves. Much puzzle material is introduced—this for a threefold reason: to make the material enjoyable, to provide a means for recreation now and later, to remove some of the undeserved mystery clinging about puzzles mathematical. Fourth, there is much emphasis on *logical development*. Not only in the chapter on logic but throughout the book an effort is made to cultivate logical statement and reasoning. Not many facts are stated without at least an indication of the proof. Fifth, the book is written in the belief that it should be a textbook, not a novel, and that a pencil and paper and the working of exercises are, if anything, more necessary to an understanding of mathematics than is work in a laboratory to the understanding of science. Finally, the book contains much material useful for the prospective teacher in secondary schools. In fact, the litho-printed edition has seen such use.

Chapters IX, X, and XI are introduced to show the student that there are geometries he never dreamed of, to show how certain facts can logically be found, to provide puzzle material, and to encourage the student to do some investigating on his own. It is interesting to find how much easier topics like groups and topology are for students than algebra which has been learned but poorly. They find these subjects interesting and the proofs understandable.

Teachers may notice the omission of any mention of trigonometry and calculus. The author feels that students who are not going on to further mathematics or science would find little use for these subjects.

The author wishes to express his appreciation of the work of The Macmillan Company in the preparation of this book and for their sympathetic counsel. As may be seen from the history of this book, several persons besides the author had to do with its inception, and without them and that abstract entity, the department of mathematics, this book would not have been. May they and it not be ashamed of the outcome!

B. W. J.

Contents

1. Sets

2. Logic

3. The Positive Integers and Zero

4. Mathematical Systems

5. Negative Integers and Rational and Irrational Numbers

6. Algebra

7. Graphs and Averages

8. Permutations, Combinations, and Probability

9. Mirror Geometry

10. Lorentz Geometry

11. Topology

Elementary Concepts
of Mathematics

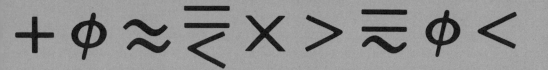

1

Sets

1. THE MEANING OF "SET"

Any occupation is characterized by the collection of materials with which one works and the means one uses to work on them. The collection of materials for a carpenter include boards, nails, and glue. A minister works with a collection of people and the scriptures of his particular religion. A mathematician works with numbers, figures, and other things or concepts which he defines as he goes along. Fundamental to all of these is the idea of a collection of objects or things. In mathematics it is customary to use the word "set" instead of "collection." In this chapter we propose to look at some of the fundamental properties of sets.

First of all, what do we mean by a "set"? We could speak of all the brick houses in a city, everything made of glass, the stars in the sky. All things in each of these categories we could say constitute a "set." We could also speak of the set of all theorems about triangles or the set of petty crimes. We would not in general use the work "flock," because that word applies only to particular kinds of sets associated with birds; we would not use the word "pair," because that again limits the kind of set we can use. But "set" seems to be a general enough term for our purposes and yet carries the idea we want.

What are the attributes of a set? First, it is a collection of "objects" or "things." Again these two words are not general enough, since they imply that what is in the set can be touched or felt, while, for example, theorems about triangles do not have this property. So we select the word "element" as being vague enough but not too vague. Thus we say that "a set is composed of its elements." If the set is that of all brick houses in a city, the elements of the set are the brick houses in the city.

A second attribute of a set is that there must be some way of telling whether or not something is in the given set. This could be determined by means of location, for example, the set of houses on a street. It could be by its use, for example, the set of all kinds of mowers. It could be by a distinctive property, for example, mammals form the set of all beasts that suckle their young. It could be described by listing: A set is composed of just you and me.

Thus for any given set and any given entity, the entity is either in the set or not in the set—it must be one and it cannot be both. Of course, we cannot always tell whether a given entity is in a given set or not. For instance, suppose X is the set of all people eating beans at this moment and c is the person in Tokyo with the most hair on his head (that is, the greatest number of hairs). It would be well-nigh impossible to find whether c is an element of X or not, but we can be sure that one or the other is the case. The "ten best makes of cars in America" is not a set because, until "best" is defined more precisely, there would be disagreement about whether a given make is in the set or not.

The above is a description of what we mean by set and is given to secure agreement on the concept. But the term is really undefined.

It should be borne in mind that a set has attributes distinct from those of any of its members. It is well known that a mob will do things which no member of the mob would think of doing. But the distinction is seen in less spectacular ways. A grain of sand is rough, but a sandy beach is smooth. The school board has an authority which no member has. The distinction exists even when the set has only one element. For instance, the subcommittee on cats of the SPCA in a small town might through natural causes be reduced to one member, Mr. Smith. Then, if all the cats in the town died, the natural thing would be to abolish the committee, but Mr. Smith might live a happy life for many years.

There are many examples of sets in mathematics, A circle is the set of points in a plane which are equidistant from a given point. A straight line is a set of points. All lines parallel to a given line are a set of lines. The odd numbers 1, 3, 5, 7, ... constitute a set of numbers. The three numbers 1, 7, and 10 constitute a set of numbers. In fact, the set consisting of the number 3 is a set of numbers containing only one element. All the letters on this page form a set.

2. ON FORMING SETS FROM SETS

First, by more restrictive description, we can describe a set which is contained in another set. A set contained in another set is called a *subset*. For instance, if X is the set of all brick houses in our city and Y is the set of all brick houses in our block, Y is a subset of X, or Y is contained in X. If X is the set of all stars in the sky and Y the set of all stars in the Milky Way, then Y is a subset of X. Our notation for the relationship of a set and subset is

$$Y \subseteq X \quad \text{or} \quad X \supseteq Y.$$

In fact, it is convenient to say that Y is a subset of X even when both X and Y are the same set. That is, *if every element of Y is an element of X, we call Y a subset of X and write $Y \subseteq X$ or $X \supseteq Y$.* The symbol[1] $Y \subset X$ is used to mean that not only is Y a subset of X but that X contains an element not in Y. In this case, we call Y a *proper subset* of X. If we represent a set by a region enclosed by a curve, we can express this relationship by a *Venn diagram*, as shown in Fig. 1:1. If A and B are each subsets of the other, they contain the same elements and we call them *equal*.

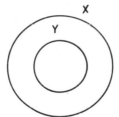

Figure 1:1

A line segment is a subset of the line on which it lies. The odd integers form a subset of the integers 1, 2, 3, 4, ... The two numbers 2, 3 constitute a subset of the set of numbers 1, 2, 3, 4.

Second, if A and B are two sets, we may consider all the elements in both A and B. These elements constitute a set because if we can tell whether an element is in A or not and in B or not, we can tell whether it is in both or not. This set we call the "intersection" (or product, or join) of A and B and write it $A \cap B$. Using a Venn diagram we could represent it as follows, where the

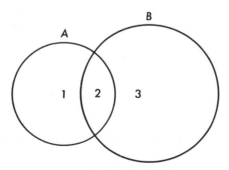

Figure 1:2

[1] In some texts the symbol $\subset$ is used in place of $\subseteq$. In such cases, no distinction is made between subsets and proper subsets.

regions numbered 1 and 2 represent the set A and the regions numbered 3 and 2 represent B. Then their intersection will be represented by the region numbered 2. For example, if A is the set of all women and B the set of all Negroes, then $A \cap B$ is the set of all Negro women (see Fig. 1:2).

Now, it might be that the two sets A and B would be defined so that they have no elements in common. For instance, A might be the set of women living on our street and B the set of men living on our street. These two sets would have no elements in common. For this purpose we define the "null set" or "empty set" and denote it by $\varnothing$. That is, if A and B have no elements in common, we would write $A \cap B = \varnothing$ and say that A and B are *disjoint*. The set of all women having wives is a null set. We make the agreement that the null set is a subset of every set, since $\varnothing$ having no elements implies that all the elements of $\varnothing$ are in any given set.

Suppose A and B stand for two lines. Each of these is a set of points. If they intersect, then $A \cap B$ is their point of intersection. If they do not intersect, then $A \cap B = \varnothing$. Suppose C and D stand for two circles; then $C \cap D$ can be two points, one point, or the null set.

Third, if we have two sets A and B, we may consider all elements in A or in B. Here we immediately are faced with an ambiguity. Is "or" the inclusive "or" which includes "both" or do we exclude "both"? Here it is not a question of which *must* it mean. "When *I* use a word," Humpty Dumpty said in a rather scornful tone, "it means just what I choose it to mean—neither more nor less." We make the agreement in this book that "or" shall mean "either or both"—the legal "and/or." When we wish to exclude "both" we shall do so explicitly. With this meaning of "or" we call the set of elements in A or in B (or both) the *union* (or sum) of A and B and write it

$$A \cup B.$$

Thus, in the Venn diagram of Fig. 1:2, the union of A and B consists of the sets numbered 1, 2, and 3.

A triangle is the union of three points, not on a straight line, and the three line segments joining them in pairs. A sphere is the union in three dimensions of all circles with a given center and radius. The union of the sets $\{1, 2, 5, 7\}$ and $\{1, 5, 6, 9\}$ is the set $\{1, 2, 5, 6, 7, 9\}$.

EXERCISES

1. Give three examples of sets from mathematics different from those given above.
2. Give an example of subset, intersection, and union in mathematics.
3. Which of the following define sets? Give reasons for your answers.
    a.   All even integers greater than 77.
    b.   The ten most important people in the United States.

    c. All days containing 25 hours.

    d. All those persons who will sing in the moonlight during the next ten years.

    e. The ten most beautiful señoritas in Mexico.

    f. All odd integers greater than 3 and less than 5.

    g. Every man who will be murdered in the next ten years.

    h. All the devils in Heaven.

4. In the state of Nebraska, for instance, state legislation applying to specific cities, say Omaha, is prohibited. Faced with this fact, how might the Nebraska state legislature legally pass a law which pertains only to the city of Omaha? What bearing does this have on the discussion above?

    In the following, use Venn diagrams or other intuitive methods to answer the questions for sets A and B.

5. What conclusions may be drawn from $A \cup B = \emptyset$?

6. Under what conditions will $A \cup B$ be the same as A?

7. Under what conditions will $A \cap B$ be the same as A?

8. Under what conditions will $(A \cap B) \cap C$ be the same as A for sets A, B, and C?

9. Since every set contains the null set, why does it follow that there is only one null set?

10. Suppose a line is thought of as a set of points and let A and B be two lines in a plane. What possibilities are there for $A \cap B$?

3. PROPERTIES OF THE SYMBOLS ∪, ∩, AND ∅

Our intuitive notions of a set assure us of certain properties having to do with the given symbols. Let A and B stand for two sets and consider the sets $A \cup B$ and $B \cup A$. If c is an element of the former, it is an element of the latter, and vice versa. In other words, the two sets $A \cup B$ and $B \cup A$ have the same elements. We express this fact by writing

$$A \cup B = B \cup A.$$

Again the set of elements in A and B as well as in C is the same as the set of elements in A as well as in B and C. We have

$$(A \cap B) \cap C = A \cap (B \cap C),$$

since either notation describes the elements common to A, B, and C. In this fashion, properties (1) through (3) below may be seen to hold. It is not a matter of proving them but merely seeing that they follow from our intuitive notions about sets A and B.

(1) Closure:
    $A \cup B$ is a set; $A \cap B$ is a set.

(2) Commutativity:
    $A \cup B = B \cup A$; $A \cap B = B \cap A$.

(3)   Associativity:
$$(A \cup B) \cup C = A \cup (B \cup C); \qquad (A \cap B) \cap C = A \cap (B \cap C).$$

Notice that each property above for union has its mate for intersection. In other words, the two symbols may be interchanged without changing the total set of properties above. This interchangeability we call "duality." We shall see that with modifications it applies to other properties as well. This is true, in particular, of the following properties involving both symbols and called the *distributive properties*.

(4a) $A \cup (B \cap C) = (A \cup B) \cap (A \cup C).$
(4b) $A \cap (B \cup C) = (A \cap B) \cup (A \cap C).$

Since this notation is not as simple as that in the previous properties, suppose we show (4a) in two ways. First we show it without use of diagrams by a method which we shall often find useful. The equality (4a) will be established if we can show the following:

(i) If d is an element of $A \cup (B \cap C)$, then it is an element of $(A \cup B) \cap (A \cup C)$.
(ii) If d is an element of $(A \cup B) \cap (A \cup C)$, then it is an element of $A \cup (B \cap C)$.

To show (i) notice that d is an element of A or of both B and C (where the "or," recall, means the legal "and/or"). If, on the one hand, d is an element of both B and C, then it is an element of $(A \cup B)$ and $(A \cup C)$. If, on the other hand, d is an element of A, then it is also an element of both $(A \cup B)$ and $(A \cup C)$. Hence in both cases (i) holds.

To show (ii) see that d is taken to be an element of both $(A \cup B)$ and $(A \cup C)$. That is, it is either in A, in which case it is in $A \cup (B \cap C)$, or it is in both B and C, in which case it is also in $A \cup (B \cap C)$. Thus (ii) is established.

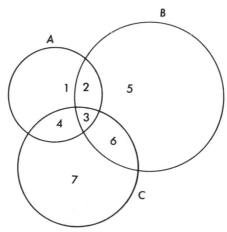

Figure 1:3

Now we show (4a) by recourse to a Venn diagram (Fig. 1:3) with numbered regions. Let *1* denote the region designated by 1 in the figure, *2* that designated by 2, and so on. Then $B \cap C = 3 \cup 6$, while $A = 1 \cup 2 \cup 3 \cup 4$. Hence

$$A \cup (B \cap C) = 1 \cup 2 \cup 3 \cup 4 \cup 6.$$

On the other hand, $A \cup B = 1 \cup 2 \cup 3 \cup 4 \cup 5 \cup 6$ and $A \cup C = 1 \cup 2 \cup 3 \cup 4 \cup 6 \cup 7$, which implies that

$$(A \cup B) \cap (A \cup C) = 1 \cup 2 \cup 3 \cup 4 \cup 6.$$

Since the same set of regions occurs for both, (4a) is established.

Notice that (4b) can be obtained from (4a) by interchanging $\cup$ and $\cap$. The proof of (4b) is left as an exercise.

Suppose we consider an example outside of mathematics of the equation (4a): Phyllis's requirements in a man are that (1) either he drive a sports car or that (2) he is both 6 feet tall and a good dancer. (Of course, if he has all three attributes, so much the better. But one should not be *too* particular.) Another way of expressing it, using (4a), would be that he has *both* of the following two attributes: (1) he drives a sports car or is 6 feet tall; (2) he drives a sports car or is a good dancer.

The final property which we consider in this section is:

(5)   If any one of the following holds, the other two also hold:

   (i) $A \subseteq B$,   (ii) $A \cup B = B$,   (iii) $A \cap B = A$.

First, if $A \subseteq B$, then (ii) holds, since taking the union of A and B adds no element to B. Conversely, if (ii) holds, there can be no elements in A which are not in B, which means that A is a subset of B. We leave the rest of the proof of (5) as an exercise.

Two special cases of (5) are of interest:

(5a) $B \cup B = B$ and $B \cap B = B$,

taking $A = B$ in (5), since then $A \subseteq B$.

(5b) $\varnothing \cup B = B$ and $\varnothing \cap B = \varnothing$,

taking $A = \varnothing$ in (5), since then $A \subseteq B$.

The property (5a) is sometimes called the *idempotent* property. Both (5a) and (5b) can be represented in two little tables:

$\cup$	$\varnothing$	B		$\cap$	$\varnothing$	B
$\varnothing$	$\varnothing$	B		$\varnothing$	$\varnothing$	$\varnothing$
B	B	B		B	$\varnothing$	B

These properties are important partially because they describe to a certain extent how sets behave and partially because they are also closely related to some properties of logic and our number system, as we shall see in later chapters.

EXERCISES

1. Show from intuitive notions of sets (Venn diagrams if you like) why each of properties (1) through (3) holds.
2. Show property (4b) using a Venn diagram.
3. Give a mathematical and a nonmathematical example of $(A \cap B) \cap C$.
4. Give a mathematical and a nonmathematical example of $(A \cup B) \cup C$.
5. If $\cup$ and $\cap$ as well as $\varnothing$ and B are interchanged in the first table after (5b), show that the second table is obtained.
6. Suppose in property (5) that (i) is replaced by $A \supseteq B$. What will the other two parts be?
7. Without using a diagram show property (4b).
8. Show from intuitive notions of sets that if (i) in (5) holds, then (iii) holds; also that if (iii) holds then (i) holds. Since we showed that if (i) holds then (ii) holds, and conversely why do these three results establish the statement of (5)?
9. Another way to show that any part of property (5) implies the other is to show the following:
    a. If (i) holds, then (ii) holds.
    b. If (ii) holds, then (iii) holds.
    c. If (iii) holds, then (i) holds.

    Why is this so? Part a was shown. Show parts b and c directly.
10. Show from intuitive notions of sets that if $A \cap B = B$ and $A \cup B = B$, then $A = B$.
11. Establish the result of Exercise 10 using only properties (1) through (5).
12. Show intuitively that if $B \cup A = B$ and $B \cap A = \varnothing$, then $A = \varnothing$.
13. Show the above by use of properties (1) through (5).
14. If Phyllis had not been so careful in stating her requirements in a man she might have said, "He must drive a sports car or be 6 feet tall and be a good dancer." This could be interpreted in two different ways, one of which is given in the text. What is the other way? Are the two sets of requirements the same? Let C be the attribute of driving a sports car, T the attribute of being 6 feet tall, and D the attribute of being a good dancer. Express both interpretations using these letters and the symbols for union and intersection.
15. Find an example outside mathematics of (4b).
16. Assuming that (i) $C \subseteq D$ implies $D \cup C = D$ for all sets C and D, and (ii) $A \cup B = B$, we may prove (iii) of (5) as follows:
    a. $A \cap B = (A \cup A) \cap (A \cup B)$, since $A \cup A = A$ by (i) and $(A \cup B) = B$ by (ii).
    b. $(A \cup A) \cap (A \cup B) = A \cup (A \cap B)$ by (4a).
    c. $A \cup (A \cap B) = A$, since $A \cap B$ is a subset of A and we can use (i) with $C = A \cap B$ and $D = A$.
    d. Hence $A \cap B = A$.

    Assuming (i) above and (iii) of (5), prove in similar fashion that (ii) of (5) holds.

17. Let $A \mathbin{\uplus} B$ denote the set of elements in A or B but not both. In properties (1) through (5) replace $\cup$ by $\mathbin{\uplus}$ and state which of these properties still hold. This is called "the symmetric difference"; we shall have another notation for this later.

4. UNIVERSAL AND COMPLEMENTARY SETS

It is sometimes convenient to consider "universal sets." One could describe this loosely as a set (or the smallest set) which contains all sets that we happen to be talking about at the time. If we were making a statement about houses on a street, the universal set might be all houses or all houses in a certain city. If we were concerned with cars, the universal set might be taken to be the set of all cars. More generally, if we are considering various sets of elements, we would choose a universal set which includes them all.

If we denote the universal set by U, we would then have, for any set B in U, in contrast to equations (5b) of Section 3,

$$U \cup B = U \qquad \text{and} \qquad U \cap B = B,$$

and we have the tables

$\cup$	U	B
U	U	U
B	U	B

$\cap$	U	B
U	U	B
B	B	B

By means of the concept of universal set we can also define a "complementary set." For instance, if A is the set of all houses on Main Street and U is the set of all houses in the city, $U - A$, called the "complementary set," would designate the set of all houses of the city not on Main Street. Thus U consists of two subsets: those houses on Main Street, designated by A, and those not on Main Street, designated by A' or $U - A$. The notation A' is simpler, but $U - A$ is a notation which has the advantage of acknowledging the dependence on what is taken to be the universal set.

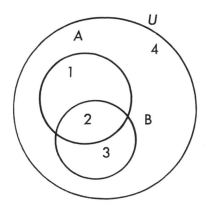

Figure 1:4

One pitfall associated with the notation $U - A$ should be noted. If $U - A = B$, it does follow that $A \cup B = U$, but $A \cup B = U$ does not imply that $B = U - A$ unless B and A are disjoint sets. For instance, in the Venn diagram of Fig. 1:4 (p. 9), $U - A$ is the set inside U but outside A; that is, $3 \cup 4$. The set 2 is in B but not in $U - A$, and the set 3 is in B and in $U - A$.

EXERCISES

1. Suppose that A and B are two sets in a universal set U and use the prime to designate complementary set in U; that is, $A' = U - A$, and so on. Using Fig. 1:4, find the sets $(A' \cap B')$ and $(A \cup B)'$ and show that they are the same.

2. Using the same notation as in Exercise 1, find the sets $(A' \cup B')$ and $(A \cap B)'$ and show that they are the same. (*Note:* The results of this and Exercise 1 are called de Morgan's laws.)

3. Using Fig. 1:4 designate $(A \cap B) \cup (A' \cap B')$.

4. Using Fig. 1:4 show that $(U - A) - B = U - (A \cup B) = (A \cup B)'$.

5. Extend the notation we have used so that for any two sets A and B, $(A - B)$ means those elements of A which are not elements of B and $(B - A)$ those elements of B which are not elements of A. Using a Venn diagram show that

$$(A - B) \cup (B - A) = (A \cup B) - (A \cap B).$$

Note that this is what we designated by $A \uplus B$ in Exercise 17 of Section 3.

6. Using the notation of Exercise 5, find $(A - B) \cap (B - A)$.

7. Under what circumstances is $A - B = A$?

8. Can $A - B$ be equal to B? Why or why not?

9. For sets A, B, and C is the following true? $A \cap (B - C) = (A \cap B) - (A \cap C)$. If the answer is no, give an example in which the equality does not hold. If the answer is yes, show it by a Venn diagram or otherwise.

10. The same as Exercise 9 with $\cap$ replaced by $\cup$.

5. PROPERTIES OF THE RELATIONSHIP "SUBSET OF"

In Section 2 we mentioned the relationship $A \subseteq B$, which is our symbol for "A is a subset of B." Recall that this means that every element of A is an element of B. Here let us list three simple properties of this relationship:

(1) For all sets A, $A \subseteq A$ (the reflexive property).
(2) If $A \subseteq B$ and $B \subseteq A$, then $A = B$ (the *antisymmetric property*).
(3) If $A \subseteq B$ and $B \subseteq C$, then $A \subseteq C$ (the *transitive property*).

The first merely states that every element of A is an element of A. The second states that if every element of A is an element of B and if every element of B is an element of A, then A and B are the same set—that is, contain exactly the same elements; this is what we mean by two sets being equal. The third statement is that if every element of A is an element of B, and every element of B is an element of C, then every element of A, being an element of B, must also be an element of C.

There are other relationships not associated with sets which have these same properties. For instance, A, B, and C might stand for three things which have hardness and $\subseteq$ might stand for the relationship "is no harder than." Then (1) is the true statement: A is no harder than A; (2) is: If A is no harder than B and B is no harder than A, then the hardness of both are the same; (3) is: If A is no harder than B and B is no harder than C, then A is no harder than C.

There are, however, many relationships which do not have this property. For instance, if A, B, and C stand for three persons and $\subseteq$ stands for the relationship "is the father of," neither property (1) nor property (3) holds. In a way, (2) might be said to hold, since the hypothesis is impossible. We shall meet these three properties again in later chapters.

6. SUMMARY

Notice that throughout this chapter we have been concerned with properties of sets which are independent of what their elements may be. For example, consider the relationship $A \cup B = B \cup A$. The elements of each of these sets might be cows. The equality for this set would merely affirm that if you combine herd A with herd B, you get the same herd as if you combine herd B with herd A. This is a pretty trivial result but it holds for all sets. A and B could stand for sets of theorems or sets of pebbles on a beach. There are many properties which a set of cows has which a set of theorems do not have —you cannot milk a theorem or prove a cow. But we here have dealt with properties of the sets themselves regardless of the character of their elements. This is typical of mathematics and we shall see many examples of this.

Although the idea of set is an abstract one, it has a very definite connection with reality. The abstraction consists merely in eliminating irrelevancies. We look at the relevant properties and make a short list from our intuitive ideas of what sets are like. Then these properties in themselves, without any further use of intuition, may be used to derive other properties to which intuition alone might not have led us. Then when these are particularized, they give us information on such practical matters as a proper phrasing of Phyllis's requirements in a man, for instance.

It is suggested that the student look for similar patterns as he continues reading this book.

EXERCISES

1. Give an example of a relationship which has the three properties ascribed above to $\subseteq$.

2. In property (5) of Section 3 we showed that the statement $A \subseteq B$ means the same thing as $B \cap A = A$. Using this fact and properties (1) through (5) of Section 3, show that $\subseteq$ has the three properties listed above.

3. In property (5) of Section 3 we showed that $A \subseteq B$ means the same thing as $B \cup A = B$. Using this fact and properties (1) through (5) of Section 3, show that $\subseteq$ has the three properties listed in Section 5.

4. Let G be the set of all people who are googlish, B the set of all people who are beemish, and P the set of all people. Suppose that the null set is the union of the following two sets: people who are googlish but not beemish, people who are beemish but not googlish. Show that every googlish person is beemish and vice versa, from intuitive notions of sets.

5. Let C be the set of all cats, F the set of cats which are fank, and G the set of cats which are gank. Now every gank cat is either fank and not gank, or gank and not fank. Show that there are no fank gank cats, by use of Venn diagrams or other means.

6. What conclusions can be drawn from the following?

$$A \subseteq B, B \subseteq C, \text{ and } C \subseteq A.$$

7. Give an example of a relationship on a set of elements which has
   a. Properties (1) and (2) but not (3) of Section 5.
   b. Properties (1) and (3) but not (2).
   c. Properties (2) and (3) but not (1).
   (Note these three results show that the three properties are independent, that is, that no one can be derived from the other two.)

7. TOPICS FOR FURTHER STUDY

Brief treatments are given in the following references: **2*** (Chap. 1), **13** (supplement to Chap. 2), **39**, and **40** (Chap. 2).

Longer treatments occur in references **12**, **23** (Chaps. 6–10), and **68**.

The algebra of sets is discussed from a somewhat more sophisticated point of view in reference **10** (Chaps. 11 and 12).

---

* Boldface numbers refer to the Bibliography on p. 328.

$$+ \; \phi \; \approx \; \overset{=}{<} \; \times \; > \; \overset{=}{\approx} \; \phi \; <$$

2

Logic

"Contrariwise," continued Tweedledee, "if it was so, it might be; and if were so, it would be; but as it isn't, it ain't. That's logic."

1. STATEMENTS

In the first chapter we were concerned with sets, which are what we work with in mathematics, as well as in many other activities. In this chapter we look at our tools. The tools of a carpenter are his hammer, saw, plane, rule, and all the contents of his tool kit; the tools of a chemist are his chemicals, bunsen burner, test tubes, and so forth; the tools of a mathematician are books, a pencil, paper, and logic.[1] Just as it behooves a carpenter to learn something about his hammer, before he sets himself up as a carpenter even though carpentry will increase his proficiency in the use of this tool, so anyone about to do some mathematics should be somewhat familiar with logic before he begins, even though doing mathematics should give him practice in the use of logic. Thus we propose to look into the structure of logic.

Important to any logical process is the set of statements with which it deals. Just as we described but did not attempt to define a set, so "statement" is too fundamental to try to define. But we might *describe* it by saying that a

[1] L. E. Dickson maintained that the most essential tool of a mathematician is a large wastebasket.

statement is something which is either true or false, but cannot be both, whatever "true" or "false" mean. The sentences "Why am I old?" and "Stop laughing at me!" are not statements in our sense, because of their form. The sentence "Black is heavy" is not a statement, at least without further explanation, because it has no meaning. On the other hand, consider the sentences "You will die tomorrow," "When I am old, I will put away childish things," and "If two sets are subsets of each other, they are equal." Suppose they have definite meaning; for instance, assume that "you" refers to just one person, that we know what is meant by "childish things," and so forth. Then we would call them statements. Whether or not we can determine if a statement is true is beside the point. No one knows whether you will die tomorrow, but everyone will admit that it is either true or false, but not both.

In practice we may attempt to find out whether a statement is true or false. This process consists in establishing some basis of truth on which we and others will agree and upon which we can build our logical structure. The contribution of logic is to generate true statements or false statements based upon certain assumptions of truth or falsity.

Here we run into some difficulty, because we are using language to describe language. Generally when we make a statement such as "It is raining today," we are affirming that it is true. In fact, unless we are asking a question or making a command, almost every sentence we write affirms that something is true. (Look at the three sentences before this in this paragraph.) In this chapter we shall continue this practice. When we make a statement which, in our opinion, may or may not be true, we shall try to make it clear that this is the case.

2. IMPLICATIONS

Much of our reasoning in mathematics and outside of it is concerned with statements of the form "If something, then something else." Such a statement is called an *implication*.

Example 1

If plants become green, then they must have had sunshine.

Example 2

If a number is divisible by 4, then it is even (that is, divisible by 2).

Example 3

If a number is even, then it is divisible by 4.

Example 4

If $A \subseteq B$, then $A \cup B = B$. [Here A and B stand for two sets. See property (5) in Chapter 1.]

Each of these is a statement because it is either true or false. Since green plants can be raised by artificial light, the statement in Example 1 is false. The statement in Example 2 is true; that in Example 3 is false, since, for instance, 6 is even and not divisible by 4. We proved in Chapter 1 that the statement of Example 4 is true.

An implication, "If p, then q," contains two statements, p and q. We call p the hypothesis and q the conclusion of the implication and use the following notation for this implication:

$$p \rightarrow q.$$

In Example 1, "plants become green" is the hypothesis and "they must have had sunshine" is the conclusion. In Example 2, "a number is divisible by 4" is the hypothesis and "it is even" is the conclusion. The implication asserts that whenever the hypothesis is true, the conclusion is also. (Nothing is affirmed about cases in which the hypothesis is not true.) In other words, when the implication is true and the hypothesis is true, then the conclusion is true. In symbols this means, if the implication $p \rightarrow q$ is true, then whenever p is true, q must be true.

Of course, there are other ways to state an implication. That of Example 1 could be expressed as follows: "To become green, plants must have had sunshine." The implication of Example 4 could be stated: "Whenever A is a subset of B, it follows that B is the union of A and B."

Many statements not in the basic form of an implication can be put into that form. For instance, "The union of two sets is a set" could be written "If A and B are two sets, their union is a set." There are other statements, of course, which cannot be put into the form of an implication, at least without much awkwardness, such as "Five is the smallest number" or "Three is a crowd."

One implication connected with sets is especially worth considering. Suppose A and B stand for two sets. Consider the two following statements:

(1) $A \subseteq B$.

(2) If a is an element of A, then it is an element of B.

Another way to write (2) is

(2a) $$p_a \rightarrow q_a,$$

where p_a stands for "a is an element of A" and q_a stands for "a is an element of B." By definition, these two statements mean the same thing. Two statements are called "equivalent" if they are either both true or both false.

For example, suppose A is the set of integers divisible by 4 and B the set of integers divisible by 2. For these sets, $A \subseteq B$ is true, since A consists of the numbers

$$(\ldots, -8, -4, 0, 4, 8, \ldots),$$

B consists of the numbers

$$(\ldots, -8, -6, -4, -2, 0, 2, 4, 6, 8, 10, \ldots),$$

and the latter set contains the former. The corresponding implication is: If a is an element of the set of multiples of 4, then it is an element of the set of multiples of 2.

A briefer way to make this same statement is:

If a is a multiple of 4, it is a multiple of 2.

If we should doubt the truth of the implications and try to show them to be false, we would have to find a number in set A which is not in set B, that is, a multiple of 4 which is not a multiple of 2. This is impossible. Notice that the statement of implication cannot be shown false by considering a number such as 3, which is not a multiple of 4.

In general, it is important to realize that if there is a single instance of a statement which is false (a counterexample), then the statement is false. On the other hand, to show the truth of a statement, one must show it to be true in *all* cases. This raises the question: What do we say about an implication when the hypothesis is false? We want an implication to be a statement; that is, it should be either true or false in all cases. For this reason we follow the usual custom and say that an implication is true if the hypothesis is false, no matter what the conclusion is. Thus, by this agreement, the following implications arc truc:

(1) If the moon is made of green cheese, then I am dead.
(2) If black is white, then 2 plus 3 is 7.
(3) If $17 = 23$, then $100 = 4 \cdot 25$.

Thus we have the following table:

Hypothesis	Conclusion	Implication
True	True	True
True	False	False
False	True	True
False	False	True

The implication "If it rains, I shall get wet," will be false if there is one case when it rains and I do not get wet. The implication "If plants become green, then they must have had sunshine" can be shown false by citing one green plant which did not have sunshine. The implication "If a number is even, then it is divisible by 4" is shown false by noting that 6 is an even number which is not divisible by 4. On the other hand, "If a number is divisible by 4, then it is even" is not false because there is no example of a number divisible by 4 which is not even; in other words this implication is true.

3. CONVERSES

We have seen that there are a number of ways of expressing an implication. What happens if we interchange the hypothesis and conclusion? If $p \rightarrow q$ denotes an implication, we call $q \rightarrow p$ its *converse*. For Example 2, let us consider various ways of expressing an implication and its converse:

	Implication (Example 2)	Converse (Example 3)
In words:	If a number is a multiple of 4, it is even.	If a number is even, it is a multiple of 4.
In symbols:	$p \rightarrow q.$	$q \rightarrow p.$
Using sets:	$A \subseteq B.$	$B \subseteq A.$

(We understand here that A stands for the set of multiples of 4 and B for the set of even numbers.) For this example, the implication, whichever way we express it, is true and the converse false.

On the other hand, there are many cases in which both an implication and its converse are true. That is, whenever p is true, then q is true, and whenever q is true, then p is true; in other words, p and q are either both true or both false. In symbols

$$p \rightarrow q \quad \text{and} \quad q \rightarrow p.$$

If this is true, we say that p and q are *equivalent statements*. We write it

$$p \leftrightarrow q.$$

In terms of the example above we would have, using set notation,

$$A \subseteq B \quad \text{and} \quad B \subseteq A; \quad \text{that is, } A = B.$$

Example 5

The statements "A number is a multiple of 2" and "A number is even" are equivalent statements because, for any given number, both statements are true or both are false.

Example 6

The following two statements are equivalent because any triangle with one property has the other: (1) The base angles of a triangle have equal measures; (2) the two sides of a triangle have equal length.

Suppose B denotes the set of triangles whose base angles have equal measure and S the set of triangles whose sides are equal in length. Let b stand for "The base angles of a triangle have equal measures" and s stand for "The sides of the triangle have equal length." Then $b \to s$ is equivalent to $B \subseteq S$; $s \to b$ is equivalent to $S \subseteq B$. Both implications are true, hence $B \subseteq S$ and $S \subseteq B$. Thus $S = B$. Since the statements in Example 6 are equivalent, the sets B and S are the same.

Mathematicians often use such terms as "if and only if" and "necessary and sufficient conditions" in this connection. Although we shall not emphasize these unduly in this book, it is worth our while to explain them briefly. Here are ways of expressing "If p, then q" using these words:

(1) p only if q.
(2) q is a necessary condition for p.
(3) p is a sufficient condition for q.

Statement (2) means, in other words, "If p, it is necessary that q" or "If p, necessarily q." Statement (3) means, in other words, "For q to be true, it is sufficient that p be true."

Corresponding statements for Example 2 are:

(1) A number is a multiple of 4 only if it is a multiple of 2.
(2) Being a multiple of 4 is a sufficient condition for being a multiple of 2.
(3) Being a multiple of 2 is a necessary condition for being a multiple of 4.

If two statements p and q are equivalent, that is, if both are true or both false, mathematicians often say "p if and only if q" or "p is a necessary and sufficient condition for q." Part of property (5) in Chapter 1 could be expressed as follows: $A \subseteq B$ if and only if $A \cup B = B$. It could also be written

$$A \subseteq B \text{ is a necessary and sufficient condition for } A \cup B = B$$

or

$$A \cup B = B \text{ is a necessary and sufficient condition for } A \subseteq B.$$

EXERCISES

1. Express in three other ways each of the following statements:
   a. If the wind blows, the lake is rough.

    b. If a google is flaccid, it is rambunctious.

    c. If a number is a multiple of 6, it is a multiple of 3.

    d. I cannot sleep unless you call to tell me you are safe.

    e. I shall be surprised if he remembers.

    f. Whenever the wind blows from the east we can smell the stockyards.

2. Express each of the following statements first as an implication, second using "only if," third using "necessary condition."

    a. Drinking coffee in the evening keeps me awake at night.

    b. Every equilateral triangle is isosceles.

    c. "Sweet are the uses of adversity."

    d. Rolling moss gathers no stones.

    e. If two lines are parallel, they do not meet.

    f. If a number is divisible by 15, then it is divisible by 3 and 5.

3. Write the converses of each of the implications in Exercises 1 and 2.

4. **a.** Find in Chapter 1 two examples of implications which are true but whose converses are not true.

    b. Find in Chapter 1 two examples of true implications whose converses are true.

5. Given the implication "If a casa is licht, it is subliminal." Assume that each part is a statement and find which of the following are equivalent to the given implication.

    a. All licht casas are subliminal.

    b. A necessary condition for a casa to be licht is that it be subliminal.

    c. A casa is subliminal only if it is licht.

    d. A casa is subliminal if it is licht.

6. Give the implication "If a brillig gyres, it glebes." Assume that each part is a statement. Which of the following are equivalent to it:

    a. A brillig gyres if it glebes.

    b. A brillig glebes if it gyres.

    c. A brillig glebes only if it gyres.

    d. A necessary condition that a brillig gyres is that it glebes.

7. *Suppose* we had agreed that an implication is *false* except when the hypothesis and conclusion are both true. Then what would be the table of the four possibilities corresponding to that at the end of Section 2? What would be some other possible agreements?

8. Give an implication. What is the converse of its converse? Give reasons for your answer.

9. For each of the following implications or statements, indicate how it could be shown not true.

    a. It never rains when the sun is shining.

    b. "And nobody
        KNOWS-tiddely-pom,
        How cold my
        TOES-tiddely-pom
        Are
        Growing."

   c. If $A \cup B = \varnothing$, then A is the null set.
   d. If $A \cup B = \varnothing$, then A is not the null set.
   e. If an implication is true, its converse is also true.
   f. "I'm never, never sick at sea."
   g. If $A \subseteq B$ and $A \cup B = A$, then $A = B$.
   h. If the conclusion of an implication is true, so is the implication.

4. TRUTH TABLES

In Section 2 we made a table for implications. Let us find tables for other statements. Here it is convenient to adapt a notation we used in Chapter 1. The symbol $A \cap B$ was used to mean the set of elements in both A *and* B. If p and q are two statements, we write $p \wedge q$ to mean "p and q." Also, the symbol $A \cup B$ was used to mean the elements in A or in B, or in both. So we use the notation $p \vee q$ to mean "p or q," where the word "or" is understood to include the possibility of both. Also let p' denote "not p." Then we can construct the following table, where T stands for true and F for false:

p	q	p'	$p \to q$	$p \vee q$	$p \wedge q$	$p' \vee q$
T	T	F	T	T	T	T
T	F	F	F	T	F	F
F	T	T	T	T	F	T
F	F	T	T	F	F	T

For instance, "p and q" is true only when p and q are both true; hence in the column $p \wedge q$ there are F's except in the first row, when both p and q are true. In the column p', when p' is T, then p is F, and vice versa. This table is called a "truth table." (It could just as well be called a "falsity table.")

Notice the fourth and last columns. Every time a T occurs in the fourth column, the corresponding entry in the last column is a T also. Whenever an F occurs in the fourth column, the corresponding entry in the last column is F. In other words, the two statements $p \to q$ and $p' \vee q$ are equivalent statements. That is, the implication "If p, then q" is true if and only if "(not p) or q" is true. In Example 2 we had the implication "If a number is a multiple of 4, it is even." An equivalent statement of this implication is "A number is either not a multiple of 4 or it is even (or both)."

This alternative way of expressing an implication is not as fruitful as one other, which we show in the following truth table:

p	q	p'	q'	$p \to q$	$q' \to p'$
F	F	T	T	T	T
F	T	T	F	T	T
T	F	F	T	F	F
T	T	F	F	T	T

Notice that the two implications are equivalent; that is, one is true when the other is true, and vice versa. If an implication is $p \to q$, then we call the implication $q' \to p'$ its *contrapositive*. An implication and its contrapositive are both true or both false.

For example, we have seen that "If a number is a multiple of 4, it is a multiple of 2" is a true implication. Its contrapositive is "If a number is not a multiple of 2, then it is not a multiple of 4." Of course, the contrapositive is true as well. If an implication is true, its converse need not be true, but its contrapositive must be.

One can also show this without recourse to truth tables. Consider the implication $p \to q$. We can show it equivalent to $q' \to p'$ by showing the following:

(1) If $p \to q$ is true, then $q' \to p'$ is true.
(2) If $q' \to p'$ is true, then $p \to q$ is true.

To show (1) we notice that $p \to q$ is equivalent to "If q is false, then p cannot be true." That is, if q is false, p must be false, which is just the second part of statement (1). We can use the same argument for statement (2). Another way to show statement (2) is to replace p by q' and q by p' in statement (1) to have

If $q' \to p'$ is true, then $(p')' \to (q')'$ is true.

But $(p')'$ is equivalent to p and $(q')'$ is equivalent to q, thus demonstrating statement (2).

Using the contrapositive is often helpful in establishing the truth of an implication.

Example 1

Suppose we know that if someone has left the house, there would be tracks in the snow. This means that if there are no tracks in the snow, then no one has left the house. So a lack of tracks will tell us that no one has left the house.

Example 2

Suppose we wish to show that if a number is a multiple of 4, it is a multiple of 2. We could do this by showing the contrapositive: If a number is not a multiple of 2, it is not a multiple of 4.

A truth table is also useful in finding the *negation* of a statement. Suppose p and q are two statements. We say that one is the negation of the other if "not p" is equivalent to "q." In other words, q is the negation of p (and p is the negation of q) is both of the following are true:

(1) If p is true, then q is false.
(2) If p is false, then q is true.

This means that if we construct a truth table for p and for q, in no instance are the truth values the same.

Suppose we find the negation of the statement "p or q." The statement "p or q" is true whenever at least one of p and q is true. That is, this statement will be false if and only if both p and q are false. Hence the negation of "p or q" is "not p *and* not q." In symbols this is

$$(p \lor q)' \leftrightarrow (p' \land q').$$

You should check this by forming the truth table for each.

We should pay special attention to the negation of an implication. To show an implication false, we need only give an example in which the hypothesis is true and the conclusion false. For instance, suppose the implication is "If he is well, he will come." This would be shown false by an instance when he is well and does not come. But the negation of an implication must be true whenever the implication is false and false whenever it is true. Hence the negation of the given implication would be "He is well and he does not come." This statement is true when and only when the implication is false. In symbols we are saying that $(p \rightarrow q)'$ and $(p \land q')$ are equivalent statements. Thus each of $p \rightarrow q$ and $p \land q'$ is the negation of the other. You may, at this point, want to review Section 4 of Chapter 1, since the idea of a complement of a set is very close to the idea of negation.

EXERCISES

1. Verify the values for the following in the truth tables of this section: $p \lor q$, $p' \lor q$, and $q' \rightarrow p'$.

2. Use the truth table to show that $(p \lor q)' \leftrightarrow (p' \land q')$.

3. Let A be the set of elements for which p is true and B the set for which q is true. Then $A' \cup B'$ corresponds to $p' \lor q'$. Find the truth table corresponding to Exercise 2 of Section 4 of Chapter 1.

4. Find a connection between Exercise 1 of Section 4 of Chapter 1 and Exercise 2 in this set.

5. Show first by translating into words, and then by use of truth tables, that $(p' \lor q)'$ is equivalent to $p \land q'$.

6. Find the negations of each of the following, first by translating the symbols into words, and second by use of truth tables.

        a. $p \land q$.   **b.** $p \lor q \lor r$.

        c. $(p \lor q) \land r$.   **d.** $(p \land q) \lor r$.

7. Find the negation of each of the statements of Exercise 2 of Section 3.

8. Write the negation of each of the following statements without using "and" in parts a and c, and without using "or" in parts b and d.

    a. The sky is blue and it is a warm day.

    b. The sky is blue or it is a warm day.
    c. The sky is not blue and it is a warm day.
    d. The sky is not blue or it is not a warm day.

  9. Write the negation of each of the following statements, without using "and" in the first two parts and without "or" in the last two parts.
    a. Today it is Wednesday and it is my birthday.
    b. Today is not Wednesday and it is my birthday.
    c. Today is not Wednesday, or it is my birthday.
    d. Today is Wednesday, or it is not my birthday.

10. Another way to write a truth table for $p \vee q$ is the following:

$\vee$	T	F
T	T	T
F	T	F

Where the T and F on the left are the truth values of p, those along the top are the truth values of q, and those in the body of the table are the corresponding truth values of $p \vee q$. Form a similar table for $p \wedge q$. Compare these two tables with those given after (5a) and (5b) in Chapter 1. What relationship is there?

11. Make a table for $\veebar$ such as that in Exercise 10, where $A \veebar B$ means "A or B but not both."

12. If an implication is written in the form "If p, then q," suppose that we call the contrapositive of its converse the "obverse" of the given implication. Write in symbols the obverse of $p \rightarrow q$. Is the converse of the contrapositive equivalent to the obverse?

13. Abbreviate contrapositive, converse, and obverse to "contra," "conv," and "obv," respectively. In the following table we have written "i" in the middle, since the converse of the converse of an implication is equivalent to the implication itself. In the middle of the first row we have written "obv,' since the contrapositive of the converse is the obverse. Fill in the rest of the table.

	Contra	Conv	Obv
Contra		Obv	
Conv		i	
Obv			

14. Using the table or by other means, find the converse of the obverse of the contrapositive of an implication.

5. ALWAYS, NEVER, AND SOMETIMES

Many statements include explicitly or implicitly one of these three words or their close relatives: "all," "none," or "some." The first two words are very stringent ones, and a statement preceded by one of these adverbs is likely to be very hard to establish. We often find ourselves in the position of Captain Corcoran of the "King's Navy," whose classic reply when challenged was, "Well, hardly ever." The reason, of course, that a statement involving "never" or "always" is difficult to make with certainty is that *one single exception suffices to prove it false*. If I made a statement that it never rains in Death Valley, it would be proved false if the oldest inhabitant could prove that one day when he was a boy it did rain. True statements involving one of these words are easier to find in mathematics than elsewhere, probably because the situation is here better controlled. I can, for example, say that no matter what number you give me I can *always* mention a number larger; this, of course, is due to the fact that any number is increased if you add 1 to it. This and many other such statements are more complex examples of such an assertion as: It is *always* true that if anything is white it is white.

It will probably be admitted that most of our "always" and "never" statements in everyday life are, and are known to be, exaggerations. They are working hypotheses which are assumed innocent until they are found guilty. You ask your neighbor, "Won't you keep my dog during my absence?" "Oh no," he replies, "he chases cats and I don't want mine chased." You say, "He doesn't chase my cat—so he won't chase yours." Just as your argument had holes in it, so had his, even though he may have seen your dog chasing a cat. It is rather a case of weighing the evidence and drawing conclusions as to probable future behavior.

These words ("always," "never," and so on) often occur implicitly. In Example 1 of Section 2 the idea was implicit that *all* plants which become green have had sunshine. This statement would be shown false if one could give *one* example of a plant which became green without having had sunshine.

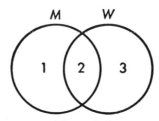

Figure 2:1

When the word "some" is used, the diagram contains overlapping circles. Suppose we have the statement "Some men are white." This statement has the diagram of Fig. 2:1, where *M* includes all men and *W* all white animals. Portion 1 includes all men who are not white, portion 2 all white men, and 3 all white animals not men, while the outside of the two circles represents all things (or animals) which are neither white nor men. Of course, it might be that a proper diagram should be one in which one circle is entirely within the other, but this would not be possible unless a stronger statement than the given one could be made.

One can represent the word "most" in a diagram by considering the relative size of the common portion. For instance, if the statement were "Most men are white animals," the larger part of circle *M* should be made to lie within *W*. However, since the statement does not imply that most white animals are men, one could not tell whether the larger part of circle *W* should lie in *M*. That would depend on the relative size of *M* and *W*. If there were fewer white animals than men, then most white animals would be men, but if there were more white animals than men, one could not tell whether the larger part of *W* were in *M* unless one had more information, such as, for instance, that *W* was at least twice as large as *M*. On the other hand, notice that the statement "*Some men are white animals*" is equivalent to "*Some white animals are men*."

EXERCISES

In each of Exercises 1 through 11, a statement is given followed by several others. Tell which of these statements are equivalent to the given one and which are not. Draw a diagram for the given statement. What kind of example, if any, would suffice to show the given statement false? Express the given statement in different words.

1. All Polynesians are brown.
   a. If a man is Polynesian, he is brown.
   b. If a man is not brown, he is Polynesian.
   c. In order to be brown, a man must be Polynesian.
   d. Whenever a man is Polynesian, he is sure to be brown.
   e. If a man is not brown, he cannot be Polynesian.
   f. If a man is not Polynesian, he is not brown.

2. All sinners are heathens.
   a. If a man is a sinner, he is a heathen.
   b. If a man is not heathen, he is a sinner.
   c. In order to be a sinner a man must be a heathen.
   d. A man is a heathen only if he is a sinner.
   e. If a man is not a sinner, he is not a heathen.
   f. A man is a sinner only if he is heathen.

3. All Americans are Canadians.
   a. Every Canadian is an American.
   b. If a man is an American, he must be a Canadian.
   c. If a man is not an American, he is not a Canadian.
   d. If a man is not a Canadian, he is surely not an American.

4. All Frenchmen revere de Gaulle.
   a. Everyone who reveres de Gaulle is a Frenchman.
   b. No Frenchman fails to revere de Gaulle.

    c. If he does not revere de Gaulle, he is not a Frenchman.

    d. If he is not a Frenchman, he does not revere de Gaulle.

5. When the sunrise is red, it is sure to rain during the day.

    a. If it is raining today, then the sunrise must have been red this morning.

    b. If it does not rain today, then the sunrise must have been red.

    c. It is does not rain today, then the sunrise must not have been red.

    d. Whenever it rains during the day, it began with a red sunrise.

6. If food is sweet, it has honey in it.

    a. Honey is sweet.

    b. If food does not have honey in it, it is not sweet.

    c. If food is not sweet, it does not have honey in it.

    d. Sweet food always contains honey.

7. There is never an earthquake in Denver.

    a. If there is an earthquake, it is not in Denver.

    b. If there is not an earthquake, it is not in Denver.

    c. Sometimes there are earthquakes in Denver.

    d. Sometimes there are earthquakes which are not in Denver.

    e. If you are in Denver, you will not experience an earthquake.

8. It never rains in the summer.

    a. If it is summer, it is not raining.

    b. If it is not raining, it is not summer.

    c. In the summer it never rains.

    d. Never in the summer does it rain.

    e. If it is raining, it is not summer.

    f. Sometimes in the summer, it does not rain.

9. Some soldiers are cruel.

    a. He is a soldier. Hence he is cruel.

    b. He is cruel. Hence he is a soldier.

    c. He is not a soldier. Hence he is not cruel.

    d. He is not cruel. Hence he is not a soldier.

10. Some women are poor drivers.

    a. A good driver can never be a woman.

    b. Sometimes women are good drivers.

    c. Sometimes good drivers are women.

    d. A poor driver is necessarily a woman.

11. Some strange people like mathematics.

    a. If a person likes mathematics, he is a strange person.

    b. If a person does not like mathematics, he is not strange.

    c. Only strange people like mathematics.

    d. Strange people like only mathematics.

12. Which of the following statements are false and which are true? Prove that your answers are correct or describe a method of establishing your conclusions.

    a. Every even number is divisible by 4.

    b. It never rains while the sky is cloudless.

    c. Every even number is divisible by 2.

    d.   A banker always has money in his pocket.

    e.   Needles and pins, needles and pins,
         When a man marries his trouble begins.

    f.   Looking at the moon makes men mad.

    g.   All professors are absent-minded.

    h.   Some wives are extravagant.

6. ARGUMENTS

When we bring together several statements and from them deduce a new one whose validity depends solely on the validity of the original statements, we have what we call an *argument*. The original statements (which may be implications) are called *premises* or *hypotheses*, the resulting one a *conclusion*, and the connecting link a *deduction*. There are two principal uses to which arguments are put. In the first place, we may wish to convince a friend (or perhaps ourselves) that something we say is true. Or we may try, from known facts, to gain additional information which we do not already know or perhaps even suspect. In the former case the conclusion is known, but it is not to be assumed, and, especially in the case of a recalcitrant friend, it is often desirable to conceal it. Our task is first to persuade him to accept some hypotheses and then, by inexorable logic, force him to the conclusion; once he accepts the hypotheses, he is lost. The situation is similar to that when one approaches a modern toll bridge—from a certain point on there is no turning back. In fact, all our reasoning must be of this character except where we lay ourselves open to the criticism of James Harvey Robinson in his *Mind in the Making*: "Most of our so-called reasoning consists in finding arguments for going on believing as we already do."

Once our friend is so careless as to admit that (1) all Pennsylvanians are Americans and (2) that all Americans are kind to animals; then he is forced to agree that all Pennsylvanians are kind to animals. In terms of sets, (1) means that the set, P, of all Pennsylvanians is a subset of the set A of all

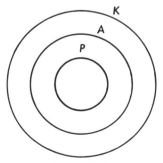

Figure 2:2

Americans; (2) means that the set A is a subset of the set K of all persons kind to animals. Figure 2:2 (p. 27) is a Venn diagram describing these relationships. It follows from the transitive property (see Section 4 of Chapter 1) that P is a subset of K.

All of this can be translated into the language of implications. Let p stand for the statement "He is a member of the set P," that is, "He is a Pennsylvanian." Let a stand for the statement "He is an American" and k for the statement "He is kind to animals." Then

(1) can be written: $p \rightarrow a$,

(2) can be written: $a \rightarrow k$,

and "all Pennsylvanians are kind to animals" because

(3) $p \rightarrow k$.

Briefly, then, $(p \rightarrow a$ and $a \rightarrow k) \rightarrow (p \rightarrow k)$.

The above is an example of a fundamental law of logic: *If there is a chain of true implications in which the hypothesis of each is the conclusion of the previous one, then it is true that the initial hypothesis implies the last conclusion and, in fact, any of the conclusions along the way.* For instance, if the first two implications below are true, the third is also:

(1) If a, then b.

(2) If b, then c.

(3) If a, then c.

Such a chain is called a *syllogism* and is the form of many arguments. In order to test an argument, it is often a good plan to try to put it into the form of a syllogism. This means that, in a true syllogism (that is, one in which each implication is true):

If the initial hypothesis is true, the conclusion is true. The contrapositive of this statement is:

If the conclusion is false, the initial hypothesis must be false.

This is the basis of the "indirect proof" or *reductio ad absurdum* arguments of high school geometry. An example of such an argument would be the following proof that there is no largest number. Suppose there were a largest number k. Then $k + 1$ would be a larger number. Thus our supposition that there is a largest number is false.

A more complex example of the "indirect proof" is the following: "Is it true that all men are brutes? If all men were brutes, then none would be good nurses. But I know a man who is a good nurse. Hence it is not true that all men are brutes." This can be dealt with more easily by putting the argument into the form of a syllogism. The second sentence of the argument affirms the truth of the following implication:

(1) If a person is a good nurse, then he is not a brute. Suppose the question "Is it true that all men are brutes?" were answered in the affirmative. Then we would have the implication.

(2) If a person is a man, then he is a brute.

In this form one cannot say that the conclusion of one implication is the hypothesis of the other. But change (1) to its contrapositive:

(3) If a person is a brute, he is not a good nurse.

Then, if (2) and (3) are both true, the following must be true:

(4) If a person is a man, then he is not a good nurse.

But the existence of a man who is a good nurse shows that the implication (4) is false. This means that one of the implications (1) and (2) must be false. Since (1) is supposedly true, it must be the implication (2) which is false. In other words, no is the answer to the question "Is it true that all men are brutes?"

Sometimes, in an argument, certain very obvious hypotheses are not stated. For instance, if a man told you, "I would rather see my daughter in the grave than see her married to you," you would probably conclude that he did not fancy you as a prospective son-in-law. The obvious and unmentioned hypothesis is that the stern parent would be very sorry to see his daughter in the grave. However, this hypothesis is very necessary to the argument.

It is interesting and instructive to note an analogy between properties of implications and set inclusion in Section 5 of Chapter 1. To emphasize this we list them side by side:

Sets	Implications
(1) $A \subseteq A$	$p \to p$
(2) If $A \subseteq B$ and $B \subseteq A$, then $A = B$.	If $p \to q$ and $q \to p$, then $p \leftrightarrow q$.
(3) If $A \subseteq B$ and $B \subseteq C$, then $A \subseteq C$.	If $p \to q$, and $q \to r$, then $p \to r$.

EXERCISES

1. What conclusions can be drawn from the truth of the following two statements? "Some men are white animals." "Some white animals are Americans." Discuss the possibilities using Venn diagrams.

2. Prepare a Venn diagram showing the relationships among the categories seagoing vessels, implements of war, and submarines. What statements would be made on the basis of your diagram? Make the list as complete as you can.

3. In Exercise 2 replace "submarine" by "airplanes" and follow the directions given.

4. Find two examples of syllogisms in Chapter 1.

In each of the following through Exercise 27, state where a syllogism can be made from the given sentences. State which of the indicated conclusions follow and which do not. If you are not sure exactly what some of the statements mean, state carefully your interpretation of their meaning. If there are any unmentioned hypotheses, point them out.

5. All Polynesians are brown.   All Balinese are Polynesians.
    a. Hence all Balinese are brown.
    b. Hence all brown people are Polynesians.
    c. Hence all brown prople are Balinese.
    d. If a man is not Balinese, he is not brown.
    e. If a man is not brown, he is not Balinese.
    f. Statements like parts d and e involving *Polynesian* and *brown.*
    g. Statements like parts d and e involving *Polynesian* and *Balinese.*

6. All rational numbers are real. If a number is an integer, it is a rational number.
    a. Hence if a number is a rational number, it is an integer.
    b. Hence if a number is an integer, it is real.
    c. Hence all real numbers are rational.
    d. Statements as in Exercise 5, parts d to g with "rational number," "real number," and "integer" taking the places of "Polynesian," "brown," and "Balinese," respectively.

7. All Americans are Canadians. If a man is a Mexican, he is an American.
    a. Hence all Mexicans are Canadians.
    b. Hence if a man is a Canadian, hc is a Mexican.
    c. Some Canadians are Mexicans.
    d. Statements analogous to Exercise 6d.

8. If any food is sweet, it has honey in it. All honey contains sugar.
    a. Hence sugar is sweet.
    b. Hence any food containing sugar is sweet.
    c. Hence any food which is sweet contains sugar.
    d. Hence honey is sweet.
    e. Statements obtained by filling in the following blanks in six possible ways with two of the three words "honey," "sugar," and "sweet." "If something is not —————, it is not —————."

9. Whenever it is going to rain my rheumatism bothers me.
    a. My rheumatism bothers me. Hence it is going to rain.
    b. My rheumatism does not bother me. Hence it is not going to rain.
    c. It is not raining. Hence my rheumatism does not bother me.

10. If you wear your sweater near that bull, he will gore you, for your sweater is red and anything red infuriates him.

11. I know I won't like Brussels sprouts, for they are just like little cabbages, and I don't like cabbage.

12. Father says, "Never do today what you can put off until tomorrow," and I act on everything father says. Hence the only things I shall do tomorrow are the things I don't need to do today.

13. All boys are bad. Most bad persons go to jail. Hence
    a. Some boys go to jail.
    b. Some persons who go to jail are boys.
    c. Not all persons who go to jail are boys.
    d. Of those who do go to jail, some are boys and some not.
    e. Some who go to jail are bad.
    f. Most persons who go to jail are bad.
    g. Most of those who go to jail are bad boys.

14. "It is easier for a camel to go through a needle's eye than for a rich man to enter into the kingdom of God."
    a. Hence if I am poor, my chances of entering His kingdom are good.
    b. Hence if I want to enter His kingdom easily, I must be poor.

15. "Blessed are ye when men shall revile you, and persecute you, and say all manner of evil against you falsely, for my sake. Rejoice, and be exceeding glad: for great is your reward in heaven: for so persecuted they the prophets which were before you." I am being persecuted. Therefore, I am to have a great reward.

16. "If any men desire to be first, the same shall be last of all, and servant of all." Now I serve everybody; hence I shall be first.

17. If you cannot do mathematics, you cannot be an engineer.
    a. Hence if you wish to be an engineer, you must learn mathematics.
    b. Hence if you learn mathematics, you can be a good engineer.

18. I know he doesn't love me, for men never tell the truth to a girl, and he told me he loved me.

19. I know he loves me, for men never tell the truth to a girl, and he has not told me he loved me.

20. I know he loves me, for men never tell the truth, and he told me he does not love me.

21. I know he doesn't love me, for men in love never tell the truth, and he told me he loved me.

22. The moon is made of green cheese. All cheese has holes in it. Therefore, the moon has holes in it.

23. "Let us start with these propositions:
    (a) It is the duty of our Government not to take any action which will diminish the opportunities for the profitable employment of the citizens of the United States. . . .
    (b) It is not in the interest of the nation to adopt any policy which makes the United States, in peace or in war, needlessly dependent upon the will of any foreign nation for any essential supply. . . . Applying these

principles to foreign trade and considering the nation as a unit, we do a useful thing if we exchange one hour of foreign labor, only if the one hour of foreign labor could not have been performed by an American. We do not, as is sometimes imagined, increase our wealth if we exchange one hour of American labor for two or more hours of foreign labor. That looks like a bargain, but it is a bad bargain, for not only do we deprive the American of his opportunity to work, but also we withdraw from the workman affected a purchasing power which reacts to the benefit of other sections of industry and of agriculture generally. Also we may weaken our self-reliance."—G. N. Peek, *Why Quit Our Own*, p. 213. D. Van Nostrand Company, Inc., Princeton, N.J., 1936.

24. I would not dare throw a spitball in class because the teacher would be sure to see it. Then she would write a note to my mother and my father would tan my hide.

25. "If the frame has been welded or straightened, most purchasers would do well not to consider the car further, for a car that shows evidence of having been in a wreck represents a pretty doubtful purchase, however low its price may seem."—*Consumers' Research*, February 1942.

26. ANTONY: "The noble Brutus
Hath told you Caesar was ambitious;
                    .   .   .
You all did see that on the Lupercal
I thrice presented him a kingly crown,
Which he did thrice refuse. Was this ambition?
Yet Brutus says he was ambitious;
And, sure, he is an honorable man."
                              —*Julius Caesar*, III, ii.

27. POLONIUS: "Neither a borrower nor a lender be;
                    For loan oft loses both itself and friend,
                    And borrowing dulls the edge of husbandry.
                    This above all: to thine own self be true,
                    And it must follow, as the night the day,
                    Thou canst not then be false to any man."
                              —*Hamlet*, I, iii.

28. Show by making a truth table that $[(p \to q) \land (q \to r)] \to [p \to r]$.

29. What conclusions can you draw from the following three implications?
          (a) p implies q.   (b) q implies r.   (c) r implies p.

30. Using the principle of Exercise 29, outline a method of proof of property (5) of Chapter 1. Then give a proof along these lines.

In each of the next six exercises there is a sequence of statements which can be put into the form of implications. In each case, form a syllogism containing all the implications in the exercise in one chain. The third is solved as a sample. The last four are taken from *Logical Nonsense* by Lewis Carroll, Blackburn and White, eds. (London: Putnam, 1934), p. 527.

31. a. On Tuesdays Susie always stops in for tea.
    b. I always get my cinnamon rolls at a little bakery around the corner.
    c. Unless I have fresh cinnamon rolls, Susie does not stop for tea.

32. **a.**   While he is ill, I cannot leave the house.
     b.   When he is not ill, I always go skiing.

33. **a.**   Puppies that will not lie still are always grateful for the loan of a skipping rope.
     b.   A lame puppy would not say "thank you" if you offered to lend it a skipping rope.
     c.   None but lame puppies ever care to do worsted work.

SOLUTION: The first can be rephrased as follows:
     a′.   If a puppy will not lie still, then it is grateful for the loan of a skipping rope.

The second statement means: If a puppy is lame, it is not grateful for the loan of a skipping rope. To make it fit with the above, we use the contrapositive to get:
     b′.   If a puppy is grateful for the loan of a skipping rope, it is not lame.
Then part c is equivalent to:
     c′.   If a puppy is not lame, it will not care to do worsted work.

Then we have a sequence of implications parts a′, b′, and c′ in which the conclusion of each is the hypothesis of the next. Hence, if the implications are true, the following one is also: If a puppy will not lie still, it will not care to do worsted work.

Another solution is the contrapositive of this implication: If a puppy cares to do worsted work, it will lie still.

34. **a.**   Nobody who really appreciates Beethoven fails to keep silence while the Moonlight Sonata is being played.
     b.   Guinea pigs are hopelessly ignorant of music.
     c.   No one who is hopelessly ignorant of music ever keeps silence while the Moonlight Sonata is being played.

35. **a.**   None of the unnoticed things met with at sea are mermaids.
     b.   Things entered in the log, as met with at sea, are sure to be worth remembering.
     c.   *I* have never met with anything worth remembering, when on a voyage.
     d.   Things met with at sea that are noticed are sure to be recorded in the log.

36. **a.**   No husband who is always giving his wife new dresses can be a cross-grained man
     b.   A methodical husband always come home for tea.
     c.   No one who hangs up his hat on a gas jet can be a man that is kept in proper order by his wife.
     d.   A good husband is always giving his wife new dresses.
     e.   No husband can fail to be cross-grained if his wife does not keep him in proper order.
     f.   An unmethodical husband always hangs up his hat on a gas jet.

7. THE METHODS OF MATHEMATICS

So far in this book we have had at least two examples of mathematical method. In each case there is something we want to learn about. We have some idea of what kinds of things we want to find out about it. So we begin by stripping our subject matter to bare essentials so that we can express it in a notation. From our knowledge of the subject matter we set up certain fundamental assumptions about what our symbols mean and rules of manipulation. Then, being careful to use only these, we try to arrive at certain conclusions involving the symbols. Finally we translate back into the original situation and, Voilà!, we have learned something.

For instance, in Chapter 1 we dealt with sets and agreed on what we meant by sets and terms such as "and" and "or." We noticed that in accordance with the meanings we had in mind for these things, there would be certain properties which we listed: (1) through (5). Then we were ready to work with the symbols using these properties and the rules of logic. We showed, for instance, by use of these properties alone, in Exercise 5b of Section 3, that if $B \cup A = B$ and $B \cap A = \varnothing$, then $A = \varnothing$. Translated this means: If all elements in set A or B are in B and if A and B have no elements in common, then A has no elements. For example, if every animal that is a fuped or a cat is a cat and if no fupeds are cats, then there are no fupeds. This is, of course, no world-shaking discovery. But it is an example of mathematical method.

Second, in this chapter we have been dealing with the rules of logic. Again we made certain agreements on the basis of experience. We eliminated irrelevancies and agreed on a notation, as well as rules of manipulation. On this basis, our truth table showed us in an exercise above that "not (p or q)" means the same thing as "(not p) and (not q)." One example of this is "It is not true that he is one or both of: 6 feet tall or a good dancer" means the same as "He is neither 6 feet tall nor a good dancer." The latter is a simpler way to express it.

We shall, of course, have many examples of mathematical method throughout this book.

8. CAUSALITY

It is one of the axioms of numbers that two things equal to the same thing are equal to each other. But that does not imply that two things which contain the same thing contain each other, or that two things which cause the same thing cause each other. As in the case of arguments, causality is much of the time a one-way road. A cold wind as well as a hot fire makes one's face glow, but the wind does not cause the fire nor the fire the wind. We could represent this situation as in Fig. 2:3, where W and F stand for the wind and fire and G stands for the glowing face. There need not be any causal connection between W and F. Only if one of the arrows were reversed could we establish such a connection: If a hot fire were caused by a glowing face, then a cold wind would indeed cause a hot fire.

A full moon makes the hounds bay the moon and lovers' hearts beat quicker, but few could claim that the hounds baying the moon made lovers' hearts beat quicker or that the latter caused the hounds to bay at the moon. Below we have Fig. 2:4, where *M*, *H*, and *L* stand for the moon, the hounds

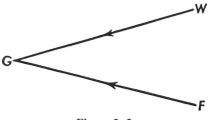

Figure 2:3

baying, and the hearts beating, respectively. There is no causal connection between *H* and *L*. Moreover, one cannot even conclude that *when* the hounds bay, lovers' hearts beat quicker, since there may be many times when the hounds bay and the moon is not full.

One has a similar situation in the common argument that to increase your pay you should get a college education, for those who have such an education as a group get higher pay than those who do not. The conclusion may be correct, but the argument is not sound, for it is possible that getting a college

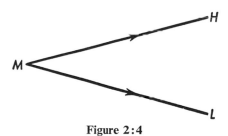

Figure 2:4

education and getting high pay may both be the result of having parents of a higher class and with larger income; also, there is *some* evidence to the effect that only the more intelligent can get such an education. In other words, the following two statements are not in the least equivalent.

(1) Those who go to college draw higher salaries than those who do not.
(2) Those who have gone to college are drawing higher salaries than they would have drawn had they not gone to college.

Furthermore, even if the latter statements held for college graduates as a whole, it might fail to hold for many individuals.

It is even possible for two things which occur together to have no common cause. For instance, my clock strikes twelve when the noonday whistle blows but the clock strikes then because it is a good clock and I wind it, while the whistle blows because someone blows it.

If A always causes B, it is true that if A occurs, then B must occur. Also, if A usually causes B, it is true that when A occurs, then B usually occurs. But the converse of neither of these statements is true. It is at this point that we have one of the commonest pitfalls of everyday reasoning. Even if hounds bayed at the moon only when it was full, they might not cause lovers' hearts to beat quicker. The sun unblocked by clouds causes the day to be bright, and whenever the day is bright there is the sun unblocked by clouds; but no one would claim that the bright day causes the sun to shine. If one of a man's eyes is blue, it is usually true that the other is also, but the blueness of the one is not caused by the blueness of the other. You may have believed in your childhood that angleworms came down from the sky because after a rain they appeared in great numbers on the pavement. Of course, it is true that if the angleworms did come down in the rain, then they would appear under the conditions observed, but that lends no support to the superstition.

A cause is one of many reasons. The former, strictly speaking, does not often occur in mathematics, for our reasons are likely to be in the nature of relationships or arguments. The answer to the question "Why is $x(x + 1)$ always even when x is an integer?" is: If x is even it is even, if it is odd, $x + 1$ is even, and in both cases we have a product of two integers one of which is even; such a product is always even. Moreover, outside mathematics our answer to "why?" is often to give a reason which is not connected with cause even though our answer begins with "because." We have, "Why is he brown? Because he is a Malayan, and all Malayans are brown."

It is very hard to pin down this causal relationship. We can agree that if A causes B, then whenever A occurs, B must occur, but we mean more than that. Usually cause precedes effect (it certainly never follows it), but many times it fails to have this property, for example, the sun causing glare on the water. We have seen that if it is true that whenever A occurs, B occurs, it does not follow that A causes B. However, in scientific practice such occurrence is assumed to be evidence of cause. For instance, to investigate the truth of the statement "Sunshine makes plants become green," the scientist might take two identical plants, give one sunshine over a period of time, and deprive the other of it but keep all other factors the same. If, at the end of the time, the former were greener than the latter, and if the results of many experiments were the same, he then would conclude that sunshine caused plants to become green. One could perform a similar experiment on individuals with regard to a college education. If a set of persons of the same background and ability (and here is the rub in the experiment) were divided into two parts and one sent to college and one not, then if the former part received higher pay than the latter, the causal relationship would be more definitely established.

EXERCISES

In each of Exercises 1 through 11, point out which conclusions follow logically and which do not. Also mention possible common causes for two or more things which occur together.

1. A depression occurred during Hoover's administration.
    a.   Hence he was one of those who brought it about.
    b.   If we feel a depression coming on, we should see to it that he is not in office.

2. Countless charts show that prices have varied with the amount of gold in this country.
    a.   Hence to regulate prices, regulate the gold content of the dollar.
    b.   Hence to regulate the gold in the country, regulate prices.

3. Statistics show that a larger proportion of graduates of Reed College go to graduate school than from any other liberal arts college. Therefore, if you wish to go to a liberal arts college and to graduate school, go to Reed College.

4. When we questioned the man about his whereabouts that particular evening he seemed very vague. If he had committed the crime, he would have been vague with regard to his whereabouts. Therefore, he was probably guilty.

5. The incidence of cancer among smokers is much greater than among non-smokers. Hence smoking causes cancer.

6. In the last ten years our main roads have been made wider and wider. And accidents have doubled.
    a.   Therefore, we should make narrower roads.
    b.   The reason for the increase in the accident rate is that on wider roads men drive faster. Hence, if we made the roads narrower, men would not drive so fast, and there would be fewer accidents.

7. One of the causes for feeblemindedness is excessive drinking, for statistics show that the two go together to a surprising degree.

8. It has been observed that during very stormy weather there are invariably sunspots.
    a.   Hence sunspots cause the storms.
    b.   Hence storms cause sunspots.
    c.   Hence when there are sunspots there are usually storms.
    d.   Hence when there are storms there are usually sunspots.

9. There is a larger proportion of criminals among those who have only elementary education than among those with extended education. Therefore, to reduce criminality, increase the age of compulsory attendance in school.

10. POLONIUS: "Costly thy habit as thy purse can buy,
                 But not express'd in fancy; rich, not gaudy;
                 For the apparel oft proclaims the man,
                 And they in France of the best rank and station
                 Are most select and generous in that."
                                             —*Hamlet*, I, iii.

11. John had never eaten yeast and on January 1 weighed 120 pounds. From then until April 1 he ate three yeast cakes a day. At the end of that time he weighed

150 pounds. Which of the following conclusions are justified and to what
extent?

a.   Eating yeast made John gain weight.

b.   Eating yeast makes people gain weight.

What additional facts would bolster these conclusions?

12.   Quote an advertisement appearing in the daily paper, magazine, or heard over
the radio—an advertisement which has to do with cause and effect. Point out
to what extent the conclusions are justified.

9. LOGICAL STRUCTURE

The usual arguments which we meet in everyday life are more complex than
those previously given, and as a result it is more difficult and usually not very
enlightening to reduce them to diagrams, although parts of an argument may
often be clarified by such a reduction. As a matter of fact, many statements
cannot be reduced to any diagram we have given, for example, the statement
"No matter what number you mention, I can always mention a bigger one."
In more complex arguments, the logical structure is best analyzed by re-
phrasing it.

Consider the following:

Although most men carry insurance continuously during most of their lifetimes,
life insurance statistics indicate that the average period for which an insurance
policy remains in force is less than ten years. This indicates that many insurance
contracts are discontinued every year, probably because they are not suited to the
needs of the policyholders. Obviously, if insurance is carried during the lifetime of
an individual, the payment of several agents' commissions and other acquisition
expenses at intervals of nine or ten years is costly. Before any insurance is pur-
chased, therefore, it should be studied carefully in order to determine whether or
not it is suitable and the best for the purpose.[1]

The argument might be rephrased as follows: (1) Although most men carry
insurance continuously during most of their lifetimes, life insurance statistics
indicate that the average period for which an insurance policy remains in force
is less than ten years. (2) Hence, many men discontinue policies and take out
new ones. (3) Under these circumstances extra commissions and other expenses are paid. (4) This is costly. (5) Hence one way to avoid extra cost is to
avoid situation (2) and hence to eliminate as many causes of the situation as
possible. (6) Probably in many cases the cause is that the policy is not suitable.
(7) One of the reasons for the unsuitability is lack of care in selecting a policy.
(8) Hence one way to save extra expense is to use care in selecting a policy.

Statements (1) and (2) are not really part of the main argument; their purpose is to show that what follows is a frequent consideration. The logic seems
sound but there are several by-arguments that apply. Statement (2) follows

---

[1] From E. C. Harwood and B. H. Francis, *Life Insurance from the Buyers' Point of
View*, p. 102. American Institute for Economic Research, Cambridge, Mass., 1940.

(1) because the span of manhood is more than ten years. Statement (5) is almost axiomatic but stems from the fact that if you find all the things that produce a result and eliminate the things, you eliminate the result. Statement (6) derives strength from the following: If a man found he could not meet the premiums he would not be able to take out another policy; neither would he do this if he did not approve of life insurance; the only other cause would be the unsuitability of the policy or the company. Notice that the argument does not show that if one uses care in selecting the policy, extra expense will necessarily be avoided—the authors do not, in fact, claim that. Other factors, such as a change of circumstances of the policyholder, may render a policy unsuitable no matter what care was taken in selecting the policy. It might even be that a change in circumstances would make a policy selected with care less suitable than one not so selected.

There is one kind of diagram which is perhaps useful in this situation. We could represent the sequence of causes by

$$L \to U \to D \to E,$$

where L is lack of care in selection, U is unsuitability of policy, D is discontinuance of policy, and E is expenses to be paid. Statement (1) is used to show D occurs, and the conclusion is that a way to avoid E is to avoid L.

We have not mentioned some important factors which play a vital role in everyday reasoning. One of these is experience, which tells us what is relevant and what is not—how much and what kind of evidence one needs before arriving at a reasonably sure conclusion. For instance, one sees the sign:

WE HAVE EATING SPACE FOR 150 PEOPLE
SO YOU CAN BE SURE OF PROMPT SERVICE

One with experience could say:

(1) I have been in little holes-in-the-wall where service was quick as a flash and huge dining halls where it took me a year to get my soup.
(2) The service depends rather on how many waiters there are, how well they are trained, and how quickly they in turn can be served in the kitchen.

Statement (1) is made to show the falsity of the conclusion. The first part of this statement is somewhat beside the point because good service in a small place would not imply poor service in a large one. The second part is more to the point, for it shows that not all large places have quick service. On the other hand, statement (2) shows what is relevant. Both statements depend on experience with restaurants.

Another important factor in any argument is one's degree of self-discipline. If anyone says of our labors, "That was a sloppy piece of work," our immediate inclination is to prove by fair means or foul that he is wrong, one of the most telling "arguments" being that *his* work was "sloppier." To rule our

emotions and to force ourselves to see and take account of relevant but un-
pleasant factors is no mean achievement. But this again is a matter which we
cannot consider here.

Finally, we wish merely to mention the field of mathematical thought,
which makes it its business to set up a precise language for the classification
of logical structure: symbolic logic. A very interesting and useful little book
on this subject is Quine's *Elementary Logic* (reference **51** in the Bibliography).
Here the machinery for examining a logical argument is set up though it is
not the author's purpose to show many of its applications. For our purposes
in everyday life the old Aristotelian logic is sufficient, and it is this logic with
which we are concerned in this chapter. However, that it not always suffices
is shown by the following example. Suppose a man makes the statement "I
am not telling the truth." There are then two possibilities. First, he is telling
the truth, in which case he must be not telling the truth because he said "I am
not telling the truth." Second, he is not telling the truth, which is exactly what
he said he was doing, which implies that he was telling the truth. We should
then be forced to conclude that at that time he was neither telling the truth
nor not telling the truth.[1]

EXERCISES

Analyze the logical structure of the following 14 arguments somewhat along the
lines of the analysis of the argument at the beginning of this section. Use diagrams
where it seems desirable.

1. Bertrand Russell describes the following "reasoning": "Two of my ser-
    vants were born in March, and it happens that both of them suffer from
    corns. By the method of simple incomplete enumeration they have decided
    that all people born in that month have bad feet, and that therefore theirs
    is a fate against which it is useless to struggle."

2. "Internationalism does not and cannot make for peace because it seeks to
    keep intact all of those elements which have always made for war. Fortu-
    nately that is now becoming apparent. The League of Nations, had it been
    able to carry out its ludicrous peace policy of enforcing economic sanctions
    against Italy, would have started a general war in the name of peace!"—
    G. N. Peek, *Why Quit Our Own*, p. 348. D. Van Nostrand Company, Inc.,
    Princeton, N.J., 1936.

3. "As an economic policy, it is impossible, by influencing the extent of plant-
    ing or breeding, to determine in advance the volume of farm production.
    As a national policy, it is unwise to attempt to reduce toward the danger
    point the volume of the production of essential foods and raw materials.
    The problem, therefore, is control of supply rather than control of produc-
    tion."—*Ibid.*, p. 46.

4. "A man should learn to detect and watch that gleam of light which flashes

[1] See reference **38**, pp. 213 ff.

across his mind from within, more than the lustre of the firmament of bards and sages. Yet he dismisses without notice his thoughts, because it is his. In every work of genius we recognize our own rejected thoughts: they come back to us with a certain alienated majesty. Great works of art have no more affecting lesson for us than this. They teach us to abide by our spontaneous impression with good-humored inflexibility then most when the whole cry of voices is on the other side. Else, tomorrow a stranger will say with masterly good sense precisely what we have thought and felt all the time, and we shall be forced to take with shame our own opinion from another.''—Emerson, essay on "Self-Reliance."

5. "You must love them [the authors of good books], and show your love in these two following ways. 1. First, by a true desire to be taught by them, and to enter into their thoughts. To enter into theirs, observe; not to find your own expressed by them. If the person who wrote the book is not wiser than you, you need not read it; if he be, he will think differently from you in many respects.

    "Very ready we are to say of a book, 'How good that is—that's exactly what I think.' But the right feeling is, 'How strange that is! I never thought of that before, and yet I see it is true; or if I do not now, I hope I shall, some day.''—John Ruskin, "Of King's Treasuries.'"

    Compare Ruskin's argument with that of Exercise 4. Are they compatible? Give the reasons for your conclusions.

6. A larger proportion of the population of Swarthmore, Pennsylvania, is in *Who's Who* than in any other city of the United States. Hence if you want to get your name in *Who's Who* come to Swarthmore to live.

7. Most people who die, die in bed. Hence, in order to live long, stay out of bed.

8. "Europe has a set of primary interests, which to us have none, or a very remote relation.—Hence she must be engaged in frequent controversies, the causes of which are essentially foreign to our concerns.—Hence, therefore, it must be unwise in us to implicate ourselves, by artificial ties in the ordinary vicissitudes of her politics, or the ordinary combinations and collisions of her friendships, or enmities.''—Washington's Farewell Address.

9. "The cruelest lies are often told in silence. A man may have sat in a room for hours and not opened his teeth, and yet come out of that room a disloyal friend or a vile calumniator. And how many loves have perished because, from pride, or spite, or diffidence, or that unmanly shame which withholds a man from daring to betray emotion, a lover, at the critical point of the relation, has hung his head and held his tongue?''—R. L. Stevenson, "Truth of Intercourse" in *Virginibus Puerisque*.

10. "In one half hour I can walk off to some portion of the earth's surface where a man does not stand from one year's end to another, and there, consequently, politics are not, for they are but as the cigar-smoke of a man.''—Thoreau, "Walking."

11. "Due to the low temperatures obtained in the chimney when gas heat is applied in highly efficient boilers, there is a possibility that water vapor in

the stack gases will be cooled below its condensation temperature and deposit on the walls of the chimney. The condensed moisture contains small amounts of acid. If the chimney is not tight, the liquid may leak through the chimney walls and discolor the interior walls; it is likely, of course, to corrode the smoke pipe, more so than occurs with other fuels, as a rule . . . A tile lining inside the chimney will help greatly to prevent this trouble."—*Consumer's Research.*

12.   PORTIA: "The quality of mercy is not strain'd.
          It droppeth as the gentle rain from heaven
          Upon the place beneath. It is twice blest:
          It blesseth him that gives and him that takes.

             .   .   .

          It is enthroned in the hearts of kings;
          It is an attribute to God himself;
          And earthly power doth then show likest God's
          When mercy seasons justice. Therefore, Jew,
          Though justice be thy plea, consider this,
          That, in the course of justice, none of us
          Should see salvation. We do pray for mercy,
          And that same prayer doth teach us all to render
          The deeds of mercy."
                    —*The Merchant of Venice*, IV, i.

13.   "To have clear understanding of the war, Americans must see how the fighting goes—the battlefields, soldiers, weapons, geography, people, cities, industries. To have a satisfactory goal, Americans must see what the fighting is *for*, all the manifestations of democracy and our way of life. *Life* shows these things with all the *reality, simplicity,* and *scope* of life itself—so that every issue is eagerly and easily read by 21,900,000 people."—A *Life* advertisement.

14.   " Makes no difference which of us is driving, our Tilt and Telescope Wheel can be quickly positioned to put steering comfort within easy reach. That's because it has seven different up-and-down settings. Moves in and out as well."

15.   A kidnapper wrote to a bereft parent: "I will give you back your child if you will tell me truly what will happen to it." What should the parent say? Give your argument.

16.   In a certain small town, the barber shaves all those men who do not shave themselves and shaves no one else. Show that the barber is not a man.

*17.   In a strange kingdom (very strange it must have been) there was a conviction that all nobles tell the truth and that all slaves lie. Once three men, all dressed alike, appeared before the king. The first one mumbled something. The king said to the second man, "What did he say?" The second said, "He says he's a slave." The king then said to the third, "What is the second man?" The third said, "The second one is a prince." Were there two nobles and one slave or two slaves and one noble? Carefully give your reasons.

*18. There was a vacancy in the department of philosophy in a large university. The head of the department brought together three well-known logicians, *A*, *B*, and *C*, who had applied for the position, to select one to fill the vacancy. Finding them all qualified, and unable to make a selection, he gave them a final test. He blindfolded them, telling them that he had some soot on a finger. He said, "I will rub a finger over the forehead of each of you. I may leave a smudge on one, two, or three of you, or perhaps on no one. When the blindfolds are removed I want you to give one tap on the floor if you see either one or two smudges. And I want to know which one of you knows that he himself is marked." The blindfolds were applied and the professor left a smudge on each of the three foreheads. Immediately the bandages were removed all three tapped. After a few moments of silence, Mr. A announced, "I know that I am marked." How did he know? Carefully give your reasons.

*19. Solve the following: There are three men, John, Jack, and Joe, each of whom is engaged in two occupations. Their occupations classify each of them as two of the following: chauffeur, bootlegger, musician, painter, gardener, and barber. From the following facts find in what two occupations each man is engaged:
    1. The chauffeur offended the musician by laughing at his long hair.
    2. Both the musician and the gardener used to go fishing with John.
    3. The painter bought a quart of gin from the bootlegger.
    4. The chauffeur courted the painter's sister.
    5. Jack owed the gardener $5.
    6. Joe beat both Jack and the painter at quoits.
    [See reference **38** (p. 189) in the Bibliography.]

*20. There were three men who simultaneously found they needed hats. One was blind and the other two could see. They went to a hatter who, being a sporting man, made this bargain with them: "I see you all wear the same size of hat. I will blindfold the two of you who can see and put a hat on each of your heads. At least one of these hats will be a black one; the others will be grey. Then I will remove the blindfolds and the first one of you who, without looking at his own hat, will tell me what color hat he wears, will be given a hat free." The three men agreed and it was done as arranged. After the blindfolds were removed, the first man who could see said "I do not know what color hat I wear." Then the second man said likewise. Finally the blind man said, "I know that my hat is black." He was correct. How did he come to that conclusion?

10. APPLICATIONS TO SWITCHING CIRCUITS

There is, strange as it may seem, a close connection between truth tables and electrical circuits. We first show how a notation can assist us in dealing with such circuits, and in Section 11 we shall relate it to truth tables and sets.

_____

* Starred problems (and sections) are thought to be more difficult than the others.

A simple electrical switch is either open, which we designate by 0, or closed, which we designate 1. Consider Fig. 2:5, where the lines denote electrical wires and the "doors" switches. In the figure, both switches are open. Notice that the current will flow from A to B if and only if one (or both)

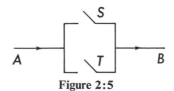

Figure 2:5

of the switches is closed. This is known as a connection "in parallel." If we enlarge our use of 0 and 1 slightly to mean that 0 indicates that the current does not flow through and 1 that it does, we have the following table, where the column headings denote the state of switch S and the row headings the state of switch T.

+	1	0
1	1	1
0	1	0

For switching circuits, it is customary to use a + sign for such a circuit in parallel. Notice that this is not the sum in the ordinary sense, since here $1 + 1 = 1$. In fact, here the symbols 1 and 0 do not stand for numbers but are merely a convenient way to designate the state of the switch. If s is the number (0 or 1) indicating the state of switch S and t is the number for T, we have

$$(s + t)\begin{cases} = 0 & \text{if both } s \text{ and } t \text{ are zero,} \\ = 1 & \text{if one or both of } s \text{ and } t \text{ are 1.} \end{cases}$$

That is; the circuit is closed if s *or* t is 1.

In contrast consider Fig. 2:6. Here, current will flow from A to B if and only

$$A \longrightarrow \quad \diagdown S \quad \diagdown T \longrightarrow B$$

Figure 2:6

if both switches are closed. Here the connection is said to be "in series." The table is

×	1	0
1	1	0
0	0	0

Thus the circuit is closed in this case if s *and* t are 1, and not otherwise.

Here it is customary to use the symbol $\times$ for multiplication in the case of a circuit in series. Again this is not multiplication, although in this case, in contrast to that for $+$, if 0 and 1 *were* numbers and $\times$ *did* designate multiplication, the table would be the same.

Comparing the two we see that a connection in parallel is associated with the word "or" and one in series with "and." (The "or" is, of course, the inclusive "or.")

Now consider the two circuits indicated in Fig. 2:7. In the second one S

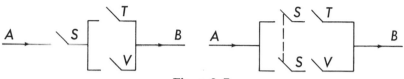

Figure 2:7

and S_1 are connected so that they are both open or both closed, and hence we consider them to be the same switch. Using our notation for "or" and "and" we see that the first diagram corresponds to $s \times (t + v)$ and the second to $(s \times t) + (s \times v)$. It is fairly apparent that these two accomplish the same thing. In either case the circuit is closed if switch S and one of T and V is closed. The first is simpler. One advantage of the notation is that we can sometimes by this means simplify the circuit.

At times it is more convenient to use a two-way switch. Suppose instead of merely being open or closed, switch S has two positions, while V and T are, as in the previous cases, open or closed. Then consider the situation in which the current flows through if S is in position 1 and V is closed or if S is in position 0 and T is closed. Then the diagram is as in Fig. 2:8. In symbols this circuit could be represented by $(s \times v) + (s' \times t)$.

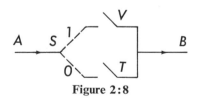

Figure 2:8

EXERCISES

1. Write out expressions for each of the circuits in Fig. 2:9, on page 46.
2. Write out expressions for each of the circuits in Fig. 2:10, on page 47.
3. Draw a diagram for each of the following:
    a. $s + (t + v)$.   **b.** $(s \times t) + (v \times w)$.   **c.** $(s \times t) + (s + v)$.
    d. $(s \times t) + (s' \times v)$.   **e.** $(s + t) \times (s + v)$.

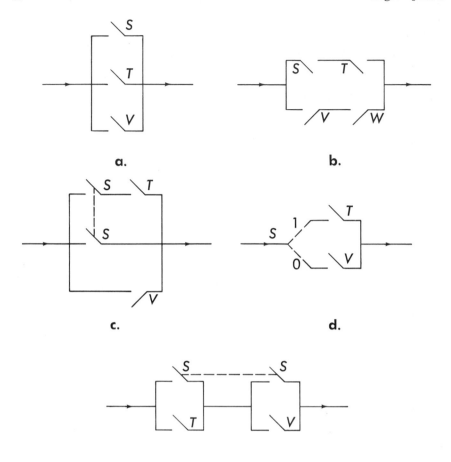

Figure 2:9

4. Draw a diagram for each of the following:

    a. $s \times (t \times v)$.   **b.** $(s + t) \times (v + w)$.   **c.** $(s + t) \times (s \times v)$.

    d. $(s \times v) + (s' + t)$. **e.** $s + (t \times v)$.

5. Which of the circuits in Exercise 3 can be simplified? Give reasons.

6. Which of the circuits in Exercise 4 can be simplified? Give reasons.

*7. Give a diagram of the wiring for the following situation. There is a stairway with a light in the middle and switches at the top and the bottom. It is to be connected so that one may turn the light on if it is off, or off if it is on, from either the top or the bottom of the stair. How can this be done? (Two-way switches are useful here.)

*8. Consider the following modification of Exercise 7. Let the stairway connect three floors, with one light in the middle and switches at each of the floors What should the connection be?

*9. A condemned man is led to a chamber where there are two chairs which look

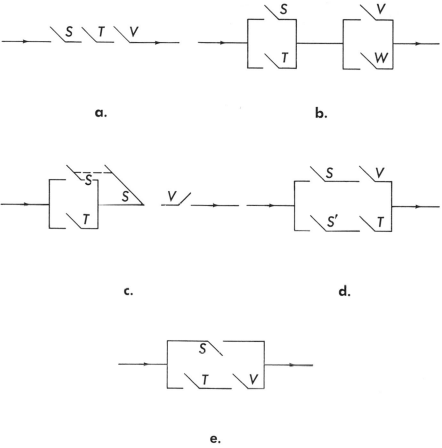

a.   b.

c.   d.

e.

Figure 2:10

exactly alike, but one is wired so that anyone sitting in it will be killed, while the other is harmless. There is a guard who knows which chair is lethal and he is an intelligent person. The prisoner is allowed to ask the guard one question to be answered yes or no, but he does not know whether the guard lies or tells the truth. The guard will answer him. After receiving the answer he must sit in one of the chairs. If he picks the nonlethal one he can go free. What question should he ask?

There is a variation of the above problem which is slightly easier to solve. For this, there are two guards instead of one. They both know which chair is lethal. The prisoner and the guards know that one of the guards always lies and that one always tells the truth, but the prisoner does not know which is which. He may ask one question of one guard to obtain his freedom; it need not have a yes or no answer. What question should he ask?

11. TRUTH TABLES, SETS, AND SWITCHING CIRCUITS

Although the reader must have noticed various analogies between the material of Section 10 and others in this and Chapter 1, it should be helpful to point out some of them more explicitly. Perhaps the most obvious one is seen by the following (see Section 4):

+	1	0
1	1	1
0	1	0

Parallel connection:

∨	T	F
T	T	T
F	T	F

Truth table:

×	1	0
1	1	0
0	0	0

Series connection:

∧	T	F
T	T	F
F	F	F

Truth table:

In each case if we replace 1 by T and 0 by F in the left-hand table, we get the right-hand one. We can say that the tables are the same except for the notation. The technical term in mathematics for such a relationship is "isomorphism." (We shall meet this idea again in Chapter 4.) Does the parallel connection correspond to any addition table? Is there an isomorphism between the tables for parallel and series connections? These questions are asked in exercises below.

There is also a connection with the negation of statements. The following occurs in Section 4:

$$(p \lor q)' \leftrightarrow (p' \land q').$$

This means that if we replace p and q by their negatives and $\lor$ by $\land$, we have the negative of $p \lor q$. In terms of the table for the connections above, if we interchange 1 and 0 in the values of p and q and in the body of the table for the parallel connection, we get

	0	1
0	0	0
1	0	1

which is the table for the series connection.

In Section 10 we noted that the following two expressions are equivalent:

$$s \land (t \lor v), \qquad (s \land t) \lor (s \land v),$$

this is, except for notation, the same as equation (4b) in Section 3 of Chapter 1. It is a distributive property which we shall be considering in more detail in Chapter 3. This is perhaps more apparent if we replace $\land$ by $\times$ and $\lor$ by $+$ to get

$$s \times (t + v) = (s \times t) + (s \times v).$$

From these considerations it is not hard to imagine that there can be developed an "algebra" which can serve with variations in notation for sets, logic, or switching circuits. Such an algebra is called a Boolean algebra after the man who first developed it, George Boole. It does not seem appropriate to develop it further here, but anyone interested in pursuing the subject can use the references given in Section 12.

EXERCISES

1. Write the switching circuits corresponding to the two sides of equation (4a) in Section 3 of Chapter 1. If $\cap$ is replaced by $\times$ and $\cup$ by $+$, will the equality hold for numbers?

2. Does the parallel connection of this section correspond to an addition table for numbers? Is there an isomorphism between the tables for parallel and series connections?

3. We have used the notation $A \uplus B$ for "A or B but not both," that is, the exclusive "or." What would be a switching circuit corresponding to this?

4. Find a connection between the answer to Exercise 3 and that for Exercise 5 in Section 4 of Chapter 1.

5. Consider the table for $\cup$ at the end of Section 3 of Chapter 1. Is this isomorphic to any table in this section?

6. Consider the table for $\cap$ at the end of Section 3 of Chapter 1. Is this isomorphic to any table of this section?

7. What circuit would correspond to $B \cup B = B$ of Section 3 of Chapter 1?

8. What circuit would correspond to $B \cap B = B$ of Section 3 of Chapter 1?

9. What in electrical circuits would correspond to the null set?

10. What in electrical circuits would correspond to the universal set?

11. There are two bridges, P and Q, across a river. Call one side of the river 0 and the other side 1. Also let the value 1 for P mean that one crosses bridge P and 0 means that one does not cross it; similarly for Q. One could make a table for where one is after crossing the bridges if he starts on side 0. That is, if one crosses both bridges P and Q we could write $1 + 1 = 0$, since the walker ends on the side he started on. If he crosses bridge P but not Q, we would have $1 + 0 = 1$, since he ends on the opposite side. Make a table for all four possibilities and compare it with other tables of this and previous sections. What would be an electrical circuit corresponding to it?

12. Work out the problem of the previous exercise for three bridges, P, Q, and R.

12. TOPICS FOR FURTHER STUDY

Brief treatments are given in the following references: **2** (Chap. 1) and **40** (Chap. 1). Somewhat longer discussions occur in references **23** (Chaps. 1–4), **54**, and **62**.

The algebra of sets and logic is called Boolean algebra; see reference **10** (Chap. 11). Applications of Boolean algebra to switching circuits is dealt with in references **31** and **53**.

There is much attention paid to mathematical method in reference **52**.

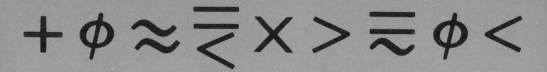

3

The Positive Integers and Zero

1. COUNTING

A man entering a theater alone says, "*one*, please" to the girl selling tickets, and the number 1 has served a useful purpose. Some persons believe that *two* can live more cheaply than *one*. A tripod (if it stands up) has *three* legs. An ordinary city block (of course, there are queer blocks in Pittsburgh) has *four* corners, although the same, alas, cannot be said for a city square. Most persons who stay at reasonable distances from buzz saws have *five* fingers on each hand. Every snowflake favors the number *six*. The crapshooter prays for "*seven* come *eleven*," and when we pay for a dozen rolls we expect to have *twelve*. To preserve our luck we refrain from continuing.

The words "one," "two," "three," and so on, are, of course, only our names for ideas which are not by any means peculiar to English-speaking peoples. We say "three," the French say "trois," the Germans "drei"—there is a name for it in almost every language. There is even an international sign language for it: the most primitive savage and the headwaiter in a fashionable hotel alike would hold up three fingers and every intelligent being, regardless of race, would understand him; that is, he would comprehend to this extent: Whatever the things he was talking about, there would be "just as many" things as there were upraised fingers. The meaning of the phrase "just as many" or its counterpart in another language might be explained in any one of various ways; one could say that it would be possible, given enough string, to connect by strings each thing with an upraised finger in such a manner that each thing would be connected with exactly one finger (but not a specified finger) and each finger would be connected with exactly one thing. One could even dispense with the string and talk abstractly about a "one-to-one correspondence" between the fingers held up and the things. This would mean that by some unspecified method each thing would be associated with or correspond to exactly one finger and each finger to exactly one thing. Similarly, if in an assemblage each man had a hat on, we should know that there were just as many hats as men (no strings are necessary except perhaps to hold the hats on). The fact that we do not then necessarily know either the number of hats or the numbers of men (although knowledge of one gives knowledge of the other) shows that there is a little more to this process we call *counting* than establishing a one-to-one correspondence. We must have established such a correspondence with some known standard—the notches on a stick, our fingers, or a set of figures learned in a certain order—before we know "how many." Our fingers or a set of numbers merely form a common yardstick by which we can measure how many things there are in any given set of things. We are particularly fortunate that, when it comes to counting, we need not refer to a standard yardstick in the National Bureau of Standards but carry around with us as a standard a memorized sequence of numbers which can, when the occasion demands it, be reduced to a sign language which all the world understands.

Perhaps the greatest usefulness of counting is the resulting ability to compare the quantity of one thing with the quantity of another thing or perhaps the quantity of one thing at two different times. Given two sets of five people we should know at a glance[1] that the number of people in one set is the same as the number in another set. But probably with even as few as ten people, and certainly with a larger collection, it would in general be easier to count each set and compare the numbers (that is, measure the quantities against a common standard) than to try to establish a direct one-to-one correspondence between the members of one set and the members of the other. Certainly the easiest way to compare the number of hens in a coop at one time with the number at another is to count them at the two times.

Possibly the most fundamental property of numbers is that *the number of objects is the same whatever the order in which (or the method by which) the one-to-one correspondence is established.*[2] If the pigeons in a cage were counted by catching each in turn and labeling them with successive numbers, the count would be the same whatever were the order in which they were caught. As a matter of fact, it is precisely this fact which underlies our counting of a set of objects in two different orders (or without regard to the order), as a means of checking one count against the other. It can, of course, be pointed out that

[1] See the story of the crow, reference **16**.
[2] See reference **38**, Chap. II.

when we are counting we do not distinguish between one object and another—they are all alike as far as the enumeration is concerned—and hence the order makes no difference, but that is not really a proof of the fundamental property. To attempt to prove it would then require the proving of something else:

> Great fleas have little fleas upon their backs to bite 'em,
> And little fleas have lesser fleas, and so *ad infinitum*.
>
> —De Morgan

One might just as well make the best of the first flea and not run the risk of calling up all "his sisters and his cousins and his aunts." We might, for instance, be foolish enough to attempt to prove that the sun will rise tomorrow. We could quote evidence to the effect that the sun has risen for the last few days and assert that there have been no signs of the earth slowing down or of the approach of a threatening heavenly body. But by the time we had completed our argument we should have made many assumptions much harder to believe or understand than is the simple assumption that the sun will rise tomorrow. So we accept without attempt at proof the statement that *the number of a set of objects is independent of their order*. It would be a queer world if this were not so.

This property of number is a very unusual one. In most respects the order in which things occur makes a great difference. To exchange the quarterback and center on a football team would usually alter the team considerably, although it would still contain eleven men. The batting order of a baseball team is a vital matter. The coat and shirt in either order are *two* articles of clothing, but a certain order is to be recommended in putting them on. Persons not under the influence of alcohol usually undress *before* taking a bath. On the other hand, in throwing a seven with two dice, it is of no particular importance which die hits the table (or the ground) first. Nor does it matter in a game of bridge whether the first or the thirteenth card dealt you is your lone ace. Indeed, it is almost paradoxical that, while a number system is an ordered set of marks, its chief use is in counting, which, in large degree, is itself independent of order.

These numbers which we get by counting things and write as 1, 2, 3, 4, . . ., are called *positive integers* (they are sometimes called *natural numbers*). (Three dots mean "and so forth"—a dot for each word.) The necessity for such a name instead of the mere appellation "number" is that later on we shall want to designate as numbers ideas which cannot be obtained by counting things, although they share many properties with the positive integers. Why we call them "positive integers" instead of "glumpers" or any other name can best be explained when we come to discuss other kinds of numbers.

Indeed, there is one new "number" not obtainable by counting, which it is convenient to introduce immediately: the number *zero*, whose symbol is 0. In our younger days we quite often read 0 as "nothing," and in many circumstances that is quite a proper meaning. (Reading it "naught" is merely another way of saying the same thing.) As a matter of fact, this is the only sense in which, for a long time, the number 0 was used. Just as I can tell you how many troubles I have, if I have some, by mentioning a number which is a positive integer, I find it convenient to use the number zero if I have none. As far as counting goes, zero means "none," and although we shall find later that this idea of zero is not adequate in some situations, we shall climb that hill when we come to it. One cannot, strictly speaking, get zero by counting things and hence it is not a positive integer; but it is a number with many of the same properties, as we shall see. In place of the long phrase "positive integer or zero" we shall often use the word "number" in this chapter; but we shall be careful to make no statements about "numbers" which are not true for all numbers we consider in this book, even though our proofs and discussions are solely in terms of positive integers and zero.[1]

The elements in many sets may be counted. We might find that there were twenty people in a room. Then the number 20 would be associated with the set of people in the room. If we let P stand for the set of people in the room, we shall denote the number of people in P by $n(P)$. We call this "the number of P." If H stood for the number of heads in the same room, we would have $n(P) = n(H)$—that is, the numbers of the two sets are the same. This does not mean that the set of heads is the same as the set of people but merely that the number associated with each set is the same. *The number of a set is another property of the set which does not depend on the kinds of things which its elements are nor on the order in which we look at them.*

Some sets which we cannot literally count have a number. If S stands for the set of grains of sand in the sea, we have no way of determining $n(S)$, but we know there is a number which describes how many elements are in S. A set is called a *finite set* if it is has a number, that is, if there is a natural number $n(S)$ which describes how many elements are in S. We might agree that $0 = n(\varnothing)$; that is, zero is the number of the null set. But there are sets which have no number associated with them and which are therefore called *infinite sets*. One such set is the set, I, of all integers. The set of points on a line is an infinite set. In fact, the points on any line segment, no matter how short, form an infinite set.

In this connection it is interesting to see that there is a one-to-one correspondence between the points of a line and the points (exclusive of the end points) of a semicircle. In Fig. 3:1 we have a semicircle with center at O and end points A and B so that the line AB is parallel to the line l. We set up the

Figure 3:1

[1] See topics 1, 2, and 3 at the end of the chapter.

one-to-one correspondence in this way: If P is any point on the line, we draw the line OP and where it intersects the semicircle is the point P', which corresponds to P. Also if P' is any point on the semicircle except A and B, the line OP' determines a point P on the line. Hence, except for the two points A and B, we have a one-to-one correspondence between the points on the semicircle and the points on the line.

EXERCISES

1. In which of the following does order play an important part, and in which is it a matter of indifference? If in any case the answer is "it depends," argue the pros and cons: the courses of a meal, the drum major and his band, the items of a shopping list, the names in a telephone directory, marks in the six tests of a course, painting the various parts of a floor, writing to one's three sweethearts, the hen and the egg, putting gasoline and water into a car, mixing the ingredients of a cake.

2. Give two not too obvious examples in which order does, and two in which it does not, make a difference.

3. The word "pair" is not applied to *any* two objects but to two which are more or less alike—a cow and a horse would not constitute a pair nor would even a road horse and a bay horse. The term "sextet" is generally associated with music. Give five other examples of such restricted number names.

4. It is stated above that 0 does not always mean "nothing." Give three examples to support this statement.

5. If the number of people in one set is less than the number in another, what precisely does this mean in terms of a one-to-one correspondence?

6. Using the results of Exercise 5 show that if the number of people in set A is less than the number in set B, and if the number in B is less than the number in C, then the number in set A is less than the number in set C.

7. What kind of correspondence describes the phrase "twice as many" in a manner analogous to the way in which one-to-one correspondence describes "just as many"? State precisely what you mean by this new kind of correspondence and give two examples of it.

8. If $n(P)$ stands for the number of people in a room and $n(E)$ for the number of eyes in the room, what will be the relationship between the two numbers?

9. Suppose we call two sets "equivalent" if they are finite sets and have the same number. Replace $\subseteq$ by "equivalent to" in the three properties listed in Section 5 of Chapter 1 and find which of these properties hold.

10. Give two examples of infinite sets not given above.

11. Explain why the union of two finite sets is itself a finite set. Must the intersection of two finite sets be a finite set?

12. Give an example of a case where the intersection of two infinite sets is a finite set. Can the union of two infinite sets be a finite set?

13.   Can the union of a finite and infinite set be finite in some cases and infinite in others?

14.   Can the intersection of a finite and an infinite set be finite in some cases and infinite in others?

15.   Show that there is a one-to-one correspondence between the even integers (0, 2, 4, 6, 8, . . .) and the integers (0, 1, 2, 3, 4, . . .).

16.   In Exercise 15 we have an example of a one-to-one correspondence between the elements of a set I, the set of integers and zero, and a proper subset E of I. Here I and E are infinite sets. Can there be a one-to-one correspondence between the elements of a finite set and a proper subset?

2. THE COMMUTATIVE AND ASSOCIATIVE PROPERTIES OF ADDITION

One of the first things we found after learning to count, and probably knew in a practical way before we could count very far, was that if you have one apple and I have two, then we both together have three; and the same is true if we have donkeys or elephants. That was the beginning of that onerous but so useful process of addition. One of its properties that came into our consciousness later but probably without much effort was the fact that it makes no difference in which order the addition is performed. One and two have the same sum as two and one. This fact follows immediately from the fundamental property mentioned in the last section. If we let dots represent the objects as follows:

•   • •

since the order of counting makes no difference, the result of counting from the left (one and two) is the same as that of counting from the right (two and one). In other words, the set is the same whether you look at it from left to right or right to left. There is, to be sure, something a little new in grouping the objects, but fundamentally it is this old question of order again. The same device can be used for any two sets of objects to establish the property that *the order of addition of two numbers makes no difference to the sum.* We express this briefly by saying: *Addition is commutative.* In symbols this amounts to writing

$$a + b = b + a$$

when a and b are any positive integers. This law or property holds even if one or both of the numbers of the sum is zero, since, for example, if you have three dogs and I have none, we both have three and $3 + 0 = 3$; and the total number of dogs is the same if I have three dogs and you have none. Also, if each has none we both together have none.

What do we really mean by addition? It can be described in terms of sets. For example, $3 + 5 = 8$, because if we have one set with three elements and another with five and if they have no elements in common, then together they have eight elements. In general, if we have two finite sets S and T whose intersection is the null set and if we denote the number of elements in S by $n(S)$ and similarly for the other sets, we have

$$n(S) + n(T) = n(S \cup T) \qquad \text{when } S \cap T = \varnothing.$$

We can regard this as a *definition of addition of integers*. Also

$$n(T) + n(S) = n(T \cup S).$$

But we know from the properties of sets that $S \cup T$ and $T \cup S$ are the same set and hence $n(T \cup S) = n(S \cup T)$, which implies

$$n(S) + n(T) = n(T) + n(S)$$

for all finite sets S and T. This shows the commutative property.

As soon as we begin to count objects in more than two sets we have another property, which again has something to do with order but in a different way from the commutative property considered above. If we have dots in the sets

•   • •   •   •   •

we can count them first by associating the first two sets whose combined number is three and, considering them as one set of three, add this set of three to the remaining one of four and get the total number seven. If, on the other hand, we had first associated the last two sets and considered them as one set of six dots, then, adding this number to the number in the first set we have the sum seven—the same number as before. This is not particularly surprising, but it is an important property or law of addition. So, we say, in brief, that *addition is associative*. This property may be expressed as follows:

$$(a + b) + c = a + (b + c),$$

where the left side means: Add a and b and then add c to the result. The right side means: Add b and c and then add this sum to a. The parentheses are used to indicate what is done first. Notice that the order in which the letters appear is the same on both sides of the equation, the sole difference lying in the placing of the parentheses.

This is the property which enables us to find by addition the total number of dots in four sets, for it reduces the work to computing three sums of two numbers each. It is probably true that we accept this property a little more easily than the commutative property—perhaps because we do not come upon it until a later stage in our development. Nevertheless, associativity is not by any means a universal property. Thus, in making iced tea we "add together" tea leaves, hot water, and ice, but (tea leaves) + (hot water) = tea and (tea) + (ice) = (iced tea), while (hot water) + (ice) = tepid water and (tea leaves) + (tepid water) is not very good. But in addition, altering the order of association of the parts does not change the results. Fundamentally we can never add more than two numbers at a time, although, with practice, we can sometimes almost immediately get the sum of three or more. This property which we have just illustrated gives sense to the question: What is one and two and four? While the question may mean one of two things (is it one added to the sum of two and four, or the sum of one and two added to four?), there is no ambiguity in the result. Notice that, in contrast to commutativity, which has to do with interchanging the order of addition of *two* numbers, associativity had to do with selecting a pair from a set of *three or more* numbers.

There are two properties of addition of integers which are so obvious that we would not need to mention them except that we shall be meeting them later in less familiar surroundings. The first is the property: *If a and b are in the set of positive integers and zero, then a + b is in the same set.* This is called the *closure property.* You are asked to show this in an exercise below.

We also know that

$$0 + a = a + 0 = a$$

whatever positive integer *a* is or, in fact, even if *a* is zero. Since the number 0 has this property, we call it the *additive identity.*

EXERCISES

1. Form an addition table of all pairs of numbers from 1 to 10 making full use of the fact that addition is commutative.

2. Which of the following are commutative? If the answer is "it depends," state the pros and cons: taking out insurance and having an accident, shooting the right and left barrels of a shotgun, the cart and the horse, the locking of the barn and the coming of a thief, dusting the furniture and mopping the floor, clipping the inside or outside of a hedge, putting a stamp on an envelope and sealing it? Give reasons.

3. Show that the set of positive integers and zero has the closure property described above.

4. If each of *a* and *b* is zero or a positive integer what can you conclude from $a + b = 0$? Give reasons.

5. Show the associative property of addition of natural numbers using sets.

6. Using sets, show that $0 + a = a + 0 = a$.

7. When we add a column of numbers from top to bottom and check our addition by adding from the bottom upward, what properties of addition are we assuming? Why?

8. Show that if *S* and *T* are two finite sets, then

$$n(S \cup T) = n(S) + n(T) - n(S \cap T).$$

9. In a certain room there are 25 blond persons, 16 girls, and 10 blond girls. Everyone in the room is either blond or a girl or both. How many persons does the room contain?

*10. Find an expression analogous to that in Exercise 8 for

$$n(S \cup T \cup U).$$

*11. (Reference 10, p. 325) Would you believe a market investigator who reports
      that, of 1000 people 816 like candy, 723 ice cream, 645 cake, while 562 like
      both candy and ice cream, 463 both candy and cake, 470 both ice cream and
      cake, and only 310 like all three?

3. THE COMMUTATIVE AND ASSOCIATIVE PROPERTIES OF MULTIPLICATION

First let us look at some of the notation we have used and will find useful.
Parentheses work from the inside out like the ripples from a pebble thrown
into a quiet millpond. For instance,

$$[(a + b) + c] + d$$

means: Add a and b, add this result to c, add that result to d. Here, owing to
the associative property of addition, it makes no difference which is done first,
but we shall soon find cases in which it does make a difference.

Sometimes, multiplication is indicated by an $\times$ as follows: 3×2. Since
this sign might be confused with the letter "x" we shall usually in this book
use the raised dot to indicate multiplication, as in $3 \cdot 2$. Sometimes we can
omit the dot without ambiguity. For instance, instead of $3 \cdot (2 + 4)$ we write
$3(2 + 4)$; instead of $3 \cdot b$ we write $3b$; instead of $a \cdot b$ we write ab. However,
$(7 \cdot 8)(2 + 4)$ is quite different from $(78)(2 + 4)$.

At the outset, notice that if each of three persons has five peanuts we could
find by counting that altogether they had fifteen peanuts. When the three pool
their resources, the five peanuts are multiplied by three. This is a concrete
instance of the abstract statement that three times five, or five *multiplied by*
three, is fifteen. We write it $3 \cdot 5 = 15$. When we learned the multiplication
table we merely memorized numerous results of such multiplications so that
once we knew that each of three persons had five peanuts, we deduced
immediately the fact that there were fifteen peanuts in all; counting in this
case became unnecessary. We also learned that five times three is fifteen. But,
since five men with three peanuts each look different from three men with
five peanuts each, it was probably some time before we realized that it was any
more than a coincidence that the total number of peanuts in one case was the
same as that in the other. We begin to see that this is indeed more than a
coincidence when we consider the diagram

$$
\begin{array}{ccccc}
\bullet & \bullet & \bullet & \bullet & \bullet \\
\bullet & \bullet & \bullet & \bullet & \bullet \\
\bullet & \bullet & \bullet & \bullet & \bullet \\
\end{array}
$$

There are three rows of five dots each—three times five; looking at it another
way, there are five columns of three each—five times three. The number of

dots is the same whichever way you look at it. We could do the same thing if there were 20 rows of 16 dots each. In fact, if we let *a* stand for some positive integer and *b* for some other (or the same) positive integer, *a* rows of *b* dots each present the same picture as *b* columns of *a* dots each; in other words, *a* times *b* is equal to *b* times *a* no matter what positive integers *a* and *b* are. Hence we say that *multiplication of positive integers is commutative*. Briefly, we may then write

$$ab = ba,$$

where *ab* is understood to mean *a* times *b* and *ba* is *b* times *a*. The fact that multiplication by zero is also commutative will appear below.

It should be seen that our diagram of dots does something which could be accomplished by no amount of comparing the product of a pair of numbers in one order with that in the other order. No matter how many examples one gives to show that one number times another number is equal to the other times the one, one could not be sure but that some untried pair might fail to have this property. But we can establish the conclusion for *every* pair of numbers by seeing that *no matter what pair* we have, we can, in imagination at least, make a diagram of dots to fit these numbers and by counting the dots in two different orders establish our result. A similar situation outside of mathematics is the following. Suppose I wished to prove that I can *always* distinguish between alcohol and water. You might try my knowledge on samples of one or the other until doomsday without *proving* that I could distinguish between them, although every successful example would lead you nearer to admitting my powers of discrimination. But if I told you that my test would be trying to light the liquid with a match, you would then know that *whatever* sample was presented to me, I could tell whether it was alcohol or water; furthermore, this could be established without a trial. Thus, while we often use examples to lead one to suspect an underlying reason, it is the latter and not the examples which proves that something *always* holds.

It is interesting to notice that the number 1 plays the same role for multiplication that the number 0 plays for addition, for, no matter what number *b* is,

$$b + 0 = b \quad \text{and} \quad b \cdot 1 = b$$

are both true. Just as 0 is called the identity element for addition (the additive identity), so 1 is called the identity element for multiplication (the multiplicative identity).

Multiplication of positive integers is also associative. Imagine three trays of glasses and suppose in each tray the glasses are arranged in the following pattern:

Consider the trays piled vertically. To find the total number of glasses we can count those in one tray by multiplying 4 by 5, getting 20; thus we have 20 stacks of glasses with 3 in each stack. This method of getting the total number could be indicated by $(4 \cdot 5)3 = 60$. The parentheses indicate that it is the product of 4 and 5 which is multiplied by 3; that is, we multiply 4 by 5 and then multiply the result by 3. There is another way in which we could count the glasses. Looking at the pile from the front we should see 5 columns of 3 glasses each—hence 15 glasses. From our knowledge of the piles we should know that there were 4 such "slices," hence 4 fifteens. This method could be indicated by $4(5 \cdot 3) = 60$. Again we are merely associating numbers in a different way. It is interesting to notice, by the way, that we can write the numbers in the same order but arrange the parentheses in a different way. Looking at the pile from the side would give us a combination of the commutative and associative property: $(4 \cdot 3)5 = 60$.

Briefly, the associative property of multiplication states that

$$(ab)c = a(bc).$$

That this product is also equal to $(ac)b$ results from the associative and commutative properties. This property of associativity does for multiplication just what it did for addition: It allows us to multiply as many numbers as we please by finding the product of any two, then the product of this result with another, and so on. Hence there is no ambiguity so far as the result is concerned in writing $4 \cdot 5 \cdot 3 = 60$ without specifying which pair is multiplied first. Therefore, in the product of three or more numbers we need no parentheses.

It is sometimes convenient to use braces { } and brackets [ ] for, although there is no gain in clarity in the expression $(5 + 6)[7 + 8]$ over $(5 + 6)(7 + 8)$, some confusion is likely to result when parentheses occur within parentheses. For instance, in the expression

$$N = 4((3 \cdot (5 + 6) + 7) \cdot 2 + 6)$$

it is a little awkward to pair the parentheses correctly, although, if one works from the inside out, there is no real ambiguity. It is, however, clearer to write it in the form

$$N = 4 \cdot \{[3 \cdot (5 + 6) + 7] \cdot 2 + 6\}.$$

We then work from the inside out, as follows:

$$\begin{aligned} N &= 4\{[3 \cdot 11 + 7] \cdot 2 + 6\} \\ &= 4\{40 \cdot 2 + 6\} \\ &= 4 \cdot 86 \\ &= 344. \end{aligned}$$

Also note that the product of two positive integers is a positive integer. We say: The set of positive integers is *closed under multiplication.* The number 0 plays a very special role with regard to multiplication. Just as the union of any number of null sets is still the null set, so

$$0 + 0 + \cdots + 0 = 0$$

no matter how many zeros there are. Hence *define* multiplication by zero as follows:

$$a \cdot 0 = 0 \cdot a = 0$$

for every number *a*, including zero. In view of this definition, the set of positive integers and zero is also closed under multiplication.

EXERCISES

1. Find the value of $[(2 \cdot 3)(4 + 5)] + 6$.
2. Find the values of
   a. $\{(2 + 5)(6 \cdot 2)\} + 1$.
   b. $[3(2 + 5)] + [(3 + 6) + 7]$.
   c. $\{(2 + 6)(7 + 5)\} + 3$.
3. Find the value of
   a. $[1 \cdot (2 + 3) + 1] \cdot 0 + 7 + 2 \cdot 5$.
   b. $2 \cdot 4 + 3 \cdot 4 + 5 \cdot (4 + 7)$.
   c. $\{1 \cdot 0 + 0\} \cdot \{0 + 7 \cdot 6\}$.
4. Show that $5(7 + 6) = (5 \cdot 7) + (5 \cdot 6)$
5. Express in words the following:
   a. $[(a + b) + c] + e$.   b. $[c(ab)]d$.   c. $[(xy) + z]t$.
6. Write the following as an equation using parentheses: *a* times the sum of *b* and *c* is equal to the sum of *a* times *b* and *a* times *c*.
7. What parentheses in Exercises 2 and 5 does the associative property of addition and multiplication render unnecessary?
8. If $a + c = a$, what must *c* be?
9. Show that if *a* and *b* are positive integers, then *ab* is not zero.
10. Suppose each of *a* and *b* is zero or a positive integer; what can you conclude from $ab = 0$?
11. If *a*, *b*, and *c* are numbers in the set of positive integers and zero and if $(a + b)c = 0$, what information does this give about *a*, *b*, and *c*?
12. In the discussion of properties of multiplication, no mention was made of sets. Why would the use of sets be awkward in this connection?

4. THE DISTRIBUTIVE PROPERTY

So far, the properties which we have considered have applied equally well to addition or to multiplication but to only one at a time. One must use care in combining addition and multiplication. For instance, $(4 \cdot 5) + 3$ is not equal to 4(5 + 3), as can be seen by performing the two operations. One cannot, then, combine multiplication and addition in a kind of associative law. But there is a property of numbers which involves both addition and multiplication. If we add 2 and 3 and multiply the sum by 6, we get the same result as if we had multiplied 2 by 6 and multiplied 3 by 6 and added the two products. We can write this as follows:

(1) $(2 + 3)6 = (2 \cdot 6) + (3 \cdot 6).$

This can be shown diagrammatically, as were the other properties, for any three numbers, and this is left to the reader. Such a diagram is important in seeing that such a process works for any three numbers—this is very different from finding for any like expression that the result of the work on one side of the equality for the particular numbers used is the same as the result of the work on the other side. This property is described by saying that *multiplication of numbers is distributive with respect to addition*, and it means that for any three numbers *a*, *b*, and *c*,

(2) $a(b + c) = (ab) + (ac).$

The reason for the terminology of the phrase in italics is that there is a kind of distribution of the multiplication throughout the addition. This may seem to be a rather awkward way of expressing the property, but it is doubtful whether there is any really shorter way of accurately describing it. Even though we briefly refer to this as the *distributive property*, the longer designation is important, for it emphasizes the relative roles of addition and multiplication. If addition were distributive with respect to multiplication we could interchange + and · in equation (1), which would give us

$$(2 \cdot 3) + 6 = (2 + 6)(3 + 6),$$

that is,

$$12 = 72,$$

which is false. In fact, *addition is usually not distributive with respect to multiplication.*

Here is an example outside of mathematics for which the distributive property does not hold. Suppose we let "touching" play the role of multiplication and "combining chemically" the role of addition. Corresponding to the left-hand side of equation (2), we might then have touching a lighted match to a combination of hydrogen and oxygen, that is, to water; corresponding to the right-hand side we should have touching a lighted match to hydrogen, touching it to oxygen, and combining chemically the results. The end result in the two cases would be very different.

This property is the hardest of the three to comprehend, since it does not appear on the surface of our everyday operations with numbers. That it really underlies our mechanical processes can be seen by examining our process of multiplying 536 by 7. Dissected, it might look like this, with the reasons indicated briefly in the right-hand column.

(i)	$536 \cdot 7 = (500 + 30 + 6)7$	Decimal notation
(ii)	$= (500 \cdot 7) + (30 \cdot 7) + (6 \cdot 7)$	Distributive property
(iii)	$= (500 \cdot 7) + (30 \cdot 7) + (40 + 2)$	Multiplication
		Decimal notation
(iv)	$= (500 \cdot 7) + [(30 \cdot 7) + 40] + 2$	Addition associative
(v)	$= 3500 + 250 + 2$	Multiplication and addition
(vi)	$= 3500 + 200 + 52$	Addition associative
(vii)	$= 3700 + 52$	Addition associative
(viii)	$= 3752$	Addition

Notice that the "carrying" part of the process consists in getting equations (v) and (vii). Strictly speaking, the associative property of addition enters in all but the last step above, since otherwise the sums would be ambiguous.

At this point let us introduce a little convention which saves writing. To make sure that there is no ambiguity we have written $(3 \cdot 5) + 2$ to certify that 3 is multiplied by 5 and the result added to 2. This has, of course, an entirely different result from that of $3(5 + 2)$, in which we add 5 to 2 first and then multiply by 3. By agreeing always to give multiplication precedence over addition—that is, always multiplying before adding whenever there is no indication to the contrary—we can write $3 \cdot 5 + 2$ to mean $(3 \cdot 5) + 2$. *It is now all the more important that we retain the parenthses when we mean that the addition should be made first.* Using this convention would enable us to omit all the parentheses in (ii), (iii), and (iv) but not in (i).

The distributive property is the source of much of the difficulty in algebra. On this account it is especially important that we establish it firmly in terms of numbers. Notice that it may now be written

$$a(b + c) = ab + ac.$$

EXERCISES

1. Use parentheses to express the associative property of addition for a particular numerical example.

2. In adding a sum of ten numbers, how many sums of two numbers would it be necessary to find? What would be your answer for a corresponding problem in multiplication?

3. If each of a, b, c is a positive integer or zero, show that $(ab)c = a(bc)$, assuming equality for positive integers.

4. Remove all the parentheses you can from the following expressions without making them ambiguous or changing their meaning, and find their values:
    a. $(3 \cdot 5) + (4 + 5) + 3(4 + 5) + (4 + 5)(3 + 4) + (3 \cdot 5)(6 + 2)$.
    b. $(2 \cdot 3) + 2(1 + 7) + 3 + 7 \cdot 6 + 4(3 + 2) + (7 + 5)(6 + 3)$.

5. In the expressions of Exercise 4, replace every multiplication by addition and every addition by multiplication. Then do for these expressions what was required in Exercise 4.

6. In the following expression insert one pair of parentheses in as many ways as possible (there are 15 ways) and evaluate the resulting expressions:

$$1 \cdot 2 + 3 \cdot 4 \cdot 5.$$

How many different results are there? Which are equal to the expression as written without parentheses?

7. Do Exercise 6 with addition and multiplication interchanged.

8. Do Exercise 6 inserting two pairs of parentheses instead of one.

9. In the equation $a(b + c) = ab + ac$ just preceding the exercises, could the parentheses be omitted? Explain the reasons for your answer.

10. Dissect the process of multiplying 827 by 3, giving the reasons for each step and including only necessary parentheses, as was done above.

11. Show that $(2 + 4)(3 + 5) = 2 \cdot 3 + 4 \cdot 3 + 2 \cdot 5 + 4 \cdot 5$. What will be the corresponding expression for $(a + b)(x + y)$?

12. Show that $5 \cdot 3 + 5 \cdot 2 + 5 \cdot 6 = 5(3 + 2 + 6)$ and find the corresponding right-hand side of the following equation:

$$ax + ay + az =$$

13. Pick out all correct right-hand sides for each of the following equations and give reasons for your answer:

(1) $$ax + ay + bx + by = \begin{cases} (a + b)(x + y) \\ a(x + y) + (x + y)b \\ a(y + b) + x(a + y) \end{cases}$$

(2) $$ar + 2a = \begin{cases} r(a + 2) \\ a(r + 2) \\ a(r + 1) + a \end{cases}$$

(3) $$ab + ac + bc = \begin{cases} a(b + c) + bc \\ b(a + c) + ac \\ a(b + c + a) \end{cases}$$

(4) $$P + Pi = \begin{cases} P(1 + i) \\ i(1 + P) \end{cases}$$

(5) $$P(1 + i)^2 + P(1 + i)^2 i = \begin{cases} P(1 + i)^3 \\ P(1 + i)(1 + i^2) \\ Pi + P(1 + i)^3 \end{cases}$$

14. $(3 + 4)8 = 8 \cdot 3 + 8 \cdot 4$. That a like expression would hold for any three numbers would follow from what properties of numbers?

15. Is $1 \cdot 5 + 3 = 1(5 + 3)$? Will this still hold if the 1 is replaced by a 4? For what positive integers a, b, c is $ab + c = a(b + c)$?

16. Use dots to prove the distributive law.

17. Suppose a student computed $3(4 \cdot 5)$ as follows: 3 times 4 is 12, 3 times 5 is 15; hence the desired product is 12 times 15, which is 180. This is wrong. What do you think would have been the source of his confusion?

18. In the first four properties of sets in Chapter 1, replace $\cup$ by $+$ and $\cap$ by $\cdot$

(where · indicates multiplication), and suppose the letters stand for positive integers or zero. Which of the properties would hold for all such numbers? If some property does not hold for all such numbers, state for which numbers it does hold.

19. For what numbers do (ii) and (iii) of property (5) of Chapter 1 hold, when the substitutions indicated in Exercise 18 are made?

20. In the tables below properties (5a) and (5b) of Section in Chapter 1, replace ∪ by +, ∩ by ·, ∅ by 0, and B by 1. To what extent do the tables now hold?

5. THE NUMBER SYSTEM TO BASE TEN

"A rose by any other name would smell as sweet," but not nearly so many people would know what you were talking about if you called it a *Rosa spinosissima*, and to some sensitive souls, a euphonious name even contributes something to the smell. Many will admit that a really good name is always partially descriptive. No matter what you call the numbers, they have the properties we have talked about but it would be incongruous, not to mention other objections, if the names of numbers—the essence of order—were themselves without reason (rhyme is not necessary). One could devise one hundred random names for the numbers from 1 to 100, but it would certainly retard the education of our youth; worse, our conception of how many a hundred men are would be very vague. We can visualize twenty men fairly well and to say that there are five sets of twenty men each, whether or not they are actually bunched in that way, gives a pretty good mental picture of how many a hundred men are. Furthermore, such a grouping would make it much easier to count them. An efficient bank teller does not count your handful of pennies one by one but in sets of five. He thereby not only works more quickly but is surer of his result. There is a disadvantage in having too small a set if one has a large number to count, because then the number of sets mounts. On the other hand, it is awkward to have the set too large. Grouping by tens is natural to us who can wiggle our fingers. The fact that we count by tens and not by twenties may be an argument for the antievolutionists, or it may merely indicate that when an ape had more than ten children she could not be bothered keeping track of them all. However, the appearance of the word "score" is damning and the French come in for their share of suspicion with their *vingt* and *quatre-vingt*. At any rate we can be proud of our ancestors in that they hit on the best way to extend the notion of number by talking of tens of tens and later of thousands of thousands.

Although we think of the method of counting by tens in terms of the numerals we use, there were many other systems of numeration which were used to count by tens. The Romans and others before them counted by tens. It is usually pointed out that their numerals are awkward because it is too difficult to calculate with them, but the Romans probably did not use them for that purpose except for the simplest calculations. For such purposes they used the abacus, that harp of the merchant, the origins of which go back to the fifth and sixth centuries B.C. However, the Hindu-Arabic numerals, which most of the civilized world uses today, have two real advantages over the Roman system. The first is the greater compactness of the symbols for the numbers from 1 to 9. The second, the giving of place value, would be applicable in the case of any nine symbols with zero and came relatively late in the development of our system of notation. In many of the early notations there was no relation between the symbol for the number 7 and that for 70. The Greeks and the Jews wrote for a time 700.80.9 for the number we would write 789. We are so accustomed to this notation of ours that it requires some-times a little effort to recall that the right-hand *digit*,[1] 9, gives the number of units, the one to the left of it, 8, the number of tens, and the one to the left of it the number of hundreds. The symbol 789 is a very elegant way of writing $700 + 80 + 9$. In our notation the use of zero plays an important part in denoting the absence of tens, for instance, in the number 304. This use of zero seems to have been very hard for the human race to invent.

Perhaps we shall appreciate the advantages of our notation more fully if we dissect this process of addition to see how it is made. To find the sum of $12 + 34 + 56$ we write it $10 + 2 + 30 + 4 + 50 + 6$ using the associative property of addition. Since addition is associative and commutative, we can rearrange the terms to have $10 + 30 + 50 + 2 + 4 + 6$. The sum of the last three is 12 or $10 + 2$, and our sum becomes $10 + 30 + 50 + 10 + 2$. Adding the first four numbers we have $1 + 3 + 5 + 1$ tens, which is 10 tens, or 100. Adding this to 2 we have our result, 102. This whole process is, of course, facilitated by writing the numbers in a column:

$$12$$
$$34$$
$$56$$

By reading down (or up) the right-hand line of numbers and then down the left-hand line we have the rearrangement necessary to adding the tens and the units. We add the right line first because it may contribute to the left line. "Carrying the one" is acknowledging this contribution. This is perhaps more clearly brought out by writing the sum in the form which one often uses when adding along a column of figures in order to make checking easier:

$$12$$
$$34$$
$$56$$
$$\overline{}$$
$$12$$
$$9$$
$$\overline{}$$
$$102$$

We say, for example, that the number 3572 has the "digits" 3, 5, 7, and 2.

Here the sum of the right-hand line appears explicitly. The 9, of course, is really a 90, but its position makes the writing of the 0 unnecessary.

To find the product of 4 and 126, we have 4(100 + 20 + 6), which, by the distributive property, is $4 \cdot 100 + 4 \cdot 20 + 4 \cdot 6 = 400 + 80 + 24$, which, by the process of addition, is 504. The connection between this and the usual process can be emphasized by writing it in the following manner:

$$
\begin{array}{r}
126 \\
4 \\
\hline
24 \\
8 \\
4 \\
\hline
504
\end{array}
$$

Zeros are not necessary to indicate the positions of 8 and 4 in the fourth and fifth lines of the scheme and hence are omitted. The process of "carrying the 2" and then the 1 is, of course, a means of using one's head instead of the paper and pencil.

6. THE NUMBER SYSTEM TO BASE TWELVE[1]

Without seeking to change the habit of counting by tens, that is, the use of the *decimal system* (a *number system to the base ten*, as it is sometimes called), we can consider the advantages, from certain points of view, in other systems. It is convenient to have 12 inches in 1 foot, for then one quarter of a foot is 3 inches. This is much more comforting than the situation in the metric system of measurement, for a quarter of a decimeter is $2\frac{1}{2}$ centimeters and we like to avoid fractions. For a similar reason, dividing the hours into 60 minutes is very helpful, for although a division into 100 minutes would give a whole number of minutes for one quarter of an hour, it would not for one third. With 60 minutes in 1 hour, there is a whole number of minutes in each of the following parts: one half, one third, one fourth, one fifth, and one sixth; and who wants to divide an hour into seven parts anyway? There are dozens of times when we find counting by twelve useful. The horses in *Gulliver's Travels*, having three toes on a foot, would perhaps have counted by twelves.

What alterations would be necessary if we had twelve fingers instead of ten? To begin with, we could manufacture an entirely new set of symbols, but that would require much mental labor in devising them and in memorizing the results as well as being a trial to the printer. And, after all, the numbers from one to nine can just as well be represented by the same symbols as before. However, if we are to take advantage of place value and indicate our twelves and twelves of twelves by the position of the symbols, we do need new

[1] See topic 4 at the end of the chapter.

symbols for the numbers ten and eleven. Let us call them t and e. Then our numbers would be:

Base 10: 1, 2, . . . , 8, 9, 10, 11, 12, 13, 14, . . . , 21, 22, 23, 24, 25, . . . ,
Base 12: 1, 2, . . . , 8, 9, t,   e,   10, 11, 12, . . . , 19, 1t, 1e, 20, 21, . . . ,

where now 10 stands for twelve, 11 for thirteen, 20 for twenty-four (being two twelves), and so forth. We could call this a *dozal* (pronounced "duzal") system [1] instead of a decimal system, since in it we count by dozens instead of tens. To avoid confusion between the two systems and to emphasize the connection of the names with the notation, let us call the above list of numbers to the base twelve

one, two, . . . , eight, nine, ten, eleven, dozen, doza-one, doza-two, . . . , doza-nine, doza-ten, doza-eleven, two dozen, two doza-one, ...

In the dozal system 100 is not one hundred but a *gross*, and 1000 is a *great gross*. We shall not attempt to give additional names.

Regarding the dozal system as a new language, it is apparent that we need some means of translating into and from this system. We illustrate the method by the following.

Example 1

How is the number 7te in the dozal system written to the base ten?

Solution. 7te means seven gross, ten doza-eleven. This to the base ten is

$$
\begin{array}{rcl}
7 \cdot 144 & = & 1008 \\
10 \cdot 12 & = & 120 \\
11 & = & 11 \\
\hline
& & 1139
\end{array}
$$

Example 2

The number which is written 1437 in the number system to the base ten is what in the dozal system?

Solution. Just as 1437 means $1 \cdot 10^3 + 4 \cdot 10^2 + 3 \cdot 10 + 7$ (where 10^3 means $10 \cdot 10 \cdot 10$, and so on) when written to the base ten, so if we write it in the form

$$a \cdot 12^3 + b \cdot 12^2 + c \cdot 12 + d,$$

it becomes *abcd* to the base twelve, where *a*, *b*, *c*, and *d* are properly chosen numbers from 0, 1, 2, . . . , 9, t, e. Now $12^3 = 1728$, which is greater than 1437 and which shows that $a = 0$. The greatest multiple of 144 less than 1437 is

---

[1] A more erudite term is "duodecimal system."

$1296 = 9 \cdot 144$. Hence $b = 9$. We then have $1437 = 9 \cdot 144 + c \cdot 12 + d$. Hence c is the number e, d is 9, and the given number is 9e9 when written to the base twelve. This process may be abbreviated as follows:

$$
\begin{array}{ll}
1437 & \\
1296 & \quad 9 \\
\hline
141 & \\
132 & \quad e \\
\hline
9 & \quad 9
\end{array}
$$

There are two methods of addition in the dozal system and they are about equal in difficulty. To add 9 and 8 first we could say, "9 plus 8 is seventeen, which is 5 more than twelve and hence is doza-five"; second, we could say, "9 lacks 3 of being a dozen and 8 minus 3 is five, which implies that the sum is doza-five." On the other hand, in multiplication we must either learn a new multiplication table or translate back and forth each time between the two number systems. Unless one is going to perform many calculations in the dozal system, the latter is the easier alternative. Then, for example, 9 times 8 is seventy-two, which is six dozen.

Let us see how addition in the dozal system goes.

$$
\begin{array}{l}
t7 \\
6e \\
et \\
\hline
\end{array}
$$

Seven and eleven are doza-six and ten is two-doza-four (ten being six and four). At the foot of the right-hand column put a 4. Carry two and ten is dozen and six is doza-six and eleven is two doza-five. Hence the sum is 254; two gross, five doza-four. To check our result we could translate the numbers to the decimal system and add as follows:

$$
\begin{array}{r}
127 \\
83 \\
142 \\
\hline
352
\end{array}
$$

$352 = 2 \cdot 144 + 5 \cdot 12 + 4$, which checks with our result in the dozal system. One easily discovers certain aids analogous to aids in the decimal system. For instance, to add eleven to any number increases the dozas by one and decreases the units by one; also it is convenient to group numbers whose sum is dozen whenever possible.

Multiplication is a little more difficult. For example,

$$
\begin{array}{r}
5e8 \\
7 \\
\hline
35^6 9^4 8
\end{array}
$$

Seven times eight is four doza-eight, seven times eleven plus four is six doza-nine, seven times five plus the six (carried) is three doza-five in the first two places.

7. NUMBERS AND NUMERALS

We have seen from the above, and we shall have more examples below, of different ways of writing a number. Sometimes it is important to distinguish between the symbol and the number it represents. In that case one often uses the word "numeral" for the symbol. For instance, 12 is the numeral for twelve in the decimal system and 10 is the numeral for twelve in the number system to the base twelve. The numeral 12 is composed of two numerals: 1 and 2. A digit is a numeral which contains just one symbol. In most cases, however, it is clear from the context whether it is the symbol or the number which is being considered. When we speak of "the number 6" we of course mean "the number which the symbol 6 represents," but there is no real ambiguity in the shorter expression.

Strictly speaking, we should write "numeral system" instead of "number system," since in the system to the base twelve, for instance, the numbers are the same; we merely have a different way of writing them. For instance, 68 in the decimal system represents the same *number* as 58 in the system to the base twelve. On the other hand, the integers are not the same *numbers* as the rational numbers; they form different *number* systems. But with this warning we continue to use the term "number system" for both, because it should cause no confusion.

EXERCISES

1. Write out the multiplication of 234 · 56 in such a way as to emphasize the properties of numbers used in deriving the result

2. Are the properties or laws of addition and multiplication mentioned in Sections 2, 3, and 4 the same for the dozal system as for the number system to the base ten? Give reasons.

3. The following are numbers written in the dozal system. Write their expressions to the base ten and check you results.
   - **a.** eee.   **b.** t5e.   **c.** tete.

4. Do what is asked in Exercise 3 for
   - **a.** ttt.   **b.** 3et.   **c.** 7e7e.

5. The following are numbers written to the base ten. Write them to the base twelve and check your results.
   - **a.** 327.   **b.** 4576.   **c.** 17560.

6. Do what is asked in Exercise 5 for
   - **a.** 531.   **b.** 7432.   **c.** 10576.

7. Each of the numbers in the following sums are written in the dozal system. Find the value of each sum in this system and check by translating the given numbers and the answer into the decimal system.

	a.	**b.**	**c.**
	t6e	7te	95e
	420	80	671
	6e1	t19	eee
	—	—	—

8. Do what is asked in Exercise 7 for

	a.	**b.**	**c.**
	ttt	3e4	516
	42e	90	t3t
	91t	1201	3t3
	—	—	—

9. Find the following products of numbers written in the dozal system. Check as in Exercise 8.

    a. 3t4 · 23.   **b.** eee · te.   **c.** 6et · 51t.

10. Do what is asked in Exercise 9 for

    a. 315 · te.   **b.** ttt · 13.   **c.** 70t · ee.

11. What is the greatest integer in the dozal system less than a great gross? Find its value in the decimal system.

*12. In the decimal system, every number which is a perfect square and whose last digit is 0 has 00 as its last two digits. Find a number in the dozal system which is a perfect square, whose last digit is 0, but yet whose next-to-last digit is not 0.

8. NUMBER SYSTEMS TO VARIOUS BASES[1]

The number system in which addition and multiplication are easiest is the *binary* or *dyadic system*, or number system to the base two. The multiplication is as follows:

$$0 \cdot 0 = 0 \cdot 1 = 0, \qquad 1 \cdot 1 = 1,$$

and addition becomes

$$0 + 0 = 0, \qquad 0 + 1 = 1, \qquad \text{and} \qquad 1 + 1 = 10.$$

Since

$$19 = 1 \cdot 2^4 + 0 \cdot 2^3 + 0 \cdot 2^2 + 1 \cdot 2 + 1,$$

this number to the base two is written 10011. Similarly, 9 is 1001. Adding, we have

$$\begin{array}{r} 10011 \\ 1001 \\ \hline 11100 \end{array}$$

[1] See topic 4 at the end of the chapter.

which we may check by translating back to the decimal system. Then our number becomes

$$0 + 0 \cdot 2 + 1 \cdot 2^2 + 1 \cdot 2^3 + 1 \cdot 2^4 = 28.$$

Multiplication is equally simple

$$
\begin{array}{r}
10011 \\
1001 \\
\hline
10011 \\
10011 \\
\hline
10101011
\end{array}
$$

which may be checked by translation. We, of course, pay for all this ease by increasing the writing involved in exhibiting the numbers. The seemingly prodigious result of our multiplication is only 171 when written to the base ten.

At the other extreme, 1 million is written 100 in the number system to the base one thousand, but a multiplication table with $999 \cdot 999$ entries is enough to make the stoutest hearts quail. Our number system is a reasonable compromise: The multiplication table is not too long and small numbers do not take too much writing.

The number system to the base two enters very usefully into numerous games and tricks. In Section 9 we shall discuss in detail one such game. A rather interesting trick involving the binary system is the following.[1] Construct four cards as follows:

1				2				4				8			
1	3	5	7	2	3	6	7	4	5	6	7	8	9	10	11
9	11	13	15	10	11	14	15	12	13	14	15	12	13	14	15

You as the "magician" ask a friend to think of a number from 1 to 15 and pick out the cards on which it occurs. You then tell him what number he is thinking of. You perform this miracle merely by adding the numbers at the top of the chosen cards. For instance, if he presents you with the cards headed 1, 2, and 4, the number which he was thinking of must have been $7 = 1 + 2 + 4$. The secret of the construction of this table can be seen by writing the numbers 1 to 15 in the binary system.

EXERCISES

1. Write 345 in the number system to the base two, to the base five, and to the base fifteen.

[1] See topic 5 at the end of the chapter.

2. Write 627 in the number systems to the bases two, five, and fifteen.

3. Extend the trick given in this section to five cards and the numbers 1 to 31.

4. What is the connection between the choice of numbers on the cards and writing them to the base two?

5. What in the number system to the base seven would be the number 51 in the decimal system? What would it be in a system to the base two?

6. Write the numbers 36 and 26 of the decimal system in the number system to the base two, add them in this number system, and check your result by expressing the sum in the decimal system. Do the same for the product.

7. Give a test for divisibility by 8 for a number in the binary system.

8. Give a test for divisibility by 3 in the number system to the base nine.

9. The numbers 55 and 26 are written in the number system to the base seven. Find their squares in the same system.

10. What would be some advantages and some disadvantages in a number system to the base eight?

11. What in the decimal system is the number 4210 as written in the number system to the base five?

12. Notice that $34^2 - 12^2 = (34 + 12)(34 - 12)$, where every number is in the decimal system. Would the same result hold if these are numbers written in the system to the base five? Why or why not?

13. Notice that $3^2 + 4^2 = 5^2$, where every number is in the decimal system. Would the same result hold if these numbers are written in the system to the base twelve? Answer the same question for $5^2 + 12^2 = 13^2$. Give reasons for your answers.

14. $(17 - 11)^2 = 17^2 - 2 \cdot 11 \cdot 17 + 11^2$, where every number is in the decimal system. Would the same result hold if these numbers were written in the system to the base twelve? Why or why not?

15. The following table indicates a method of converting 101 from the decimal system to the binary system. In the first line we give the quotients for successive divisions by 2 and in the second line the remainders. The conclusion is that 101 written in the binary system is 1100101. Explain why this works:

Quotient	50	25	12	6	3	1	0
Remainder	1	0	1	0	0	1	1

16. Use the method of Exercise 15 to write 101 in the number system to the base three.

17. Devise a trick analogous to that using the four cards above, where now the number system to the base three is involved instead of the binary system and the cards serve to select some number from 1 to 26.

18. Another interesting application of the number system to the base two is contained in the following device, which reduces all multiplication to multiplication and division by 2. It is a method used by some Russian peasants in very recent times. For example, to find $37 \cdot 13$, write

13	37
6	74 ×
3	148
1	296
	481

The left column is obtained by dividing by 2, disregarding the remainder, and the right by multiplying by 2. In forming the sum, 74 is omitted because it occurs opposite the even number 6. Why will this process always work?

19. Determine a set of six weights which make possible weighing loads from 1 to 63 pounds if the weights are to be placed in just one pan of the scales.

*20.[1] Prove that each number can be represented in exactly one way in the following form:

$$a + 3b + 9c + 27d + \cdots,$$

where each of a, b, c, and d is 1, 0, or -1.

*21. Determine a set of four weights which make possible weighing loads from 1 to 40 pounds if weights can be placed in either of the pans. How can you generalize this problem?

9. NIM: HOW TO PLAY IT

An interesting application of the number system to the base two is the system for winning a game called Nim. The theory of this game is complete, in that if one knows the system he can always win unless his opponent (either by accident or design) also follows the system, in which case who wins depends solely upon who plays first. The story is told of a well-known intellectual of doubtful morals who, on meeting on shipboard an unsuspecting sucker in the form of one who thought he knew all about the game, used his knowledge so artfully as to completely finance his trip abroad. Let the reader take warning.

In the simplest form of the game, a certain number of counters are put into three piles or rows. To play, one draws as many counters as he wishes from a pile; they must all be from the same pile and he must draw at least one. There are two contestants, they play alternately, and the winner is the one who draws the last of the counters. By way of illustration, suppose at a stage in the game there remain only two piles of two counters each. If it is A's turn to play he loses for if he draws a complete pile, B will draw the remaining pile; if he draws one counter from one pile, B can draw one from the other pile and in his following turn A will have to draw one leaving just one for B. For the sake of brevity we can call 0 2 2 a *losing combination*, since whoever is faced with that array will lose if his opponent plays correctly. On the other hand, 1 1 1 is a *winning combination*. In general, if it is a contestant's turn to play and the situation is such that he can win by playing properly, no matter how his

---

[1] See topic 6 at the end of the chapter.

opponent plays, we say that he is faced with a *winning combination*; if, on the other hand, it is true that, whatever his play, his opponent can win by playing correctly, we call it a *losing combination*. It will turn out that every combination is one or the other of these. The best way to begin the investigation of this game is to play it.

EXERCISES

1. Which of the following are winning combinations: 1 2 3, 1 4 5, 0 4 4, 3 4 5, 2 1 1?
2. Can any of the above be generalized into certain types of winning combinations?

10. NIM: HOW TO WIN IT

We shall first give the device by which it may be determined, whatever the numbers in the piles, whether any given combination is "winning" or "losing." After that we shall prove that our statement is correct. The method of testing which combination it is is as follows. Write the numbers in the piles in the number system to the base two and put them in the position for adding; if the sum of every column is even, it is a losing combination—otherwise it is a winning combination. For instance,

$$
\begin{array}{ccc}
3 & \text{is} & 11 \\
4 & \text{is} & 100 \\
5 & \text{is} & 101
\end{array}
$$

The sums of the first and third columns are 2, but the sum of the second is 1, which is odd. Hence, when the numbers of counters in the piles are 3, 4, and 5, respectively, we have a winning combination. But if 5 were replaced by 7, the last row would be 111 and the test shows that 3, 4, 7 is a losing combination. A direct analysis of the possible plays will show that this is indeed the case. Notice that we *do not add the three numbers to the base two*; we merely *add each column*.

By way of saving circumlocution, let us call it an *even combination* when, after writing the numbers of the counters in the piles to the base two, the sum of every column is even; otherwise, call it an *odd combination*. We then wish to prove that **an odd combination is winning and an even combination losing**. Notice first that a player can win in one play if there remains just one pile, and only under those circumstances can he win in one play. In that case, the number of counters in that pile written to the base two will have at least one 1 in it, and since the other numbers are 0, it will be an odd combination. Hence a player certainly cannot win in one play unless he is faced with an odd combination. Thus, if a player can so manage that *every* time after he has played the combination is even, his opponent cannot win; and someone must win, since there cannot be more plays than the total number of counters. It, therefore, remains for us to show (1) that a play always changes an even combination into an odd one, and (2) that faced with an odd combination one can always so play that the result is an even combination. To show the first it is only necessary to see that taking any number of counters from a pile changes the number in that pile written to the base two; that is, at least one 1 will be changed to a 0 or at least one 0 to a 1—then the sum of that column, if even before the play, must be odd after the play. To show the second, pick out the first column, counting from the left, whose sum is odd. At least one of the numbers written to the base two must have a 1 in this column. Call one such number R. It will look like this:

$$\underline{\hspace{4cm}}1\underline{\hspace{4cm}}$$

where the dashes indicate digits (to the base two) that we do not know about. Find the next column (again counting from the left) whose sum is odd. The digit of R in this column may be 0 or 1. To be noncommittal, call it a. And so proceed until we have R looking like this:

$$\underline{\hspace{2.5cm}}1\underline{\hspace{1cm}}a\underline{\hspace{2.5cm}}b\underline{\hspace{1cm}}c\ldots$$

where $1, a, b, c, \ldots$ are the digits in the columns whose sums are odd. Now write the number T

$$\underline{\hspace{2.5cm}}0\underline{\hspace{1cm}}A\underline{\hspace{2.5cm}}B\underline{\hspace{1cm}}C\ldots$$

where A is 0 if a is 1 but 1 if a is 0, similarly for B, C, ... In other words, T is the number obtained from R by changing from 0 to 1 or 1 to 0 all the digits of R which occur in columns whose sum is odd. Now T is less than R and hence we can take enough counters from the pile containing R counters to have T left. Then the sum of each column will be even and we have completed our proof.

We illustrate the process for the following example. Suppose we are faced with 58, 51, and 30 counters in the respective piles. We first write these numbers in the binary system,

$$
\begin{array}{rcl}
58 & \text{is} & 111010 \\
51 & \text{is} & 110011 \\
30 & \text{is} & 11110 \\
   &   & *\ ***
\end{array}
$$

where we have placed a star under each column whose sum is odd. R, the pile from which we draw, will be a number which has a 1 in the first starred column counting from the left. In this case all three numbers have this property, so that we may choose R to be any one of the three numbers. Choose it to be the top number, 58. We then change every digit in the top line which occurs in a starred column. The top line becomes 101101, which is 45 in the decimal system and every column will then have an even sum. Hence, to win, take 13 counters from the pile having 58. We could also win by taking 15 from the second or 21 from the third. Notice that if one of the numbers had a 0 in the first starred column counting from the left, we could not draw from that pile to win, for replacing the first 0 by a 1 would increase the size of the number.[1]

EXERCISES

1. Determine which of the following combinations are odd and state in each case of an odd combination how you would play to leave an even combination.
   a. 3 7 9.   **b.** 4 11 29.   **c.** 2 4 8.   **d.** 4 10 14.
2. Prove that if there are 6, 5, and 4 counters in three piles, respectively, one can play in three different ways to leave an even combination.
3. Under what conditions will it be possible to play in three different ways to leave an even combination? Will it ever be possible to play in more than three different ways to leave an even combination? Will it ever be possible to play in exactly two different ways to leave an even combination?
*4. Sometimes the game of Nim is played so that the *loser* is he who takes the last counter. What would then be the system of winning?

11. SUBTRACTION

You give the cashier one dollar to pay for your thirty-five-cent meal and she drops the coins into your eager hand, trying to make it seem as if you had not spent a thing by counting: "thirty-five, forty, fifty, a dollar." But you, wise as you are, know that she started counting at thirty-five and that another dollar is well on its way toward leaving you forever. It might be, if you are especially intelligent, that you had already guarded yourself against her sleight-of-hand by answering for yourself the question, "How much change should I get?" As a matter of fact, to ambitious souls, 3 + 7 = 10 is occasionally interesting, because it is useful in answering the question: "I have 3; how many more do I need to get in order to have 10?" 3 + ? = 10. It is the process of answering such a question that we call *subtraction*. Assuming we are all ambitious, then, we fortify ourselves by saying that subtraction is a convenient means of finding how much more we have to get to achieve our ambition. It goes something like this: I have three dollars and want ten—I am thus three on my way toward having ten and have three less than ten dollars to go. Subtraction is thus reduced to our former method of counting, with the difference that we omit some of the objects from our count or, what amounts to the same thing,

---

[1] See topic 7 at the end of the chapter.

remove some of the objects before we start to count. Subtracting 3 from 10 is accomplished by taking away three objects from ten and counting what is left. Employing the usual sign for subtraction we have $10 - 3 = 7$ as another way of saying $3 + 7 = 10$.

Subtraction is just the opposite process from addition; for example, if you subtract three and then add three you are just where you started; you merely replace the three you took away. We therefore call *subtraction the inverse of addition*.

We have seen that if a and b are positive integers, than $a + b$ is one also; that is, the answer to $a + b = ?$ is always a positive integer. But $a + ? = b$ does not always have an answer which is a positive integer. For instance, there is no positive integer which can be added to 11 to get 6. When there is a positive integer that can be added to a number a to get a number b, we say "a is less than b" or "b is greater than a." This is written

$$a < b \qquad \text{or} \qquad b > a.$$

Since $a + x = b$ means the same as $x = b - a$, we can also say that $b - a$ is a positive integer if and only if $b > a$.

This relationship can be shown graphically if we arrange the symbols for the positive integers on a line as follows:

$$\begin{array}{ccccccccc} . & . & . & . & . & . & . & . & . \; . \; . \\ \hline 0 & 1 & 2 & 3 & 4 & 5 & 6 & 7 & . \; . \; . \end{array}$$

Here the numbers indicated increase as we move from the left to the right along the line. That is, if the symbol for one number occurs to the left of the symbol for another number, the former number is less than the latter. If the symbol a occurs to the left of the symbol b, we may find the number $b - a$ by "counting off" from a to b along the line.

There is another symbol closely associated with the above which we use when we wish to include the possibility of equality. If a is greater than or equal to b, we write $a \geq b$. This may also be read: a is not less than b. An equivalent statement is: b is less than or equal to a, written $b \leq a$.

Notice that this symbol has the same three properties which the symbol $\subseteq$ for inclusion of sets had in Section 5 of Chapter 1 and the symbol $\rightarrow$ for implication in Section 6 of Chapter 2.

The property of having or not having an inverse has many parallels outside of mathematics. The inverse process is often much more difficult and many times impossible. To put a spoonful of sugar into a cup of coffee is the simplest of acts but to remove it is, if possible at all, a much more difficult process. This is, of course, a kind of addition and subtraction. There is also the idea of opposite or inverse direction. The turnstiles through which one leaves a subway station are expressly built to allow no inverse path. Although the pole vaulter must be careful how he descends, it is not nearly so difficult a feat as propelling himself up and over the bar. It is easier to get into jail than out.

The converse of a statement is a kind of inverse if one thinks of it in terms of the direction of implication. For example, consider the two statements; That her title is Mrs. implies that she is a woman; that she is a woman implies that her title is Mrs. Only one of these is true. On the other hand, the following proposition and its converse, or inverse, both hold: That the sum of the squares of the lengths of two sides of a triangle is the square of the length of the third side implies that the triangle is a right triangle. Devotees of *Alice in Wonderland* will remember the following passage:

"Then you should say what you mean," the March Hare went on.

"I do," Alice hastily replied; "at least—as least I mean what I say—that's the same thing, you know."

"Not the same thing a bit!" said the Hatter. "Why, you might just as well say that 'I see what I eat' is the same thing as 'I eat what I see'!"

"You might just as well say," added the March Hare, "that 'I like what I get' is the same thing as 'I get what I like'!"

"You might just as well say," added the Dormouse, who seemed to be talking in his sleep, "that 'I breathe when I sleep' is the same thing as 'I sleep when I breathe'!"

"It *is* the same thing with you," said the Hatter, and here the conversation dropped. . . .

Apparently the last proposition did have a true inverse for the case considered.

Our definition of "inverse" applies to all these examples. To put a spoonful of sugar into a cup of coffee and then take it out (without doing anything else) is to maintain the status quo. The pole vaulter starts on the ground and ends on it (although it must be admitted that the exact location is slightly different). If the first proposition and its inverse are combined we get the triviality: That her title is Mrs. implies that her title is Mrs. It should be pointed out that while there is always an inverse proposition there is not always a *true* inverse proposition; similarly, one can always describe an inverse action or process, but it is not always possible to perform it.

12. DIVISION

Multiplication also has an inverse process. In this case we call the inverse process "division." Sometimes we can divide and sometimes we cannot, just as in the set of positive integers and zero, we could sometimes subtract and sometimes not. Consider the question $5 \cdot ? = 30$. In words, this is: What can I multiply 5 by to get 30? That is, if I have five sets with the same number of elements, and if the total number of elements in the five sets is 30, how many elements are in each set? We call the answer, if it exists, the quotient of 30 and 5. If the answer exists we say that 30 is *divisible by* 5, or 5 is a *factor* of 30. The number 30 is not divisible by 7, for instance, because there is no integer which, when multiplied by 7, gives 30.

In general, the integer b is *divisible* by a if there is an integer c such that $ac = b$. We can also express this property by saying that a is a *factor* of b. Notice that 1 is a factor of every integer and that every integer is a factor of itself.

The factors of 30 are 1, 2, 3, 5, 6, 10, 15, and 30. Thus we can arrange 30 elements of a set into two subsets of 15 each, three subsets of 10 each, ten subsets of 3 each, and so on. But since 7 is not a factor of 30, we cannot arrange 30 elements into seven subsets with the same number of elements in each: If we try this we will have seven subsets of 4 each and two left over. If 30 is divided by any factor, the remainder will be zero.

There are various notations for division. We can write

$$b \div c = f, \qquad b/c = f, \qquad \text{or} \qquad \frac{b}{c} = f.$$

They all mean the same thing and are equivalent to saying $b = cf$.

EXERCISES

1. Give various examples outside mathematics of processes or operations which do or do not have inverses.

2. Is subtraction commutative or associative? Is division?

3. Is division distributive with respect to addition? (Consider division on both sides.)

4. Is division distributive with respect to subtraction?

5. Is $5 + (3 - 2) = (5 + 3) - 2$? In general, is $a + (b - c) = (a + b) - c$, where a, b, and c are positive integers and $b > c$?

6. Is $7 - (3 + 2) = (7 - 3) + 2$? In general, is $a - (b + c) = (a - b) + c$, where a, b, and c are positive integers and $a > (b + c)$?

7. Is $7 - (3 - 2) = (7 - 3) - 2$? In general, is $a - (b - c) = (a - b) - c$, where a, b, and c are positive integers and $a > b > c$?

8. Is $a - (b - c) = (a + c) - b$ if a, b, and c are positive integers and $a > b > c$?

9. Point out what properties of numbers are used in subtracting 345 from 723.

10. How many possible meanings can the following ambiguous symbol have: $64/8/2$? What are the various values?

11. How many possible meanings can the following ambiguous symbol have: $256/16/8/2$? What are the various values?

12. What positive integer or integers are factors of every positive integer? Of what numbers is 0 a factor?

13. In what sense is 0 divisible by every positive integer? If b is a positive integer, to what is $0/b$ equal?

14. Is zero divisible by zero in any sense? Discuss your answer.

15. Examine the process of division of 3240 by 24.

Answer: Write

$$24\underline{)\,3240\,}\underline{\,135}$$
$$24$$
$$\overline{84}$$
$$72$$
$$\overline{120}$$
$$120$$

In the first line of the division, 2400 is found to be the greatest multiple of 2400 less than 3240. One can omit the 00 after 24 by placing it judiciously below the 32. Again, the 0 in 840 can be omitted for a similar reason, and so it goes; 3240 is found to be $24(100 + 30 + 5) = 135 \cdot 24$. Explain the process of dividing 3198 by 13.

16. If 3240 and 24 were translated into the dozal notation, would the former be divisible by the latter?

17. If 3240 and 24 are two numbers in the dozal notation (for example, if 24 is two doza-four), will the former be divisible by the latter?

*18. a, b, and c are three numbers less than 10. When will the two-digit number ab (its digits are a and b in that order) be divisible by c in the dozal notation as well as in the decimal system? Will the two answers ever be the same?

*19. The converse of an implication could be thought of as being its own inverse, because the converse of the converse is the implication itself. Could a similar remark be made about the contrapositive and obverse?

13. DIVISORS

We have previously called attention to the fact that the base twelve is convenient for certain purposes because it has, for its size, a large number of divisors. It has, including 1 and 12, six divisors (that is, 12 is divisible by six different numbers). All the numbers through 23 have less than seven divisors. Again, 24 has more divisors than any smaller number. (Such numbers are called *highly composite*.) What is the next such number? We leave the question to anyone who is interested.

Another line of inquiry which has to do with divisors concerns the so-called perfect numbers: $6 = 1 + 2 + 3$ is the sum of its divisors including 1 but not including 6; 6 is thus called a *perfect number*. The next one is $28 = 1 + 2 + 4 + 7 + 14$. Is there another one less than 50? There is a formula (see Section 14) which gives all the even perfect numbers, but no one has found an odd perfect number. In fact, whether or not there are odd perfect numbers is one of the unsolved mysteries of the mathematical world. Fame awaits him who can find an odd perfect number or prove that there are none. It is known that if P were an odd perfect number it would have at least five different prime[1] factors and be divisible by a number $p^n > 10^{12}$, where p is a prime number and n is a positive integer.

[1] See Section 14.

14. PRIME NUMBERS

From the point of view of having divisors, there are numbers that are especially peculiar: integers greater than 1 which have no divisors except themselves and 1. These are the *prime numbers*—the first five are 2, 3, 5, 7, and 11. Integers greater than 1 which are not prime numbers are called *composite numbers*. The worst of it is (although from many points of view it is very fortunate) that *there is no last prime number*. Euclid, of geometry fame, proved this a long time ago. Here is his proof essentially. Suppose the fiftieth prime number were the last one. Multiply together the fifty prime numbers and add 1 to the product. Call the result *B*. *B* cannot be divisible by any one of the fifty primes because the remainder is 1 when it is divided by any of them. Call *q* the smallest number greater than 1 which divides *B*. If *q* had a divisor greater than 1 and smaller than *q*, this divisor would divide *B* and contradict our assumption about *q*. Therefore, since *q* divides *B* and has no divisors greater than 1 and less than *q*, *q* is a prime number different from the fifty primes. Hence our assumption that there are only fifty primes is false. Notice that we have not proved that *B* or even *q* is the fifty-first prime, but we have proved that *q* is a prime which is not one of the first fifty. Furthermore, no matter how many primes we thought there were, we could always prove, by the above method, that there is one more. A short way of saying this is: *There are infinitely many prime numbers* or *the set of prime numbers is an infinite set*. This merely means that no matter what number you name, there are more than that many primes. It is like the various word battles of our childhood: to "My father gets ten dollars every day" you countered with "My father gets a hundred dollars every day." Your opponent in turn raised the ante and whoever could count the further won. An interesting thing about the phenomenon in this connection is that if you were boasting about the number of primes and I were boasting about the number of positive integers, again the victory would go to him who could count furthest, even though there are many positive integers which are not prime numbers and every prime number is a positive integer. Furthermore, if we extend our concept of "just as many" which we considered in Section 1, we can say that there are just as many prime numbers as there are positive integers because we can establish a one-to-one correspondence between the prime numbers and the positive integers: There is a tenth prime, there is a millionth prime, there is a trillionth prime, and so on.

There is an interesting theorem about primes which goes by the name *Bertrand's postulate*. It is this: Strictly between any number *A* (*A* > 1) and its double there is a prime number, where "strictly between" means "between and not equal to *A* or 2*A*," that is, not only greater than *A* but less than 2*A*. For instance, if *A* = 5, then 7 (but not 5) is strictly between 5 and 10. Although it seems not strange in the least that this theorem should be true, it is rather difficult to prove.

At this point we shall mention two statements about the occurrence of prime numbers which most persons believe to be true but which no one has yet been able to prove. Two prime numbers which differ by 2 (such as 29 and 31) are called *twin primes*. It is *believed* that there are infinitely many twin primes. The second unproved statement is this: Every even number greater than 2 is the sum of two prime numbers. This latter is called *Goldbach's conjecture*, but no one has proved it.

Although primes are peculiar from the point of view of divisibility, this very fact makes them the indivisible numbers into which numbers can be factored, for when, in the process of dividing, one comes to a prime, the division stops. *If a number is expressed in two different ways as a product of prime numbers, the two products differ at most in the order in which the factors occur*; that is, each prime factor appears the same number of times in each product. This is called the *law of unique decomposition* into prime factors. For example, consider the number 315. It has a factor 5 and $315 = 5 \cdot 63$. The number 63 is divisible by 3 and $315 = 5 \cdot 3 \cdot 21$. The number 21 is divisible by 3 and $315 = 5 \cdot 3 \cdot 3 \cdot 7$, and all the factors are prime numbers. If we had divided first by 3, the result might have been $315 = 3 \cdot 7 \cdot 5 \cdot 3$. In both of the expressions for 315 as a product of prime factors, each prime 3, 5, and 7 appears the same number of times. A convenient way of expressing the factorization is

$$315 = 3^2 \cdot 5 \cdot 7.$$

The formula for all even perfect numbers (see Section 13) is

$$2^{p-1}(2^p - 1),$$

where not only p but $(2^p - 1)$ are prime numbers.

EXERCISES

1. If B is 1 more than a product of the first n primes, consider q, the smallest prime divisor of B. Is q the $(n + 1)$st prime for any of $n = 3, 4, 5, 6$?

2. Why is 2 the only even prime number?

3. Show that if k is a composite number, it has a factor greater than one whose square is less than or equal to k.

4. Show that B as defined in Exercise 1 is a prime number for $n = 1, 2, 3, 4, 5$. (It can be shown that this is not the case for $n = 6$.)

5. Show that none of the nine consecutive integers from 212 to 220 inclusive is a prime number. Note that $212 = 2 \cdot 3 \cdot 5 \cdot 7 + 2$. Find a succession of twelve consecutive composite numbers.

6. How many prime numbers are there strictly between 11 and 22?

7. We have shown that there are just as many prime numbers as positive integers. Why does this not prove that every positive integer is a prime number?

8. Why does Bertrand's postulate imply that there are infinitely many prime numbers?

9. Prove that there are more than twice as many positive integers less than 1000 as there are prime numbers less than 1000. Would this be true if 1000 were replaced by 1,000,000?

10. Prove by use of Bertrand's postulate or by any other means the following: 1 more than the product of the first n primes is *never* the $(n + 1)$st prime unless $n = 1$.

11. In 1937, a man by the name of Vinogradoff proved the following result: There is a number M such that all odd numbers greater than M are the sum of three prime numbers (for example, $15 = 3 + 5 + 7$). Would this result imply the following theorem, or would it be implied by it? There is a number M such that every even number greater than M is the sum of two prime numbers.

*12. Show that if A and B have no factor except 1 in common, the number of divisors of AB is the product of the number of divisors of A and of B. (The number itself and 1 are counted as divisors.)

15. TOPICS FOR FURTHER STUDY

Where no reference is given, you should not conclude that there are no references but merely that you should have more profit and pleasure from investigating the topic without consulting references.

1. The history of the number concept and the number system: references **16** (Chaps. 1 and 2), **60** (Chaps. 1 and 2), and **72**.

2. Properties of positive integers and zero: references **14** (Chap. 1 and its supplement) and **18**.

3. Counting and big numbers: references **38** (Chap. 2).

4. Numbers to other bases: base twelve, reference **63**; base eight, references **33** and **64**; base two and puzzles, reference **38** (pp. 165–171).

5. The window reader (another trick like the one at the close of Section 8): reference **6** (pp. 321–323).

6. Bachet weights problem (see Exercise 19 of Section 8): reference **6** (pp. 50–53).

7. Nim: reference **64** (pp. 15–19).

8. Highly composite numbers—what ones are they? See Sections 14 and 13.

9. Perfect numbers: references **36** (pp. 56, 57) and **65** (pp. 80–82).

10. Magic squares: references **6** (Chap. 7) and **65** (pp. 159–172).

11. Puzzles: references **6** (Chap. 1), **38** (Chap. 5—see also additional references at the close of that chapter), and **42**.

12. Find every number whose square is a four-digit number having the last three digits equal, that is, of the form *baaa*.

13. The eight-queens problem (the problem is to place eight queens on a chess board so that no one will threaten any other one): references **26** and **42** (Chap. 10).

14. What mathematicians think about themselves: references **7, 11, 28**, and **70**.

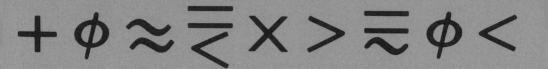

4

Mathematical Systems

1. NUMBERS ON A CIRCLE: ADDITION

"Tomorrow and tomorrow and tomorrow creeps in its petty pace from day to day." "If winter comes . . . ." Morning, noon, and night, and then another day begins. Sunday, Monday, . . . , Saturday, and Sunday comes again. "History repeats itself." "Life is a ceaseless round."

Recall the Mad Hatter's tea party with the many places set:

"It's always six o'clock now,"

A bright idea came into Alice's head. "Is that the reason so many tea things are put out here?" she asked.

"Yes, that's it," said the Hatter with a sigh; "it's always teatime, and we've no time to wash the things between whiles."

"Then you keep moving round, I suppose?" said Alice.

"Exactly so," said the Hatter; "as the things get used up."

"But when you come to the beginning again?" Alice ventured to ask.

"Suppose we change the subject," the March Hare interrupted, yawning, "I'm getting tired of this. I vote the young lady tell us a story."

Alice, of course, had mentioned one of the most important properties of moving around a table: that one must, sooner or later, if one continues in the same direction, come back to the starting point. The table was not round, for it had corners, but as far as the property which Alice mentioned was concerned it might even better have been round. For aesthetic reasons we change the setting slightly and draw a picture of a round table with the seven numerals

from 0 to 6 instead of teacups (Fig. 4:1). One could even label the divisions of
the circle with the days of the week or by increasing the number of divisions
and reversing the direction we should have the face of a clock. Motion around
the table can be used as a basis for defining a new kind of addition which it is
interesting to consider. To add 2, for instance, to any number from 0 to 6,

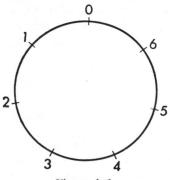

Figure 4:1

start with the point on the circle bearing the number to which 2 must be
added and move two divisions in a counterclockwise direction, that is, in the
direction opposite to that in which the hands of a clock move when it is
running properly. Thus, to get 3 + 2 we start at 3 and, moving two divisions
in a counterclockwise direction, we come to 5. Hence, even on the circle,
3 + 2 = 5. Adding by the same method, wonder of wonders, 3 + 4 = 0 and
4 + 5 = 2! We call this process *addition on the circle*. No matter what addition we perform, our result is one of the seven numbers. We subtract by
merely going in the opposite (that is, clockwise) direction. Of course we could
reverse the direction of the numbers around the circle and have the same results.

It is not hard to see that these seven numbers on the circle with the process
of addition have the following four properties:

(1) The result of every addition is one of the numbers; that is, the set of
numbers is *closed under addition*.
(2) Addition is *associative*: $(a + b) + c = a + (b + c)$.
(3) There is a number 0 such that $0 + a = a$ for each number of the set; that
is, there is an *additive identity* element.
(4) For each number a of the set there is a number b of the set such that
$a + b = 0$; that is, each number has an *additive inverse*.

The first property is apparent, because if we start at one of the points and
move a whole number of divisions in a counterclockwise direction, we must
arrive at another point. The associative property holds because the numbers

represented by each side of the equals sign correspond to moving first a spaces, then b and finally c.

The third property follows because adding zero is equivalent to not moving at all. Finally, to get the additive inverse of a number a, we merely move a steps around the circle in the clockwise direction.

Since many other sets and operations have this set of properties, it is helpful to list the abstract properties by means of which we define a *group*. Such an abstraction is useful because any property which we can show to follow from the definition of a group will also follow, without further proof, for all examples.

Definition

Given a set S of one or more elements and a process of operation (such as addition or multiplication), which, to be noncommital, we denote by $\#$. Then S is said to form a "group under $\#$" if, for every pair of elements a and b (the same or different), the following properties hold:

(1) $a \# b$ is in S: the *closure property*.
(2) $(a \# b) \# c = a \# (b \# c)$: the *associative property*.
(3) There is an element e of S, called the *identity* element, such that $a \# e = e \# a = a$, for all elements a in S.
(4) Every element a of S has an *inverse* b in S such that

$$a \# b = b \# a = e.$$

There is another property which addition of numbers on the circle has: $a + b = b + a$. This is the commutative property. Thus if

(5) $a \# b = b \# a$, we call the group *Abelian*.

The word "group" in this technical sense has much meaning packed into it, like Humpty Dumpty's "impenetrability." Said he,

"I meant by 'impenetrability' that we've had enough of that subject, and it would be just as well if you'd mention what you mean to do next, as I suppose you don't mean to stop here all the rest of your life."

"That's a great deal to make one word mean," said Alice in a thoughtful tone.

"When I make a word do a lot of work like that," said Humpty Dumpty, "I always pay it extra."

Although we referred to the circle with seven divisions in our development above, our arguments were general enough for us to affirm that the numbers $0, 1, 2, \ldots, n - 1$, with the operation of addition on a circle with n divisions, form a group. A graphic way to see these properties for the circle of seven divisions is to form the addition table:

+	0	1	2	3	4	5	6
0	0	1	2	3	4	5	6
1	1	2	3	4	5	6	0
2	2	3	4	5	6	0	1
3	3	4	5	6	0	1	2
4	4	5	6	0	1	2	3
5	5	6	0	1	2	3	4
6	6	0	1	2	3	4	5

You should notice how the closure and commutative properties are illustrated by this table. It may also be used to subtract. For instance, to find x satisfying $3 + x = 1$, we look across the row numbered 3, to the number 1 which occurs in the column headed 5. This shows that $x = 5$; that is, 5 is what must be added to 3 to get 1. What property of the table shows that subtraction is always possible? If $\#$ means addition, property (3) may be thought of as closure under subtraction.

2. NUMBERS ON A CIRCLE: MULTIPLICATION

It is also interesting to consider multiplication on a circle. To get $2 \cdot 3$ we can take two steps around the circle (numbered from 0 to 6) taking three divisions in a stride. Since we should end on the 6 we have the natural result that $2 \cdot 3 = 6$. But, using the same process, we get the astonishing result that $3 \cdot 4 = 5$. Lo, we need a new multiplication table! Here it is.

·	1	2	3	4	5	6
1	1	2	3	4	5	6
2	2	4	6	1	3	5
3	3	6	2	5	1	4
4	4	1	5	2	6	3
5	5	3	1	6	4	2
6	6	5	4	3	2	1

Notice that although this table could be obtained by counting out on a circle, it could also be derived by considering the remainders when the various products are divided by 7. For instance, $5 \cdot 3 = 15$ has the remainder 1, which is what appears in the table as the product of 5 and 3. Notice that, as in the case of addition, all our results are among the numbers from 0 to 6.

The identity for multiplication is, of course, the number 1, which is in the set. Now consider the existence of an inverse for multiplication. Suppose we wish to solve $4x = 1$. We look in row 4 and see that the number 1 is in column 2; thus 2 is the multiplicative inverse of 4.

Now let us be more systematic and consider two questions about the multiplication table on the circle of 7 divisions.

Question 1. Do the numbers 0, 1, 2, 3, 4, 5, and 6 and multiplication on the circle of 7 divisions form a group? The answer is no, because the fourth property of a group fails in the case of the equation $0 \cdot x = 1$, which has no solution.

Question 2. Do the numbers 1, 2, 3, 4, 5, and 6 and multiplication on the circle of 7 divisions form a group? To see that the answer is yes, note for property (1) that only these six numbers appear in the table, which implies closure. We shall prove the associative property later. The identity 1 is certainly in the set. Finally, $ax = 1$ is solvable, since to get the solution we look in row a for the number 1 and x is the number of the column in which the 1 occurs. Since each row contains 1, the equation is always solvable.

The answer to the second question naturally leads to another: In a circle of n divisions, do the numbers

(1) $1, 2, 3, \ldots, n - 1$

form a group under multiplication? (Why, if 0 is included, must they fail to form a group?)

Notice first that if the multiplication table of the numbers in (1) contains a zero, the closure property does not hold, since 0 is not in the set (1). But if we have two numbers r and s in (1) whose product is n, then, on the circle, $rs = 0$. For instance, if $n = 15$, $t = 3$, and $s = 5$, we have $rs = 0$ on the circle with 15 divisions. This leads to

Result 1. If n is a composite number (see Section 14 of Chapter 3), then the numbers (1) on the circle with n divisions do not form a group under multiplication.

Another way to state this result is: The numbers (1) with multiplication do not form a group unless n is a prime number. Now we want to show that if n is a prime number we do indeed have a group; that is,

Result 2. If n is a prime number, then the numbers (1) on the circle with n divisions form a group under multiplication.

We have checked all the properties except that of an inverse. To show that an inverse exists, consider the equation $bx = 1$ on the circle with n divisions. This will have a solution if the number 1 is in row b. To show this, notice that row b in the multiplication table will consists of the remainders when the following numbers are divided by n:

$$b, 2b, 3b, \ldots, (n - 1)b.$$

First, none of these remainders is zero, since n is a prime number and neither b nor any of $1, 2, 3, \ldots, n - 1$ is divisible by n. Second, no two remainders are the same, because we can show as follows that if two were the same, the row would contain a zero, contrary to the previous sentence. To show this, suppose the remainders for bx and by were the same with $n > x > y > 0$. Then $bx - by$ divisible by n implies that the remainder is zero, but[1] $bx - by = b(x - y)$ shows that this is in the bth row, which is a contradiction.

Finally, since no remainder in row b is zero, no two are the same and there are $(n - 1)$ remainders; they must be the numbers $1, 2, \ldots, n - 1$ in some order. This shows that the row b contains a 1 and $bx = 1$ has a solution on the circle. This completes the proof of result 2. We can thus combine the two results in the following statement: The numbers $1, 2, 3, \ldots, n - 1$ on a circle of n divisions form a group under multiplication if and only if n is a prime number.

3. SOME FUNDAMENTAL PROPERTIES OF GROUPS

There are three more properties of a group which are useful and which are easy to derive from the other properties:

 (i) In any group there is just one identity element.
(ii) Every element of a group has just one inverse.
(iii) In any group the equations $a \# x = b$ and $y \# a = b$ are solvable, using the sign $\#$ for the operation of the group.

We now prove (i). Suppose the group had two identity elements, e and e'. Then, since e is an identity element, $e \# e' = e'$. Since e' is an identity element, $e \# e' = e$. This shows that $e' = e$, since they are both equal to $e \# e'$.

We leave the proof of (ii) as an exercise and prove the first part of (iii). To solve $a \# x = b$, we let a' be the inverse of a and write

$$a' \# (a \# x) = a' \# b.$$

But, by the associative property: $a' \# (a \# x) = (a' \# a) \# x = e \# x$. Since e is the identity element we then have

$$x = e \# x = a' \# b.$$

This shows that if there is an element x such that $a \# x = b$, then $x = a' \# b$. If we put this value of x into the original equation we can see that the equality holds. The second equation in (iii) can be dealt with similarly.

If $\#$ is addition, we call the additive inverse of an element its *negative* or

---

[1] See Exercise 15 of Section 5.

opposite and, since in this case the identity element is 0, we write $a + (-a) = (-a) + a = 0$. If $\#$ is multiplication, we call the inverse of an element its *reciprocal* and, since 1 is the identity element, we write

$$a \cdot a^{-1} = a^{-1} \cdot a = 1 \qquad \text{or} \qquad a \cdot \frac{1}{a} = \frac{1}{a} \cdot a = 1.$$

We could prove that addition and multiplication have each of the properties (i), (ii), and (iii), but it is simpler to use the notation $\#$ and do it all at once.

EXERCISES

1. Show why the addition table for the circle with 7 divisions shows the commutative and closure properties and the existence of an inverse.

2. In a certain year, the first of April is a Monday. What day of the week in that year is the Fourth of July?

3. Did Magellan's voyage around the world prove that the earth was round?

4. What relationships can you discover between addition of hours on the face of a clock and the dozal system of notation?

5. Form the addition tables for the circles of 6 and 8 divisions. Will these numbers with addition form an Abelian group? Assume associativity.

6. Consider the numbers from 0 to 10 inclusive written on a circle and make the resulting multiplication table. Do the numbers from 1 to 10 and multiplication on the circle form an Abelian group? Assume associativity.

7. Write the numbers from 0 to 11 on a circle and consider the questions raised in Exercise 6. How does the multiplication table differ essentially from that for 7 or 11 divisions? Assume associativity.

8. For which of the following number of divisions of the circle will multiplication of nonzero numbers yield a group: 13, 16, 15, 26, 31? Assume associativity.

9. On a circle of 12 divisions, which of the following are solvable for x? When there is a solution, give all solutions. Where there is no solution, show why there is none.

    a. $x + 8 = 3$.   **b.** $5x = 7$.   **c.** $2x = 5$.   **d.** $10x = 2$.

10. Find an example in this section of an indirect proof and analyze its structure after the manner of Chapter 2.

11. Do the numbers 1, 5, 7, and 11 on the circle of 12 divisions and the process of multiplication form a group? Assume associativity.

12. Prove property (ii), using the symbol $\#$ or first using $+$ and then $\cdot$.

13. Prove the second part of property (iii).

*14. Notice that in the multiplication table given above both diagonals read the same from left to right as from right to left, the diagonals being 1 4 2 2 4 1 and 6 3 5 5 3 6. Why is this so for other circles? Can you find any other interesting properties of the multiplication tables?

*15. If b and m have no factors in common except 1, show that if $bx = by$ on the circle of m divisions, then $x = y$ on that circle. Show that this implies that the numbers $b, 2b, 3b, \ldots, (m - 1)b$ are in some order equal on the circle with m divisions to $1, 2, \ldots, m - 1$, if b and m have no factors in common.

*16. Show that the positive numbers less than m, having no factors in common with m with the process of multiplication on a circle of m divisions, form a group.

*17. A circle is divided into n parts, where n is a positive integer. Prove that if, for a given positive integer b, m is the smallest number of times one can lay b off on the circumference and have a closed polygon (that is, return to a point previously touched), then n/m is the greatest common divisor of b and n; that is, n/m is the largest number which divides both b and n.

4. FIELDS

We have now shown that if p is a prime number, then

(1) The numbers $0, 1, 2, \ldots, p - 1$ and addition on the circle with p divisions form an Abelian group.

(2) The numbers $1, 2, \ldots, p - 1$ and multiplication on the circle with p divisions form an Abelian group.

It can also be shown (see Exercise 15 of Section 5) that the distributive property holds; that is,

(3) For multiplication and addition on a circle with p divisions, $a(b + c) = ab + ac$.

In mathematics we have a special name for a number system which has these three properties—we call it a *field*. (The Germans call it a "Körper," which is their word for "body.") Neither word seems especially appropriate but mathematicians have agreed on the term and we shall use it, too. We thus have a number system with just p numbers in it and which has many properties in common with the integers. It has other properties as well. For instance, we can always divide except by zero; this is true because of statement (2) above. (Why?)

Above we defined a field in terms of groups and the numbers on a circle. It is instructive to write out a more general definition like that for a group, so that we can derive some additional properties which will hold for any *field* no matter whether it has to do with numbers on a circle or not.

Definition

A *field* consists of a set S of at least two numbers, or elements, which we designate by $a, b, c, \ldots$, and two operations called addition and multiplication, denoted by $+$ and $\cdot$, respectively, with the following properties:

(1) Closure: $a + b$ and ab are in S.
(2) Associativity: $(a + b) + c = a + (b + c)$ and $(ab)c = a(bc)$.
(3) Existence of an identity element:
    (a) For addition it is 0 and $0 + a = a + 0 = a$ for each a in S.
    (b) For multiplication it is 1 and $1 \cdot a = a \cdot 1 = a$ for each a in S.
(4) Existence of an inverse except for multiplication by 0.
    (a) Addition: For each a there is a $(-a)$ such that
$$a + (-a) = (-a) + a = 0.$$
    (b) Multiplication: For each $a \neq 0$ there is an a^{-1} such that
$$a \cdot a^{-1} = a^{-1} \cdot a = 1.$$
(5) Commutativity: $a + b = b + a$ and $ab = ba$.
(6) Distributivity: $a(b + c) = (ab) + (ac)$ and $(b + c)a = (ba) + (ca)$.

Notice that, as for numbers on a circle with p divisions, the first four properties can be described by saying that for addition the elements form a group and for multiplication the elements without zero form a group. Property (5) makes the groups Abelian and property (6) is one which involves both addition and multiplication.

Just as was the case with a group, there are some other properties of a field which follow from the definition and are useful whether the field is that of numbers on a circle or any other field. There is some advantage in proving them for all fields at once, but if you feel more comfortable about it you can think in terms of numbers on a circle in the rest of this section.

(7) $a \cdot 0 = 0 \cdot a = 0$, no matter what number a is.

To show this notice first that

$$0 \cdot a + a = 0 \cdot a + 1 \cdot a = (0 + 1) \cdot a,$$

using the property of the identity for multiplication and the distributive property. But $0 + 1 = 1$, and thus

$$0 \cdot a + a = 1 \cdot a = a.$$

Thus $0 \cdot a$ has the property that if you add it to a you get a. This is the property of the additive identity. Since in connection with a group, we showed that there is only one additive identity, we have $0 \cdot a = 0$. In view of the commutative property, $a \cdot 0 = 0$ also.

(8) $a(-b) = (-(ab))$.

Let us illustrate this first for numbers on the circle with 7 divisions and choose $a = 2$ and $b = 3$. Then $(-b) = 4$ and $a(-b) = 2 \cdot 4 = 1$ on the circle. But $ab = 6$ and $(-6) = 1$ on the circle. For this case we have shown the equality. Now, to show it in terms of letters, we notice first that

$$a[(-b) + b] = a \cdot 0 = 0.$$

Using the distributive property we have

$$a(-b) + (ab) = 0.$$

Thus when we add $a(-b)$ to ab we get zero, which means that $a(-b)$ must be the additive inverse of ab, which is written $(-(ab))$.

(9) $(-a)(-b) = ab$.

To show this we would start with

$$(-a)[(-b) + b]$$

and use the previous properties. This is left as an exercise.

Notice that for the properties (7), (8), and (9) we did not assume the existence of a multiplicative inverse. These properties hold, for instance, for numbers on a circle of 12 divisions. (See Exercise 7 below.)

(10) If $ab = 0$, then $a = 0$ or $b = 0$ or both.

To prove this, assume $a \neq 0$. Then $ab = 0$ implies $a^{-1}(ab) = 0$ by property (7) and

$$0 = (a^{-1} \cdot a)b = 1 \cdot b = b = 0$$

by the associative property and that of the identity 1. Hence if a is not zero and $ab = 0$, then b must be zero.

EXERCISES

1. Why do the numbers on the circle with 12 divisions not form a field?
2. Would the numbers on the circle with 15 divisions form a field? Why?
3. Illustrate property (9) for two numbers on the circle of 7 divisions.
4. Prove property (9).
5. Show that in a field $0 \cdot x = a$ is solvable for x only when $a = 0$.
6. On the circle of 12 divisions find two numbers a and b, neither zero, whose product is zero.

7. Which of the properties of a field hold for the numbers on a circle of 12
   divisions?

8. Are there inequalities for numbers on a circle; that is, does $a < b$ mean anything?

9. State two properties which the integers have which do not hold for numbers
   on a circle of p divisions.

10. Recall that in Chapter 1 we used the notation $A \uplus B$ to mean "A or B but
    not both." Compute the truth tables (see Section 4 of Chapter 2) for $\uplus$ and
    $\cap$. Find a relationship between them and the addition and multiplication
    tables for a circle of 2 divisions.

5. EQUIVALENCE AND CONGRUENCE

We have already used the word "equivalent" in Chapter 2. We called two
statements "equivalent" if they are both true or both false. They were not
necessarily the same statements. Similarly, we might say that two sets are
equivalent if they have the same number of elements; they would not have to
be the same sets. We could say that two cars are equivalent if they cost the
same. In general, we use the word "equivalent" both inside and outside
mathematics to mean that in some respects the objects or concepts to be
compared are the same. There are three fundamental properties which every
equivalence relation must have, using $\cong$ to mean "equivalence."

(1) $A \cong A$ (reflexive property).

(2) $A \cong B$ implies $B \cong A$ (symmetric property).

(3) $A \cong B$ and $B \cong C$ imply $A \cong C$ (transitive property).

(Notice that these are similar to but not the same as the properties of set
inclusion in Section 5 of Chapter 1.) Any relationship which has these three
properties we call an *equivalence relationship*.

Congruence of triangles is an equivalence relationship. They need not be
the same but to be congruent they must have certain properties in common.
"Living in the same house as" is another equivalence relationship.

Whenever we have an equivalence relationship we have a *classification*. If
the relationship is equal price in cars, for instance, we can include in one class
all cars of a given price and we have just as many classes as we have prices.
Paralleling the three properties of equivalence we have the following for a
classification:

(1c) A is in its own class.

(2c) If A is in the same class as B, then B is in the same class as A.

(3c) If B is in the same class as A and also in the same class as C, then A and C
     are in the same class.

A classification implies an equivalence relationship and vice versa.

There is a classification of the integers which is closely connected with the numbers on a circle considered in Sections 3 and 4. We might classify all the integers according to their remainders when divided by 3, for instance. We could then apportion the integers into three sets or classes as follows:

$$\text{Class of 0:} \qquad \ldots, -6, -3, 0, 3, 6, \ldots$$
$$\text{Class of 1:} \qquad \ldots, -5, -2, 1, 4, 7, \ldots$$
$$\text{Class of 2:} \qquad \ldots, -4, -1, 2, 5, 8, \ldots$$

Every integer would be in one of these classes and no integer would occur in two. In other words, we could call two integers *equivalent* if their remainders when divided by 3 were the same. It is customary to use the word "congruent" for this kind of equivalence. We use the notation

$$a \equiv b(\text{mod } 3)$$

to mean: a and b have the same remainders when divided by 3. The "(mod 3)" indicates that the classification is with respect to 3. We read the statement "a is congruent to $b(\text{mod } 3)$."

There is a simpler way to think of the congruence of a and b. Notice that in each class one gets each number from the preceding one by adding 3. Hence any two numbers in the same class differ by a multiple of 3. This means that instead of dividing a and b by 3 and looking at their remainders, we can determine whether they are in the same class by testing whether their difference is divisible by 3.

If we were to classify the integers according to their remainders when divided by an integer m, we would have m different classes corresponding to the remainders: $0, 1, 2, 3, \ldots, m - 1$. Then, as above,

$$a \equiv b(\text{mod } m)$$

would have two equivalent meanings:

(1) The remainders when a and b are divided by m are equal.
(2) The difference $a - b$ is divisible by m.

(We call these meanings equivalent because for any integers a, b, and m statements 1 and 2 are both true or both false.) The advantage of having two equivalent meanings is that in any given situation, we can use the one that seems easier to apply.

Is this congruence an equivalence relationship? Let us test the required properties in turn:

(1co) $a \equiv a(\text{mod } m)$, since $a - a = 0$, which is divisible by m.
(2co) If $a \equiv b(\text{mod } m)$, then $b \equiv a(\text{mod } m)$, since if $a - b$ is divisible by m, so is $b - a$, since $mq = a - b$ implies $m(-q) = b - a$.

(3co) If $a \equiv b(\text{mod } m)$ and $b \equiv c(\text{mod } m)$, then $a \equiv c(\text{mod } m)$, since if a and c have the same remainders as b when divided by m, they have the same remainders as each other.

There are two other properties of congruences which are like those of equality. We know that

(4) If $a = b$, then $a + c = b + c$.

(5) If $a = b$, then $ac = bc$.

These are the so-called "well-defined properties" of equality with respect to addition and multiplication. They also hold for congruences as follows:

(4co) If $a \equiv b(\text{mod } m)$, then $a + c \equiv b + c(\text{mod } m)$.

(5co) If $a \equiv b(\text{mod } m)$, then $ac \equiv bc(\text{mod } m)$.

The proofs are left as exercises.

Now look back at our numbers on a circle. Consider, for instance, the circle with 7 divisions. If we start at 0 and count off, say, 23 divisions we end at the mark 2, which is the remainder when we divide 23 by 7. If the circle had 11 divisions and we counted off 23 divisions, we would end at the mark 1, since 1 is the remainder when 23 is divided by 11. Writing the numbers on a circle is merely a device for picking out the remainder.

Now it is very easy to prove the associative property for multiplication on a circle. We merely notice that, by the associative property for integers, $a(bc) - (ab)c = 0$ and hence $a(bc) \equiv (ab)c(\text{mod } m)$, which shows that on the circle of m divisions the numbers $a(bc)$ and $(ab)c$ are the same, and by property (5co), any of a, b, c can be replaced by numbers in the same class without altering the truth of the congruence.

EXERCISES

1. Show property (3co) of congruences using the fact that $a \equiv b(\text{mod } m)$ means that $(a - b)$ is divisible by m.
2. Show properties (4co) and (5co).
3. Prove using only properties (1c), (2c), and (3c) that in a classification no element can be in two different classes.
4. Give three examples of equivalence relationships different from those given in this section.
5. Prove that if $a \equiv b(\text{mod } m)$ and $c \equiv d(\text{mod } m)$, then $ac \equiv bd(\text{mod } m)$.
6. Prove that if $a \equiv b(\text{mod } m)$, then $a^2 \equiv b^2(\text{mod } m)$.

7. Show that if $a^r \equiv b^r(\text{mod } m)$ and $a \equiv b(\text{mod } m)$, where r is some positive integer, then

$$a^{r+1} \equiv b^{r+1}(\text{mod } m).$$

Why does this show that if $a \equiv b(\text{mod } m)$, then $a^s \equiv b^s(\text{mod } m)$ for all positive integers s?

8. Compare the properties of an equivalence relation with those of inclusion of sets in Section 5 of Chapter 1.

9. Using the result of Exercise 7 and other properties of a congruence, show that if $a \equiv b(\text{mod } m)$, then

$$a^5 + 7a^2 + 2 \equiv b^5 + 7b^2 + 2(\text{mod } m).$$

10. Show that if $a + c \equiv b + c(\text{mod } m)$, then $a \equiv b(\text{mod } m)$. Why does this follow from one of the properties above?

11. Show that $ax \equiv b(\text{mod } m)$ is solvable if a and m have no common factor greater than 1.

12. A schedule is to be arranged for a league of seven teams such that each team plays every other team exactly once. Call the teams 0, 1, 2, 3, 4, 5, and 6. Note that three games can be played in a day and each day there must be one team which is idle (plays with itself). Number the days 0, 1, 2, 3, 4, 5 and 6. We have the following schedule for the first four days:

Day	Pairings			Idle team
0	(1, 6)	(2, 5)	(3, 4)	0
1	(0, 1)	(2, 6)	(3, 5)	4
2	(0, 2)	(3, 6)	(4, 5)	1
3	(0, 3)	(1, 2)	(4, 6)	5

Note that on day 0 we pair a and b so that $a + b \equiv 0(\text{mod } 7)$, on day 1, $a + b \equiv 1(\text{mod } 7)$, and so on. Complete the table and show why it works.

13. Could the system of scheduling used in Exercise 12 be used for a league of nine teams? Why?

14. Could the system of scheduling used in Exercise 12 be used for a league of eight teams? Why? If not, what modification could be used?

*15. Prove that on a circle of m divisions, the distributive property holds; that is, $a(b + c) = ab + ac$.

*16. Does $ac \equiv bc(\text{mod } m)$ always imply that $a \equiv b(\text{mod } m)$? If your answer is yes, prove it. If your answer is no, give an example when it does not hold and state what congruence is always implied by the given one.

*17. Give an example of a relationship which satisfies properties (2) and (3) of an equivalence relationship but not (1).

6. TESTS FOR DIVISIBILITY[1]

It is easy to tell in any case whether or not subtraction is possible merely by comparing the size of the numbers. But it is more difficult to tell whether division is possible short of actually carrying through the process to see how it comes out. However, in some cases, one can easily tell ahead of time whether or not division is possible. This is easiest for 2, 5, and 10. There also is a well-known test for divisibility by 9. To illustrate this, consider the number 145,764. Add the digits, that is, 1, 4, 5, 7, 6, and 4, to get 27. This test tells us that since this sum is divisible by 9, the number itself is divisible by 9. If the sum had not been divisible by 9, the number given would not have been divisible by 9. The test then is this: *It is true that a number is divisible by 9 if the sum of its digits is divisible by 9 and only in that case.*

It is not hard to prove this test. First let us see how it goes for the number 1457. This number can be written

$$1000 + 400 + 50 + 7 = 10^3 + 4 \cdot 10^2 + 5 \cdot 10 + 7.$$

This number will be divisible by 9 if it is congruent to 0(mod 9). Now $10 \equiv 1 \pmod 9$ and, from properties (4co) and (5co) (see Exercise 6 above), we have

$$1457 \equiv 1^3 + 4 \cdot 1^2 + 5 \cdot 1 + 7 = 1 + 4 + 5 + 7 = 17 \equiv 8 \pmod 9.$$

Since 17 is not divisible by 9, 1457 is not. We have actually shown more than we set out to do. We have shown that the remainder when 1457 is divided by 9 is 8. In this case, the remainder when the number is divided by 9 is the same as the remainder when the sum of its digits is divided by 9. (Instead of dividing 17 by 9 to get its remainder, we could have added 1 and 7.)

It is true that the remainder when any number is divided by 9 is the same as the remainder when the sum of its digits is divided by 9. Let us see how the proof goes for any number of four digits. Such a number could be written $10^3 a + 10^2 b + 10c + d$. This number is congruent to

$$1^3 a + 1^2 b + 1 \cdot c + d \equiv a + b + c + d \pmod 9.$$

This completes the proof.

This result has a useful application in the process which goes by the name "casting out the nines"[1]—an excellent and ancient way of checking addition and, for that matter, subtraction, multiplication, and division. For purposes of illustration consider the sum

$$\begin{array}{r} 137 \\ 342 \\ 890 \\ \hline 1369 \end{array}$$

---

[1] See topic 2 at the end of the chapter.

We compute the remainders when divided by 9 by adding the digits once, and perhaps twice, as follows:

$$137 \equiv 11 \equiv 2(\text{mod } 9)$$
$$342 \equiv 9 \ \ \equiv 0(\text{mod } 9)$$
$$890 \equiv 17 \equiv 8(\text{mod } 9)$$

$$\overline{1369 \equiv 19 \equiv 1(\text{mod } 9)}$$

Since $2 + 0 + 8 \equiv 1(\text{mod } 9)$, our remainders check. In practice it is, of course, not necessary to write the (mod 9), and one may drop our multiples of 9 as one goes along. For instance, in the 890 one can omit the 9, arriving at the remainder 8 directly; in 1369 we can drop out the 9 and the 3 and 6, since $3 + 6 = 9$. Thus we can "cast out the nines" wherever they occur.

It should be pointed out that the tests for divisibility we have given so far in this section depend on our number being written in the decimal system. In other systems of numeration, the tests would be different. Suppose we consider, for instance, the number system to the base seven. If we let s stand for "seven" and if 1364 were a number written to the base seven, then this would mean

$$n = 1 \cdot s^3 + 3 \cdot s^2 + 6 \cdot s + 4.$$

Then we can get the remainder after division by 6 by adding the digits, since $s \equiv 1(\text{mod } 6)$ shows that

$$n \equiv 1 + 3 + 6 + 4(\text{mod } 6).$$

Hence in this number system we can check for divisibility by six by adding the digits. The test for divisibility by seven is here very easy: A number written to the base seven will be divisible by seven if and only if its last digit is zero.

7. REMAINDERS FOR LARGE NUMBERS

This congruence notation is useful in finding remainders when certain large numbers are divided by small ones. For instance,

$$8^{20} - 1$$

is a number with 19 digits, and to find its value and divide by 7 would be a laborious task; but on the circle with 7 divisions, it has the same value as $1^{20} - 1 = 0$, since 8 has the value 1 on the circle. In congruence notation this is

$$8^{20} - 1 \equiv 1^{20} - 1 \equiv 0(\text{mod } 7).$$

This also affirms that the remainder when 8^{20} is divided by 7 is 1.

To find the remainder when the same number, $8^{20} - 1$, is divided by 5, we have the series of congruences

$$8^{20} - 1 \equiv 3^{20} - 1 \equiv (3^2)^{10} - 1 \equiv 4^{10} - 1 \equiv (4^2)^5 - 1 \equiv 1^5 - 1$$
$$\equiv 0 \pmod 5.$$

Hence $8^{20} - 1$ is divisible by 5. A shorter way to arrive at this result would be to notice that $4 + 1 \equiv 0 \pmod 5$ implies $4 + 1 + (-1) \equiv 0 + (-1) \pmod 5$ by property (4co) and thus $4 \equiv -1 \pmod 5$. Now, from property (9) of Section 4, $(-1)(-1) = (-1)^2 = 1^2 \equiv 1 \pmod 5$ and $(-1)^{10}$, being (-1) taken ten times as a product, is equal to 1. [In fact, $(-1)^u = 1$ if u is even and is equal to -1 if u is odd.] Thus

$$4^{10} - 1 \equiv (-1)^{10} - 1 = 1 - 1 \equiv 0 \pmod 5.$$

EXERCISES

1. Find the remainder when 9^{30} is divided by 10, that is, the last digit in the given number. What is the remainder when the same number is divided by 4?

2. Find the last digit in the number 7^{43}.

3. Prove that if x is an integer, then $x^3 \equiv 0, 1$ or $-1 \pmod 7$.

4. Prove that the square of any odd number is one more than a multiple of 8.

5. (From reference 21) A dog told a cat that there was a mouse in the five-hundredth barrel. "But," said the cat, "there are only five barrels there." The dog explained that you count like this:

$$
\begin{array}{ccccc}
1 & 2 & 3 & 4 & 5 \\
9 & 8 & 7 & 6 & \\
 & 10 & 11 & 12 &
\end{array}
$$

    and so on,

    counting back and forth along the line of barrels as indicated, so that the seventh barrel, for instance, would be the one marked 3, the twelfth barrel the one numbered 4. What was the number on the five-hundredth barrel?

6. Prove that for every number of five digits it is true that the remainder when it is divided by 9 is the same as the remainder when the sum of its digits is divided by 9.

7. Show that by adding the digits of the number, then adding the digits of the sum of the digits, and so on, it is true that to find the remainder when any number is divided by 9 the only division necessary is dividing 9 by 9.

8. Give a test for divisibility by 4.

9. Is there any way to tell solely by the last digit in a number whether or not the number is divisible by 9? By 7? Give your reasons.

10. What is a test for divisibility by 11? Is there any way of comparing the remainders when the number itself and the sum of its digits is divided by 11, as was done previously for 9? If not, can the sum of the digits be replaced by some other simple expression in terms of the digits to give the desired result? Prove your statement for any number of four digits.

11. In a number system to the base twelve, for what number b would the following be true: A number is divisible by b if the sum of its digits is divisible by b and only in that case.

12. In Exercise 11, would there be more than one number b for which the statement could be made?

13. In a number system to the base thirteen, for what numbers b could the statement in Exercise 11 be made?

14. Show that the remainder when $a + b$ is divided by 37 is equal to the remainder when $1000\, a + b$ is divided by 37. For example: $457 + 324$ and $457,324$ have the same remainders when divided by 37.

15. Show that the remainder when $b - a$ is divided by 7 is equal to the remainder when $b + 1000\, a$ is divided by 7. For example, $457 - 324$ and $324,457$ have the same remainders when divided by 7.

16. Can you find any other numbers which have simple tests for divisibility?

17. What is a test for divisibility by 2 in the number system to the base three?

18. Check the result of multiplying 257 by 26 by casting out the nines. Show why it works.

19. If a set of numbers is correctly added, will it always check under casting out the nines? If a sum checks by casting out the nines is the addition necessarily correct?

20. For the dozal system of notation, what would be the check corresponding to casting out the nines?

21. The result for 1457 in the second paragraph of Section 6 could have been obtained more quickly by dividing 1457 by 9 and finding the remainder. Why was this not done?

22. Show how the following trick can be performed and why it works. Also mention any possibility of failure. You select any number, form from it another number containing the same digits in a different order, subtract the smaller of the two numbers from the larger, and then tell me all but one of the digits of the result. I can then (perhaps) tell you what is the other digit of the result.

*23. If N is any number of four digits (that is, between 1000 and 10,000), show that the sum of the digits of $99N$ is 18, 27, or 36. This may be made the basis of a trick as follows. Memorize the 18th name on page 18, the 27th name on page 27, and so on, of your local telephone directory. Then announce to a friend that you have memorized the telephone directory and by way of proving it ask him to select a number unknown to you, multiply it by 100, subtract the number from the result, and add the digits of the final result. He tells you his answer and you tell him he can find such and such a name in, say, the 27th place on page 27. One variation of this trick is not to ask for the number but to find for yourself which it is by asking if it is odd or even, less than 30 or more than 30.

*24.   Can you devise any tricks similar to the one in Exercise 23?

8. OTHER GROUPS

Groups consisting of multiplication or addition on a circle are not by any means the only kinds of groups. In this section we shall discuss other groups, one of which is not Abelian.

First recall that the addition table of numbers on a circle of three divisions looks like this:

$$
\begin{array}{c|ccc}
+ & 0 & 1 & 2 \\
\hline
0 & 0 & 1 & 2 \\
1 & 1 & 2 & 0 \\
2 & 2 & 0 & 1 \\
\end{array}
$$

Call this group $A_1(3)$.

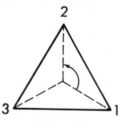

Figure 4:2

Next consider the rotations about its center of the equilateral triangle in Fig. 4:2. We confine our attention to the three rotations which take the triangle into coincidence with itself, that is, more briefly, "take the triangle into itself." These rotations we call A, B, and C, where A stands for no rotation, B a rotation of 120°, and C a rotation of 240°, all in a counterclockwise direction. We define the "sum" of two rotations to be the result of performing first one and then the other. For instance, $B + C$ means "rotate through 120° and then through 240°." This has the same result as not moving the triangle at all. Hence we say that $B + C = A$. The addition table is

$$
\begin{array}{c|ccc}
+ & A & B & C \\
\hline
A & A & B & C \\
B & B & C & A \\
C & C & A & B \\
\end{array}
$$

Let us test this for the properties of a group. Since the sum of two rotations which takes the triangle into itself takes the triangle into itself (or since only A, B, and C occur in the body of the addition table) the property of closure holds. Now A is the identity element, since,

$$A + A = A, \qquad A + B = B, \qquad A + C = C,$$

and there is an inverse, since each row contains all the letters. Verifying the associative property directly would be somewhat tedious, and we show this later. Hence we have a group. Furthermore, it may be seen that it is an Abelian group. Call it $A_2(3)$.

Although these groups had different origins, it is easy to see that if we replace A by 0, B by 1, and C by 2 in the second addition table we get the first table. Any two groups which have the same addition tables except for the letters or symbols involved are called *isomorphic*. Two isomorphic groups are essentially the same. In this case it may be seen merely by connecting by lines the successive divisions of the circle for $A_1(3)$ and replacing the 3 with a 0. Adding numbers on that circle is essentially the same as rotation through $0°$, $120°$, and $240°$. As a matter of fact, noticing such a relationship in the beginning (perhaps you did) would have eliminated the necessity of showing that in both cases the properties of a group hold.

Now we develop a third group $A_3(3)$, which, in spite of its different origin, turns out to be isomorphic with these two. Write

$$a = \begin{pmatrix} 1 & 2 & 3 \\ 1 & 2 & 3 \end{pmatrix}, \qquad b = \begin{pmatrix} 1 & 2 & 3 \\ 2 & 3 & 1 \end{pmatrix}, \qquad c = \begin{pmatrix} 1 & 2 & 3 \\ 3 & 1 & 2 \end{pmatrix},$$

where the symbol a means "move 1 into 1, 2 into 2, and 3 into 3"; b means "move 1 into 2, 2 into 3, and 3 into 1"; c means "move 1 into 3, 2 into 1, 3 into 2." Each element of the group is then a reordering, or "permutation," of the numbers 1, 2, and 3. The "product" of any two elements of this group we understand to be that permutation which results from performing first one and then the other. For instance, bc means "first do b and then[1] c." What is the result? Notice that b moves 1 into 2 and c moves 2 into 1 and hence bc moves 1 into 1; b moves 2 into 3 and c moves 3 into 2 and hence bc moves 2 into 2. Similarly, bc moves 3 into 3. This is just what permutation a does. Thus we write $bc = a$. In this fashion we can form the multiplication table

·	a	b	c
a	a	b	c
b	b	c	a
c	c	a	b

where the first line gives the products a^2, ab, ac in order, and so on.

---

[1] In some situations, the opposite interpretation is made.

We could verify in detail that this is a group, but it is simpler to compare it with the previous group and see that it is the same except that we have used lowercase letters instead of capital letters. Moreover, consider Fig. 4:2 and think of a triangle made of cardboard. We consider permutation b to mean "Move vertex 1 into position 2, vertex 2 into position 3, and vertex 3 into position 1." This is the same as rotating the triangle in a counterclockwise direction about its center through an angle of 120°. Similarly, the permutation c is equivalent to a rotation through 240°. This shows directly that $A_3(3)$ is isomorphic to the group of rotations through 0°, 120°, and 240°.

Notice now that the rotations are not the only ways of taking our equilateral triangle into itself. We might keep one vertex fixed and interchange the other two: a reflection about an altitude. This gives us three other permutations:

$$d = \begin{pmatrix} 1 & 2 & 3 \\ 1 & 3 & 2 \end{pmatrix}, \qquad e = \begin{pmatrix} 1 & 2 & 3 \\ 3 & 2 & 1 \end{pmatrix}, \qquad f = \begin{pmatrix} 1 & 2 & 3 \\ 2 & 1 & 3 \end{pmatrix}.$$

Here, if we are to consider these as things to be done to a triangle, an important caution is in order. Just as b is a rotation through 120° regardless of what the particular numbering is, so d is a reflection in the altitude through the vertex at the lower right-hand corner (the 1 position), e the reflection in the vertical altitude (the 2 position), and f the reflection in the altitude through the vertex in the lower left-hand corner (the 3 position). We illustrate this for the products db and bd (Fig. 4:3). Notice that bd and db are not the same. This

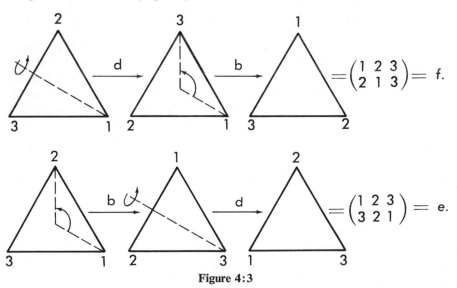

Figure 4:3

means that if we have a group it is not Abelian. By this means we can form the following multiplication table (you should check the entries):

	a	b	c	d	e	f
a	a	b	c	d	e	f
b	b	c	a	e	f	d
c	c	a	b	f	d	e
d	d	f	e	a	c	b
e	e	d	f	b	a	c
f	f	e	d	c	b	a

The table shows that we have closure, because only the six letters appear in the body of the table; the element a is the identity. Since each row contains a, there is always an inverse. The associative property is quite obvious, because $(xy)z$ and $x(yz)$ both mean: Do x, then y, and then z. So we have here a group with six elements. But it is not Abelian, since, for example, $bd \neq db$. (Remember that bd means: "Do b first and then d.") Actually this is the smallest group which is not Abelian; all groups with less than six elements are Abelian. We call this group G(3). It is called "the group of symmetries of the equilateral triangle." The term "symmetry" is not unnatural for the permutations d, e, and f, which are reflections in an altitude. The rotations are a kind of symmetry about the center of the triangle.

If we select the elements a, b, and c from G(3) we have the group $A_3(3)$, which is called a *subgroup* of G(3). The elements a and f form another subgroup of G(3). However, the elements a, b, c, and d do not form a group, because cd is not one of a, b, c, and d, and the closure property thus fails to hold. It is a property of groups which we shall not attempt to prove here—that the number of elements in any subgroup is a divisor of the number of elements in the group. Notice that 2 and 3 are divisors of 6, but 4 is not. Also, since every group contains an identity element, every subgroup must contain the identity element of the group.

It is not hard to show that every group with just three elements is isomorphic to every other group with just three elements. Let e be the identity element and r and s the other two elements of the group. Then we know the following part of the multiplication table:

	e	r	s
e	e	r	s
r	r		
s	s		

Keeping in mind the restriction that each row and column must contain e, r, and s in some order, we see that there is only one way in which to fill in the blank part of the table, for s cannot occur in the last place in the second row and hence must occur in the second with e in the last. This determines the second and hence last row and we have the multiplication table.

	e	r	s
e	e	r	s
r	r	s	e
s	s	e	r

which, except for the letters involved, is the same as the preceding ones.

It is sometimes difficult to tell when two given groups are isomorphic, but certain facts are a help. Since each has only one identity element, these must correspond to each other. Then if b is another element of one group and B of the other group, we can look at the powers of b and B. If we find, for instance, that b^2 is equal to the identity for its group and B^2 is not the identity for its group, we know that B and b cannot correspond to each other. We can also sometimes tell by the general properties of the groups. For instance, the group G(3) cannot be isomorphic to the group for addition on the circle of 6 divisions, even though the two have six elements each, because the former is not Abelian and the latter is.

Sometimes we want to find whether a subset of elements of a group itself forms a group. Let G denote the group and S a subset of its elements. We do not have to check for the associativity property because, if the whole set G has this property, any subset must have. We do have to check S for closure and it is easy to see whether or not S contains the identity element, although what follows shows that for a finite set, closure implies that the identity is in the set. The remaining property affirms that the inverse of any element in S is in S. If S is a *finite* subset, we can show that this is implied by closure. To this end, suppose S is a subset of elements of G which has the closure property. Either S consists of the identity element alone, in which case it is a trivial group, or it contains an element b which is not the identity. Then, if we take the operation of the group to be multiplication, the closure property requires that S contains

$$b, b^2, b^3, b^4, \ldots.$$

In fact, all powers of b are in S. But S being finite implies that the powers of b cannot all be different. Suppose then that $b^r = b^s$, where $0 < r < s$. Since G is a group, the inverse of b, call it b^{-1}, is in G and $b^r = b^s$ implies, if $r > 1$, that

$$b^{-1} \cdot b \cdot b^{r-1} = b^{-1} \cdot b \cdot b^{s-1}; \quad \text{that is,} \quad b^{r-1} = b^{s-1}.$$

Continuing in this fashion we see that in G, $e = b^u$ for some positive integer u. Since all powers of b are in S, then e is in S as well as b^{u-1}, the inverse of b. Thus we have shown: If S is a finite subset of a group G, which has the closure property, then it is a group itself.

EXERCISES

1. Find all the subgroups of G(3).
2. If H is the group of integers 1, 2, 3, 4, 5, 6(mod 7) with multiplication, is it isomorphic to G(3)?
3. Is the group H of Exercise 2 isomorphic to the set of numbers on the circle of six divisions with addition?
4. Prove that the elements of G(3) can be written

$$a, b, b^2, d, db, db^2$$

and that the multiplication table can be constructed from the following: $b^3 = a, d^2 = a, bd = db^2$, and the knowledge that the associative law holds. (The elements b and d are called *generators* of the group.)
5. Find the multiplication table of each of the following permutation groups:

(1) $a = \begin{pmatrix} 1 & 2 & 3 & 4 \\ 1 & 2 & 3 & 4 \end{pmatrix}$, $b = \begin{pmatrix} 1 & 2 & 3 & 4 \\ 2 & 1 & 4 & 3 \end{pmatrix}$, $c = \begin{pmatrix} 1 & 2 & 3 & 4 \\ 3 & 4 & 1 & 2 \end{pmatrix}$, $d = \begin{pmatrix} 1 & 2 & 3 & 4 \\ 4 & 3 & 2 & 1 \end{pmatrix}$.

(2) $a = \begin{pmatrix} 1 & 2 & 3 & 4 \\ 1 & 2 & 3 & 4 \end{pmatrix}$, $b = \begin{pmatrix} 1 & 2 & 3 & 4 \\ 2 & 3 & 4 & 1 \end{pmatrix}$, $c = \begin{pmatrix} 1 & 2 & 3 & 4 \\ 3 & 4 & 1 & 2 \end{pmatrix}$, $d = \begin{pmatrix} 1 & 2 & 3 & 4 \\ 4 & 1 & 2 & 3 \end{pmatrix}$.

Show that each of the groups above has the closure property and that division is always possible. Point out which are Abelian.
6. Show that all groups containing exactly four elements are isomorphic to one of the groups of Exercise 5.
7. Consider the group of rotations of a square through multiples of 90° about its center. Which group in Exercise 5 is it isomorphic to?
8. A square can be reflected in four lines through its center: horizontal, vertical, and two through pairs of opposite vertices. What permutations do these correspond to? Are any listed in Exercise 5? Do these form a group?
9. H is the group formed by the numbers of a circle with 6 divisions and the process of addition. K is the group formed by the numbers 1, 2, 4, 5, 7, and 8 on a circle with 9 divisions and the process of multiplication. Are H and K isomorphic? Is K isomorphic to G(3)?
10. Is either of the groups of Exercise 5 isomorphic with multiplication or addition on a properly chosen circle?
11. Consider four different things which may be done to an implication: i means "leave it alone," c means "take the contrapositive," v means "take the converse," and o means "take the obverse." Find the multiplication table and show that these four form a group. Which group in Exercise 5 is it isomorphic to?
12. Show that the numbers 1, 5, 7, and 11 with multiplication (mod 12) form a group. Which group in Exercise 5 is it isomorphic to?
13. Show that the numbers 1, 4, 13, and 16 with multiplication (mod 17) form a group. Which group in Exercise 5 is it isomorphic to?

*14. A group has the elements e, v, u, u^2, u^3, vu, vu^2, and vu^3, where e is the identity element, $u^4 = e = v^2$ and $vu = u^3v$. Write the complete multiplication table and point out any relationships you can find between it and the groups of Exercise 5. Is there any relationship like those indicated in Exercises 7 or 10?

9. TOPICS FOR FURTHER STUDY

Where no reference is given, you should not conclude that there are no references but merely that you should have more profit and pleasure from investigating the topic without consulting references.

1. Some properties of powers of numbers on a circle (see Section 7).
2. Tests for divisibility: reference **8**.
3. With equipment given in this chapter, any of the following may be accomplished:
   a. Prove: If a is an element of a group with a finite number of elements and t is the least power of a such that $a^t = e$, where e is the identity element, then t is a divisor of the number of elements of the group.
   b. Find all the groups (no two isomorphic) with five elements. What ones of these can be obtained by multiplication or addition on a circle?
   c. Prove: If a is a fixed element of a group, then the set of elements y of the group for which $ay = ya$ themselves form a group.
   d. Find three groups with eight elements, no two of which are isomorphic. (There are five such groups.)
4. Find a method for solving such problems as the following: What is the smallest number which is equal to 2 on the circle of 3 divisions, 5 on the circle of 7 divisions, and 8 on the circle of 11 divisions?
5. Congruences: reference **36** (Chap. 2).
6. Group: references **7** (pp. 278–283), **10** (Chap. 6), **48** (Vol. 3, part 9), and **53** (Chap. 18).

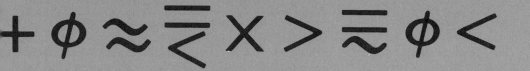

5

Negative Integers
and Rational
and Irrational Numbers

1. INVENTIONS

The numbers 0, 1, 2, 3, ... used throughout Chapter 3 are called *non-negative integers*; 0 is zero, and 1, 2, 3, 4, ... are the positive integers. These numbers, the decimal system for the representation of them, and the rules for computing with them were created by human ingenuity to meet various human needs. To assume that men have always known and been able to use the results of Chapter 3 is like assuming that they have always had motorcycles and been able to ride them. The fact is that numbers and vehicles alike are not mysterious creations of nature plucked from trees or hooked in lakes; they are "of the people, by the people, and for the people." Just as there are several kinds of vehicles (oxcart, wagon, buggy, train, steamship, motorcycle, automobile, submarine, dirigible, airplane), so there are several kinds of numbers (zero, positive integers, negative integers, primes, perfect numbers, rational numbers, algebraic numbers, irrational numbers, real numbers, imaginary numbers, complex numbers). Just as new kinds of vehicles have been invented to have new uses, new kinds of numbers have been invented to have new uses. In this chapter we first introduce negative integers and find that the set of all integers (which includes 0, the positive integers, and the negative integers) has an important property which the smaller set of positive integers fails to have: that, for every pair of integers a and b, there is an integer x such that $a + x = b$.

Alice has something to say on the subject.

"And how many hours a day did you do lessons?" said Alice, in a hurry to change the subject.

"Ten hours the first day," said the Mock Turtle; "nine the next, and so on."

"What a curious plan!" exclaimed Alice.

"That's the reason they're called lessons," the Gryphon remarked, "because they lessen from day to day."

This was quite a new idea to Alice, and she thought it over a little before she made her next remark. "Then the eleventh day must have been a holiday?"

"Of course it was," said the Mock Turtle.

"And how did you manage on the twelfth?" Alice went on eagerly.

"That's enough about lessons," the Gryphon interrupted in a very decided tone; "tell her something about the games now."

Our immediate association of negative numbers with the above quotation bears witness to the fact that we are so accustomed to various *interpretations* of negative numbers that we are apt to forget that their origin is really an *invention*. But we are getting ahead of our story.

2. NEGATIVE INTEGERS

From a mathematician's point of view, we introduce the negative integers because the positive integers and zero do not form a group under addition. The latter have the first three propertes of a group, but there is not in general an additive inverse. The negative numbers fill this gap. If we were to be rigorous we would need to show or so define the negative integers that the complete set of integers (positive, zero, and negative) have all the properties of of a group. In addition, we would want the group to be Abelian. Here we just assume that the whole set of integers forms a commutative additive group.

Here let us look carefully at some of the properties of the negative integers as part of the set of integers. In the first place we define $(-a)$ by the equation

$$(1) \qquad\qquad (-a) + a = 0$$

when a is a positive integer or zero. Since we want this to hold also when a is a negative integer, we define $(-a)$ to be the additive inverse of a no matter whether a is positive, zero, or negative. Then (1) shows that not only is $(-a)$ the additive inverse of a, but also a is the additive inverse of $(-a)$; hence

$$(2) \qquad\qquad -(-a) = a$$

for all integers a. Thus the additive inverse of the additive inverse of a number is the number itself.

3. ADDITION OF INTEGERS

Now notice that we have two different meanings of the minus sign. Here we use it to denote the additive inverse. Previously when we wrote $b - a$ we meant that number which when added to a gives b; but this had meaning only when b was greater than or equal to a. To reconcile the two meanings of the minus sign we want to have

$$(3) \qquad b - a = b + (-a) = (-a) + b,$$

no matter what integers a and b are. To justify this we need notice merely that

$$[b + (-a)] + a = b + [(-a) + a] = b + 0 = b.$$

So $b + (-a)$ has the property that if we add a to it we get b. This is just what we had agreed previously should be the meaning of $b - a$ when b is greater than or equal to a. Since previously $b - a$ had no meaning when b is less than a, we are now free to give it the meaning in (3) in all cases.

For example, $5 - 3 = 2 = 5 + (-3)$. What about $3 - 5$? In this case, $3 - 5 = (-2)$, since if we add 5 to (-2) we get 3. Notice that in this case $3 - 5 = -(5 - 3)$. Using letters we have

$$a - b + (b - a) = a + (-b) + b + (-a) = 0,$$

and thus

$$(4) \qquad a - b = -(b - a).$$

In other words, $a - b$ is the opposite or additive inverse of $b - a$, no matter what the relative values of a and b.

We leave as an exercise the proof of the following:

$$(5) \qquad (-a) + (-b) = -(a + b)$$

Notice, again, that in (1), (2), (3), (4), and (5) the letters a and b can stand for any integers, positive, zero, or negative.

Here we should take note of another convention in mathematics. The expression $18 - 10 - 4$ *could* have two meanings:

$$18 - (10 - 4) \qquad \text{or} \qquad (18 - 10) - 4.$$

We choose the latter because we want to have

$$18 - 10 - 4 = 18 + (-10) + (-4).$$

Thus $18 - 10 - 4$ is equal to 4 and not to $18 - 6 = 12$.

EXERCISES

1. Express each of the following as a single integer:
   a. $7 - 11 + (3 - 4)$.
   b. $-(18 - 25) - (13 - 21)$.
   c. $-(-22 + 18)$.
   d. $10 - 3 - 2 + 5 - 7$.

2. Express each of the following as a single integer:
   a. $9 - 15 + (6 - 7)$.
   b. $-(10 - 15) - (17 - 36)$.
   c. $-(-55 + 21)$.
   d. $13 - 6 + 7 - 2 - 3$.

3. If the temperature at 6:00 P.M. is $+5°F$ and the temperature falls $8°F$ between 6:00 and midnight, what is the temperature at midnight?

4. Show that $(2 + 7) - 8 = 2 + (7 - 8)$ and that $2 - (8 - 7) = (2 - 8) + 7$. Express in letters the properties of which these are examples.

5. Show property (5) and explain why it is really just another form of (4).

6. Give numerical examples of (3) and (4) when $a > (-b) > 0$.

4. MULTIPLICATION OF INTEGERS

We found in Chapter 3 that if a and b are positive integers, then the product ab can be visualized as the total number of objects in a sets of which each set contains b things. The question of evaluating the product ab when one (or both) of a and b is negative is more delicate. As an actual matter of fact, such products have not yet been defined in this course and they are entirely meaningless until meaning has, in some way or other, been attached to them. We are at liberty to define these products as we please, but it is convenient to define them so that certain properties hold.

First, suppose a and b are positive integers; what do we want $b(-a)$ to be? This should be

$$(-a) + (-a) + \cdots + (-a),$$

where the $(-a)$ appears b times. By our definition of the addition of negative numbers, this must be

$$-(a + a + \cdots + a),$$

where the a occurs b times; that is, $-(ba) = -(ab)$. Thus from our previous definitions it follows that $b(-a) = -(ab)$. Then since we want the commutative property to hold for such multiplication we have by definition

$$b(-a) = (-a)b = -(ab),$$

where a and b are positive integers. If either or both are zero, this also holds.

It remains to define $(-a)(-b)$ for positive integers a and b. Here the distributive property can be used to determine what the definition is to be. On the one hand,

$$(-a)[b + (-b)] = (-a) \cdot 0 = -(a \cdot 0) = -0 = 0.$$

If the distributive property is to hold,

$$0 = (-a)[b + (-b)] = (-a)b + (-a)(-b) = -(ab) + (-a)(-b).$$

This means that if $(-a)(-b)$ is added to $-(ab)$ we get zero. But, when ab is added to $-(ab)$ we get zero. Hence we define the product of negative numbers so that

$$(-a)(-b) = ab.$$

To summarize our definitions we have

(1)
$$(-a)b = b(-a) = -(ab),$$
$$(-a)(-b) = ab$$

where a and b are any positive integers or zero. Notice that we had the same results for multiplication on a circle.

Actually, equations (1) are defined to hold even when a and b do not stand for positive integers or zero. One could show this for all integers as a consequence of other properties just as we showed properties (8) and (9) of a field in Section 4 of Chapter 4. But, for our purposes it is simpler just to assume (1) for any integers a and b. For instance, if $a = (-5)$ and $b = 3$, then $(-a) = 5$ and $(-a)b = 5 \cdot 3 = 15$. This is equal to minus the product of a and b, since this product is (-15) and its opposite is 15.

It is especially important to notice here our convention for expressions involving subtraction and multiplication. For instance, $2 - 3 \cdot 5$, by our agreement, means: Multiply 3 by 5 and subtract the result from 2; that is,

$$2 - 3 \cdot 5 \text{ means } 2 - (3 \cdot 5), \qquad \text{that is, } -13.$$

This is quite a different result from that of subtracting 3 from 2 and multiplying the result by 5:

$$(2 - 3) \cdot 5 = -5.$$

EXERCISES

1. Verify equations (1) for the following pairs of values of a and b: (1) $a = 5$, $b = (-7)$; (2) $a = (-3)$, $b = (-2)$.
2. Verify equations (1) for the following pairs of values of a and b: (1) $a = (-8)$, $b = (-4)$; (2) $a = (-3)$, $b = 6$.
3. Carefully use the definitions of products to show that
$$-3(2 - 7) = (-3)2 + (-3)(-7).$$
4. In each of the following expressions insert one pair of parentheses in as many ways as possible (there are 15 ways) and evaluate the resulting expressions:

    a. $1 - 2 - 3 - 4 - 5.$ **b.** $1 - 2 \cdot 3 - 4 - 5.$

How many different results are there? Which are equal to the expression written without parentheses?

5. Pick out all correct right-hand sides for each of the following equations and give reasons for your answer:

a. $Sr - S = \begin{cases} r(S - 1) \\ S(r - 1) \\ r(S - 1) + r - S \end{cases}$

b. $ax - ay + bx - by = \begin{cases} (a - b)(x + y) \\ (a + b)(x - y) \\ (a + x)(b - y) \end{cases}$

c. $ab - ac + bc = \begin{cases} a(b - c) + bc \\ c(a - b) + ab \\ b(a - c) + ac \\ -b(-a - c) - ac \end{cases}$

6. Prove that $(-ab)c = -a(bc)$, where a, b, and c are nonnegative integers, assuming the associative property of multiplication for nonnegative integers.

7. Show that $(-ab)c = -(a(bc))$ for exactly one of a, b, and c negative (three cases), assuming equations (1) for all integers and the associative property for multiplication of nonnegative integers.

8. Show that $(-1)a = (-a)$ and $(-1)(a + (-b)) = (-a) + b$.

9. Without assuming the results of Exercise 8, but assuming the distributive property for integers and (1) above, show that

$$(-a)(b - c) = -(ab) + ac = ac - ab.$$

5. INEQUALITIES AND ABSOLUTE VALUES

We all know that 2 is less than 3, 7 is less than 10, and so forth. When it comes to inequalities involving negative numbers, the relationship is not quite so clear. One way of defining inequality is with reference to the number line in Fig. 5:1, where the positive integers appear to the right of 0 on the line and the negative integers on the left. Then a number a is said to be less than a number b if the point corresponding to a is to the left of the point corresponding to b.

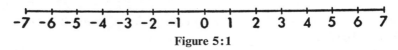

Figure 5:1

We can also define inequality arithmetically as we did in Section 11 of Chapter 3. Thus

$$a < b \text{ means } a + x = b, \quad \text{for some positive integer } x.$$

It is also convenient to have a notation to indicate how many units from 0, regardless of direction, a given number is. For instance, -3 and $+3$ are the same number of units from 0 on the number line but in opposite directions; each is 3 units from 0. We designate this by two vertical lines and call it the *absolute value*; that is, the absolute value of both -3 and 3 is 3, and we designate it by

$$|-3| = |3| = 3.$$

In general,

if a is positive, $\quad |a| = a;$
if a is negative, $\quad |a| = -a;$
if $a = 0$, $\quad\quad\quad |a| = 0.$

Thus, except for zero, the absolute value of a number is always positive.

The absolute value can be used not only to designate the distance between the point 0 and another point but also the distance between two points. To see this, consider two points whose coordinates are a and b, respectively. The distance between these two points will be the same as the distance between $a - x$ and $b - x$, no matter what number x is. If we then choose $x = b$, we see that the distance between a and b is the same as the distance between $(a - b)$ and 0, which is $|a - b|$. That is: The distance between two points with coordinates a and b is

$$|a - b|.$$

Now recall the "well-defined" properties of equality (see Section 5 of Chapter 4). These were

(1) If $a = b$, then $a + c = b + c$.

(2) If $a = b$, then $ac = bc$.

These may be interpreted in several ways. They affirm that we may add the same quantity to both sides of an equality or multiply both sides by the same quantity without altering the fact of equality. It also means that in the expressions $a + c$ and ac we may replace a by a number b which is equal to it without changing the number which the sum or product represents.

Do these properties hold for inequalities? Our first question would be for integers a, b, and c:

If $a < b$, does it follow that $a + c < b + c$?

Looking at the numbers on the line we see that $a < b$ means that the symbol a is to the left of the symbol b. If c is positive, the point corresponding to $a + c$ will be c units to the right of the point corresponding to a, and the point corresponding to $b + c$ will be c units to the right of the point corresponding to b; thus the point $a + c$ will still be to the left of $b + c$. If c is negative, the same conclusion holds.

We can also obtain the same result arithmetically: $a < b$ means that there is a positive integer x such that $x + a = b$. Then using the well-defined property for equality we may add c to both sides of the equality and have $(x + a) + c = b + c$. The associative property yields

$$x + (a + c) = b + c,$$

which proves our result.

The second question would be, for integers a, b, and c:

If $a < b$, does it follow that $ac < bc$?

Here we immediately see that some exception must be made, since if $c = 0$, $ac = bc$. For this question the number line is not as useful as the numerical treatment. So we start with $x + a = b$ for some positive integer x. By the well-defined property for equality we may multiply both sides of the equation by c and have

$$c(x + a) = cb.$$

Using the distributive property we have

$$cx + ca = cb.$$

If cx is positive, then $ca < cb$; if $cx = 0$, then $ca = cb$ and if cx is negative, then $ca > cb$. Hence, since x is positive, we have $ca < cb$ if c is positive and $ca = cb$ if $c = 0$. Now, suppose that cx is negative; that is, c is negative. Then $-(cx)$ is positive and we have

$$-(cx) + cx + ca = -(cx) + cb,$$
$$ca = -(cx) + cb.$$

This shows that ca is the sum of cb and the positive integer $-(cx)$ and hence $cb < ca$. Thus, summarizing our results, for the well-defined properties we have

(1i) If $a < b$, then $a + c < b + c$.

(2i) If $a < b$, then $ac < bc$ if c is positive,
                        $ac = bc$ if $c = 0$,
                        $ac > bc$ if c is negative.

We have assumed two other properties of inequality which, for completeness, should be listed here:

(3i) If a and b are any two integers, exactly one of the following relationships holds:

$$a < b, \qquad a = b, \qquad b < a.$$

(4i) If $a < b$ and $b < c$, then $a < c$ (the transitive property). (Compare these with the properties of set inclusion in Chapter 1, Section 5.)

EXERCISES

1. In each of the following, replace ? by $>$, $<$, or $=$.
    a. 1 ? -5.   **b.** -4 ? -7.
    c. If $a < b$, then $a - 5$? $b - 5$; $7a$? $7b$; $-8a$? $-8b$.

2. In each of the following, replace ? by $<$, $>$, or $=$.
   a. -3 ? -2.   b. -2 ? 1.
   c. If $a < b$, then $a - 3$? $b - 3$; $5a$? $5b$; $-3a$? $-3b$.

3. Suppose in (1i) and (2i) that $a < b$ were replaced by $a > b$; how should the conclusions be altered?

4. Suppose in (1i) and (2i) that $a < b$ were replaced by $a \leq b$ (meaning "a is less than or equal to b"); how should the conclusions be altered? In this case show that (2i) can be stated with only two parts instead of three.

5. Suppose in (1i) and (2i) that $<$ were replaced by $\geq$; how should the conclusions be altered?

6. If $a \geq b \geq c$ or $c \geq b \geq a$, show that $|a - b| + |b - c| = |a - c|$.

7. Prove the converse of Exercise 6.

8. If $ac < bc$, under what conditions will it follow that $a < b$? That $a > b$? Is $a = b$ a possibility also? Prove your results.

9. In Exercise 8 we considered division. Should we also consider subtraction? That is, does $a + c < b + c$ imply $a < b$? Why or why not?

10. If $a < b$, does it follow that $a^2 < b^2$? Discuss all possibilities for integers a and b.

11. If $|a| < |b|$, does it always follow that $|ac| < |bc|$? State any exceptions and prove your results.

12. What conclusions can be drawn from $|a| = |b|$?

13. In property (2) of Section 5 in Chapter 1, replace $\subseteq$ by $\leq$ and suppose that A and B denote numbers. Prove, using the four properties of inequalities listed above that

$$A \leq B \quad \text{and} \quad B \leq A \quad \text{implies } A = B.$$

14. We call an integer b "between" the integers c and d if either $c < b < d$ or $d < b < c$. Show that if the integer b is between c and d, and if a is an integer different from zero, then ab is between ac and ad, no matter whether a is positive or negative.

15. If, in properties (3i) and (4i), a and b stand for sets and $<$ is replaced by $\subseteq$, do these properties hold?

*16. Show that $|a + b| \leq |a| + |b|$.

*17. Prove that $|ab| = |a| \cdot |b|$.

*18. If $a < b$, does it follow that $a^3 < b^3$? Discuss all possibilities.

6. INTRODUCTION OF RATIONAL NUMBERS[1]

The mathematical need for an extension of the set of integers and the practical need extend from the same problem: having to divide something into a number of equal parts. Dividing a pie into five equal parts is equivalent in mathematical notation to solving the equation $5x = 1$. Just as in Section 2

[1] See reference 53, Chap. 3.

we created the additive inverse, $(-a)$, of each positive integer a, so we now create the multiplicative inverse, a^{-1} or $1/a$, of each integer a different from zero. These numbers have, by definition, the property that

$$a^{-1}a = 1 = a \cdot a^{-1}; \quad \text{that is, } (1/a)a = 1 = a(1/a).$$

To emphasize this relationship we use the following table:

	Addition	Multiplication
Identity element:	0, since	1, since
	$a + 0 = 0 + a = a$	$a \cdot 1 = 1 \cdot a = a$
Inverse element:	$(-a)$, since	$1/a$, since
	$(-a) + a = a + (-a) = 0$	$(1/a)a = a(1/a) = 1$

When we were considering division of integers, we used the notation b/a to denote that integer, if one exists, such that $(b/a)a = b$; that is, b/a is the solution of the equation $ax = b$. Just as in our development of the negative integers we extended the notation $a - b$ to cases in which $a < b$, so here we extend the notation b/a to cases in which b is not divisible by a. Notice that

$$[b(1/a)]a = b[(1/a)a] = b \cdot 1 = b.$$

Hence, if we identify b/a with $b(1/a)$ we have a solution of every equation of the form $ax = b$, where a and b are integers and $a \neq 0$. These numbers we call the rational numbers. That is: The solutions of all equations of the form $ax = b$, where a and b are integers and $a \neq 0$, constitute the rational numbers. They are designated by b/a or $\dfrac{b}{a}$. These numbers are called rational because they are ratios of integers—not because they have attributes opposite to those of inmates of asylums.

It is opportune here to point out that there are limitations on what numbers we can create. Even an omnipotent Diety cannot create a black rock which is entirely white because being white would be incompatible with the property of being black. So man "made in God's image" cannot create a *number* x such that $0 \cdot x = b$ for $b \neq 0$, since the existence of such a number would be incompatible with our conception of a number. The incompatibility between the existence of such a number and our ideas about numbers can be shown in several ways. One way is to show as follows that multiplication by such a "number" would not be associative. Suppose $0 \cdot x = b$ for some number x when b is an integer. Then, since $0 \cdot x$ and b are the same number, we can multiply them by 0 and have

$$0(0 \cdot x) = 0 \cdot b.$$

But, by the associative law, the left side is equal to $(0 \cdot 0)x = 0 \cdot x = b$, whereas, since b is an integer, we know that the right side, $0 \cdot b$, is zero. This shows that if the associative law is to hold and $0 \cdot x = b$ for some number x, then b must be 0. But if $b = 0$, any number x will do. In other words, the quotient $b/0$ cannot be a number (that is, cannot have all the properties of a number) unless b is 0; in that case $0/0$ can be given any value you please. Since we wish the quotient b/a not only to have the properties of numbers but to have just one value, *we exclude a $= 0$ as a divisor.* We assume, although it is possible to prove it, that there is only one number x such that $ax = b$, when $a \neq 0$.

We have introduced the rational numbers because with just the integers we do not have a multiplicative inverse. This is the only property of a field (see Section 4 of Chapter 4) which the integers lack. We have in this section defined the rational numbers. Now we must introduce other definitions so that they will behave as we want them to, that is, form a field.

7. EQUIVALENCE OF FRACTIONS

We have assumed that there is just one *number x* such that $ax = b$ for given integers $a \neq 0$ and b. But, of course, there may well be a number of different ways to represent this number. For instance, the number 4 can be represented as $2 + 2$, $3 + 1$, $\frac{8}{2}$, four, and so on. So there may be many different fractions which represent the same number. If $ax = b$, then x is represented by the fraction b/a, assuming $a \neq 0$. But if the well-defined property is to hold for multiplication (see Section 5 of Chapter 4),

$$ax = b \qquad \text{implies} \qquad kax = kb$$

for any integer k. This means that kb/ka and b/a must represent the same rational number. So we write

$$\frac{kb}{ka} = \frac{b}{a} \qquad \text{if } k \neq 0.$$

The equals sign here means, of course, that the numbers represented by the fractions are equal. Since the fraction is the symbol and the symbols are not the same, one should, strictly speaking, call the fractions "equivalent" when the numbers they represent are equal.

There is another way to test the equivalence of two fractions by using the above twice. Consider

$$\frac{b}{a} \simeq \frac{bc}{ac}, \quad \frac{d}{c} \simeq \frac{da}{ac},$$

where a and c are assumed to be different from zero, all four letters are integers, and we use the $\simeq$ to emphasize the equivalence. Now the fractions

bc/ac, da/ac are equivalent only if they are the same, since their denominators are the same. This means that, assuming the transitive property of equivalence

$$\frac{b}{a} \quad \text{and} \quad \frac{d}{c}$$

are equivalent fractions if and only if the fractions bc/ac, da/ac are the same; that is, $bc = da$.

Thus, if a, b, c, and d are integers with neither a nor c zero, then, by *definition*,

$$\frac{b}{a} \cong \frac{d}{c} \quad \text{means} \quad bc = ad;$$

that is, the two fractions represent the same rational number if and only if $bc = ad$. (The $\cong$ sign is used in this place, as above, to emphasize the idea of equivalence. Elsewhere we shall use the equals sign to indicate equivalent fractions, since there is no ambiguity.) For instance, to determine if $\frac{14}{23} \cong \frac{154}{299}$, we ask: Is

$$14 \cdot 299 = 23 \cdot 154?$$

The answer is no, as may be seen by multiplication or, more cleverly, by noticing that the last digits of the products are different.

We leave it as an exercise to show that this definition is an equivalence relationship. (See Section 5 of Chapter 4.)

8. ADDITION AND MULTIPLICATION

In Section 7 we showed how to determine whether or not two fractions represent the same rational number. It now behooves us to define addition and multiplication of rational numbers in terms of the fractions which represent them. Since these are new numbers, we have much liberty in this direction, but if we want these new numbers to have the properties we are used to, we must so define addition and multiplication that these properties hold. Suppose we list some of the properties which we wish to hold for rational numbers a, b, and c:

Addition	Multiplication
$a + b = b + a$	$ab = ba$
$(a + b) + c = a + (b + c)$	$(ab)c = a(bc)$
$a + 0 = a$	$a \cdot 1 = a$
If $a = b$, then $a + c = b + c$	If $a = b$, then $ac = bc$
$a(b + c) = ab + ac$	

Everyone should believe that

$$\tfrac{3}{17} + \tfrac{5}{17} = \tfrac{8}{17}.$$

Using the distributive property for rational numbers we can establish this result as follows:

$$\tfrac{3}{17} + \tfrac{5}{17} = 3(\tfrac{1}{17}) + 5(\tfrac{1}{17}) = (3 + 5)(\tfrac{1}{17})$$
$$= 8(\tfrac{1}{17}) = \tfrac{8}{17}.$$

Notice that we are really *defining* $\tfrac{8}{17}$ to be the sum of $\tfrac{3}{17}$ and $\tfrac{5}{17}$ in order that the distributive property will hold, or, less technically, so that seventeenths will have one of the same properties that dogs have; for example, that three dogs plus five dogs is eight dogs. The general result of which this is an example is

$$\frac{b}{a} + \frac{c}{a} = \frac{(b + c)}{a}.$$

On the other hand, no one should believe that $\tfrac{17}{3} + \tfrac{17}{5} = \tfrac{17}{8}$, but alas, some do! To find its true value, we could begin as above:

$$\tfrac{17}{3} + \tfrac{17}{5} = 17(\tfrac{1}{3}) + 17(\tfrac{1}{5}) = 17(\tfrac{1}{3} + \tfrac{1}{5}).$$

But what is $\tfrac{1}{3} + \tfrac{1}{5}$? Surely it is not $\tfrac{1}{8}$, for this is less than $\tfrac{1}{3}$ as well as $\tfrac{1}{5}$. Here a familiar trick is used. By the well-defined property for addition (the fourth in the left-hand column of the table above) we can in a sum replace any fraction by one equivalent (or equal) to it without altering the sum. We have seen that we can easily add two rational numbers if the fractions which represent them have equal denominators. Such a "common" denominator for the given sum must be both a multiple of 3 and of 5. The least such number is 15. Hence, in the sum $\tfrac{1}{3} + \tfrac{1}{5}$ we replace $\tfrac{1}{3}$ by $\tfrac{5}{15}$ and $\tfrac{1}{5}$ by $\tfrac{3}{15}$, since $b/a = kb/ka$ if $k \neq 0$, and have

$$\tfrac{1}{3} + \tfrac{1}{5} = \tfrac{5}{15} + \tfrac{3}{15} = \tfrac{8}{15},$$
$$\frac{17}{3} + \frac{17}{5} = 17\left(\frac{8}{15}\right) = \frac{(17 \cdot 8)}{15} = \frac{136}{15}.$$

We have, of course, here used a more roundabout method than is necessary. We could more expeditiously have used the trick on the original sum and would have the familiar

$$\tfrac{17}{3} + \tfrac{17}{5} = \tfrac{85}{15} + \tfrac{51}{15} = \tfrac{136}{15}.$$

In general,

$$\frac{a}{b} + \frac{c}{d} = \frac{ad}{bd} + \frac{cb}{bd} = \frac{(ad + cb)}{bd}.$$

We now look at some products of rational numbers. It is true that

$$\tfrac{2}{7} \cdot \tfrac{3}{5} = \tfrac{6}{35}.$$

Let us see how we can use the rules to obtain this result. Put $a = \frac{2}{7}$, $b = \frac{3}{5}$. Then $7a = 2$ and $5b = 3$ and therefore $(7a)(5b) = 2 \cdot 3$, $35ab = 6$, and $ab = 6/35$.

In general, we define the product of two rational numbers as follows in terms of the fractions which represent them:

$$\frac{a}{c} \cdot \frac{b}{d} = \frac{ab}{cd},$$

where c and d are different from zero.

We have shown that if the properties we want are satisfied, then we must define addition and multiplication as above. If we were going to put our new number system on a firm foundation, we would then have to proceed to show that all the given properties are indeed satisfied. But this would be somewhat tedious, and we leave this to others for the most part.

We have been rather careful to use the word "fraction" to mean the symbol as against the term "number," which is what it represents. A fraction has a numerator and denominator but a number does not. Strictly speaking, we add and multiply numbers but not fractions. But there is no real harm in referring to the addition or multiplication of fractions because it would be difficult to misunderstand what is meant by it.

Notice that if b and c are any integers different from zero, the fractions $0/b$ and $0/c$ are equivalent fractions, since $0 \cdot b = 0 \cdot c = 0$, and hence

$$\frac{a}{c} + \frac{0}{b} = \frac{a}{c} + \frac{0}{c} = \frac{a+0}{c} = \frac{a}{c}.$$

Thus, whenever b is not zero, the rational number $0/b$ has the property of the additive identity. Hence we identify $0/b$ with 0. Similarly, it can be shown that if $b \neq 0$, then b/b is the multiplicative identity and hence we write $b/b = 1$.

What is the additive inverse of a rational number a/b? In accordance with our previous notation it would be $-(a/b)$. There are two other forms in which this can be expressed. First

$$\frac{-a}{b} + \frac{a}{b} = \frac{(-a)+a}{b} = \frac{0}{b} = 0,$$

Hence $(-a)/b$ is another form of the additive inverse of a/b. Furthermore, by the equivalence property,

$$\frac{-a}{b} = \frac{-a}{b} \cdot \frac{-1}{-1} = \frac{a}{-b}.$$

Thus the three ways in which the additive inverse of a/b can be expressed are

$$\frac{-a}{b}, \frac{a}{-b}, -\frac{a}{b}.$$

We shall deal with the multiplicative inverse in Section 9.

EXERCISES

1. Find the value of each of the following:

$$\frac{5}{6} + \frac{3-4}{7}, \qquad -\frac{7-2}{7} + \frac{6}{11}.$$

2. Which of the following equalities hold whatever numbers a and b are, provided no demoninator is zero? Give the reasons for your answer.

    a. $\dfrac{3a + b}{3c} = \dfrac{a+b}{c}.$ b. $\dfrac{7a}{7b+c} = \dfrac{a}{b+c}.$

    c. $\dfrac{5a + 5b}{5c + 5d} = \dfrac{a+b}{c+d}.$

3. Which of the following equalities hold whatever numbers a and b arc, provided no denominator is zero? Give reasons for your answers.

    a. $\dfrac{a + 3b}{3d} = \dfrac{a+b.}{d}$ b. $\dfrac{3d}{a+3b} = \dfrac{d}{a+b}.$

    c. $\dfrac{2a + 6b}{2c + 6d} = \dfrac{a+b.}{c+d}$

4. Express each of the following as a single fraction:

    a. $\frac{3}{7} \cdot \frac{4}{9} \cdot \frac{7}{8},$ b. $\frac{3}{7}(\frac{4}{9} + \frac{7}{2}).$ c. $\dfrac{3+5}{6} \cdot \dfrac{4}{7}.$

    d. $\frac{4}{8} \cdot \frac{5}{9} + \frac{1}{6} \cdot \frac{5}{16}.$ e. $\frac{3}{4}(\frac{4}{2} - \frac{4}{7}).$

5. Express each of the following as a single fraction:

    a. $\dfrac{a}{b} + \dfrac{c}{d}.$ b. $\dfrac{a}{b} + \dfrac{c}{d} + \dfrac{e}{f}.$ c. $\dfrac{a}{b} - \dfrac{c}{d}\left(\dfrac{e}{f} - \dfrac{g}{h}\right).$

6. Express each of the following as a single fraction:

    a. $\dfrac{a}{b} - \dfrac{c}{d}.$ b. $a\left(\dfrac{b}{c} - \dfrac{d}{e}\right).$ c. $-\dfrac{r}{s} + \dfrac{x}{y}\left(\dfrac{a}{b} - \dfrac{c}{d}\right).$

7. Prove that if a and b are any two rational numbers and $ab = 0$, then $a = 0$ or $b = 0$.

8. Show that $(a/1) + (b/1) = (a+b)/1$ and $(a/1)(b/1) = ab/1$ and thus that $a/1$ behaves like a.

9. Prove that if $b \neq 0$, then b/b is the multiplicative identity. (Thus we call it 1.)

10. Prove that for no positive integers b and c is it true that

$$1/c + 1/b = 1/(c+b).$$

11. Show that the equivalence of fractions has the three properties of an equivalence relationship.

9. DIVISION

Is there a rational number x such that $ax = b$? The answer is "It depends." If a and b are *both* zero, the equation is satisfied by *every* rational number. We showed in Section 6 that if a is zero and b is not zero, the equation cannot be satisfied by any number which has the usual properties of numbers of our acquaintance. Thus, if $a = 0$, then the equation $ax = b$ *never* has a unique solution for x: The equation has either many solutions (when $b = 0$) or none whatever (when $b \neq 0$).

In case a and b are rational and $a \neq 0$, we can show that the equation

$$ax = b$$

has exactly one solution. If

$$a = \frac{j}{k}, \quad b = \frac{m}{n},$$

where j, k, m, and n are integers and none, except perhaps m, are zero, then

$$a \cdot \frac{k}{j} \cdot \frac{m}{n} = \frac{j}{k} \cdot \frac{k}{j} \cdot \frac{m}{n} = \frac{m}{n} = b.$$

Therefore, the equation is satisfied when

$$x = \left(\frac{k}{j}\right)\left(\frac{m}{n}\right) = \frac{km}{jn}.$$

This means: *To divide b by a, multiply b by the reciprocal of a.*

This shows that there is a rational number x such that $ax = b$ when a and b are rational and $a \neq 0$. That x is the only rational number that satisfies the equation can be shown as follows: Let $ax = b$ and $ay = b$. This implies that $ax = ay$. Since a is not zero, we can find a rational number r such that $ra = 1$ (that is, $r - 1/a$). Multiply both sides of $ax = ay$ by r and get $rax = ray$; that is, $x = y$. The multiplicative inverse of b/a is thus a/b, because

$$\frac{b}{a} \cdot \frac{a}{b} = \frac{ba}{ab} = \frac{1}{1} = 1.$$

We have defined equality, addition, and multiplication so that we can always add, subtract, multiply, and divide except by zero. The rational numbers have all the required properties of a field (see Section 4 of Chapter 4).

10. INEQUALITIES OF RATIONAL NUMBERS

A rational number a/b is said to be "positive" if a and b are positive integers: In that case $-(a/b)$ will be negative. Since, by the equivalence property,

$$\frac{a}{b} = \frac{(-1)a}{(-1)b} = \frac{(-a)}{(-b)},$$

we see that a/b is also positive when a and b are both negative. Thus

(1) The fraction a/b represents a positive rational number if a and b are both positive or both negative.

On the other hand, we have seen that $-(a/b) = (-a)/b = a/(-b)$. Thus

(2) The fraction a/b represents a negative rational number if one of a and b is positive and the other negative.

The rational numbers consist, thus, of the positive rational numbers, the negative rational numbers, and zero.

Then, if r and s are two rational numbers, as before, we say that $r < s$ means that there is a positive rational number x such that $r + x = s$. For instance, $\frac{1}{3} < \frac{5}{3}$, since $\frac{1}{3} + \frac{4}{3} = \frac{5}{3}$ and $\frac{4}{3}$ is a positive rational number. Also $-(\frac{3}{2}) < \frac{1}{2}$, since $-(\frac{3}{2}) + 2 = \frac{1}{2}$ and 2 is a positive rational number. In fact, we may arrange these and other rational numbers on a line as shown in Fig. 5:2.

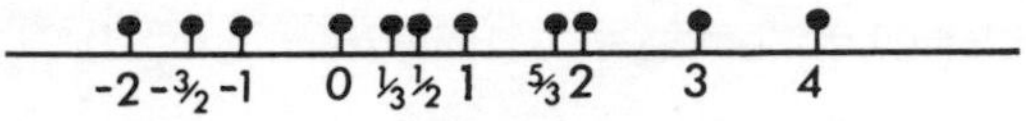

Figure 5:2

Do properties (1i) and (2i) of Section 5 hold when the letters involved there stand for rational numbers? The answer is yes; now we might carry through the proof along previous lines. But the only properties we assumed in Section 5 were the associative, commutative, and well-defined properties. If we assume, as we have done, these same properties for the rational numbers, the proofs of properties (1i) and (2i) do for rational numbers as well. This is one advantage of having kept track of just what we assumed in the proof.

One way of finding whether $r < s$, for two rational numbers, is to compute $s - r$. If the answer is positive, $r < s$, and if the answer is negative, then $r > s$. There is a simpler way. Suppose we wish to compare $\frac{3}{5}$ and $\frac{4}{7}$. We can write them as fractions with the same denominator: $\frac{21}{35}$ and $\frac{20}{35}$. Since $\frac{20}{35} + \frac{1}{35} = \frac{21}{35}$, we see that $\frac{20}{35} < \frac{21}{35}$. Thus $\frac{4}{7} < \frac{3}{5}$.

More generally, suppose we wish to compare a/b and c/d, where a, b, c, and d are integers and neither c nor d is zero. By multiplying numerator and denominator by -1 if necessary, we can take these fractions to have positive denominators. Then we write

$$\frac{a}{b} = \frac{ad}{bd} \qquad \text{and} \qquad \frac{c}{d} = \frac{cb}{bd}.$$

Then

$$\frac{c}{d} - \frac{a}{b} = \frac{cb}{bd} - \frac{ad}{bd} = \frac{cb - ad}{bd}.$$

Since bd is a positive integer, the last fraction will represent a positive rational number if and only if $cb - ad$ is positive; that is, $cb > ad$. Thus we have shown that

For bd positive, $c/d > a/b$ if and only if $cb > ad$.

The rational numbers satisfy properties (1i), (2i), (3i), and (4i) of Section 5. One may also compare rational numbers by means of their decimal values (see Section 11).

EXERCISES

1. Express each of the following as a single fraction:

   a. $\dfrac{\frac{16}{8}}{4}$.

   b. $\dfrac{5(\frac{1}{2} - \frac{3}{4})}{\frac{3}{7}}$.

   c. $\dfrac{3(\frac{4}{3} - \frac{1}{5})}{3(\frac{1}{2} - 1)}$.

   d. $\dfrac{\dfrac{x(a - b)}{c} + \dfrac{y}{d}}{\dfrac{x}{cd}}$.

2. Express each of the following as a single fraction:

   a. $\dfrac{16}{\frac{8}{4}}$.

   b. $\dfrac{7}{8} + \dfrac{2(\frac{1}{3} - \frac{1}{4})}{\frac{8}{3}}$.

   c. $\dfrac{\dfrac{a}{b}}{\dfrac{c}{d}}$.

   d. $\dfrac{\dfrac{r}{s} - \dfrac{t(x - y)}{w}}{\dfrac{x}{s} - \dfrac{y}{w}}$.

3. Arrange the following rational numbers in order from least to greatest: $(-3)/7$, $8/13$, $7/(-3)$, $5/8$.

4. Arrange the following rational numbers in order from least to greatest:

   $13/8$, $21/13$, $34/21$, $-(2/3)$, $-(3/5)$.

5. If $a > b > 0$, show that $a/b > (a + 1)/(b + 1)$. (This is a special case of Exercise 12.)

6. How many possible different meanings can the following ambiguous symbols have?

   a. $1/2/3$.   b. $x/y/z$.   c. $1/2/3/4$.   d. $t/x/y/z$.

7. If a and b are rational numbers, prove that ab is a rational number.

8. If a and b are rational numbers, prove that $a + b$ is a rational number.

9. Prove that if a and b are rational numbers and $a \neq 0$, then b/a is a rational number.

10. We showed above that if two rational numbers are written as fractions with positive and equal denominators, then one is less than the other if the numerator of one is less than the numerator of the other. Suppose a/b and a/c are two fractions with positive numerators. If $b < c$, is $a/b < a/c$? If not, what relationship is there between a/b and a/c?

11. Suppose we wished to compare the rational numbers a/b and c/d and instead of finding equivalent fractions in which the denominators were equal we found those in which the numerators were equal, as follows:

$$\frac{a}{b} = \frac{ac}{bc} \quad \text{and} \quad \frac{c}{d} = \frac{ca}{da}.$$

Could we by comparing the denominators tell which fraction is the greater?

12. If b and d are positive integers, show that the number $(a + c)/(b + d)$ always lies between a/b and c/d.

13. If b and c are positive integers and $b > c$, which of the following must hold?

$1/b > 1/c, 1/b < 1/c, 1/b = 1/c$.

14. Answer the question in Exercise 13 if b and c are negative integers.

11. DECIMALS

Since we use the decimal system of reckoning (rather than the dozal or some other system), it is as natural to represent rational numbers as sums of hundreds, tens, units, tenths, hundredths, and so on, as it is to represent positive integers as sums of units, tens, hundreds, and so on. Fractions such as $\frac{3}{10}$, $\frac{271}{100}$, and $\frac{52}{1000}$, in which the denominators are 10, 100, and 1000, occur so frequently that a special way of writing them has been invented. Just as $20 = 2(10)$, $200 = 2(100)$, $2000 = 2(1000)$, we have $2 = 2(1)$, $.2 = 2(\frac{1}{10})$ $.02 = 2(\frac{1}{100})$, $.002 = 2(\frac{1}{1000})$. Every time we divide by 10 we move the *decimal point* one place to the left, and every time we multiply by 10 we move it one place to the right, adding zeros if necessary to indicate its position.

The monetary system of the United States makes use of decimals; for example, 712 cents make 7.12 dollars; and 2.99 dollars make 299 cents. The metric system of measurement was devised to fit in with our decimal notation: 100 centimeters = 1 meter, 1 decimeter = .1 of a meter, and so on. On the other hand, the division of 1 foot into 12 inches is not easily adaptable to this system, for 1 inch = .08333 ... foot. It would, however, fit in nicely with the dozal system, for in that case .1 would mean one "dozenth" (that is, one twelfth) and we should have 1 inch = .1 foot in dozals while 1 decimeter = .124972 ... meter in the same system.

We have all had experience with multiplication and division of decimals but it may not be without profit here briefly to explore the subject to find the reasons back of familiar rules. (Sometimes, alas, they are not as familiar as they should be.) We have the following

Rule

In multiplying two decimals, the number of places to the right of the decimal point in the product is the sum of the number of places to the right of the decimal point in the numbers of the product.

The following illustration will show the reason back of the rule; suppose we multiply 1.576 by 2.32. The former is $1576(1/10^3)$ and the latter is $232(1/10^2)$. Hence the product is equal to

$$232(1576)(1/10^3)(1/10^2).$$

But

$$\frac{1}{10^3} \cdot \frac{1}{10^2} = \frac{1}{10^3 \cdot 10^2} = \frac{1}{10^5},$$

since 10 taken as a product three times multiplied by 10 times 10, is 10 taken as a product five times. Thus, to get the product of the given numbers, we multiply 232 by 1576 and place the decimal point five places from the right. If our numbers were .001576 and .0232, respectively, we would have $1/10^6$ instead of $1/10^3$, $1/10^4$ instead of $1/10^2$, and hence $1/10^{10}$ instead of $1/10^5$. Notice, by the way, that the number of digits to the right of the decimal point of any number (not counting terminal zeros) is equal to the least power of 10 by which the number need be multiplied to become an integer.

The rule for division is just the reverse of that for multiplication. Since the number of places to the right of the decimal point in the product is the sum of the number of places for the members of the product, so the number of places in the quotient is the number of places in the dividend minus the number of places in the divisor. For instance, in dividing 5.37 by 253.2 we write

$$\begin{array}{r} .02 \\ 253.2 \,\overline{)\, 5.370} \\ 5.064 \end{array}$$

The position of the decimal point in 5.064 is determined by that in 5.37, the dividend. Since 5.064 has three places to the right of the decimal point and 253.2 has one, the quotient must have $3 - 1 = 2$ places to the right of the decimal point.

Some numbers are easily expressed (or represented) in decimal form. For example, $\frac{2}{5} = .4$, $-\frac{10}{4} = -2.5$, $\frac{3}{8} = .375$. But there are other numbers, such as $\frac{1}{3}$, $\frac{5}{7}$, and $\frac{1492}{1776}$, for which representation in decimal form is not so simple. We have all written

$$\begin{array}{r} .3333333 \cdots \\ 3 \,\overline{)\, 1.0000000 \cdots} \end{array}$$

and then

$$\tfrac{1}{3} = .3333333 \cdots.$$

To examine the connection between the number $\frac{1}{3}$ and the *nonterminating decimal* $.3333333 \cdots$, let

$$\begin{aligned} s_1 &= .3 \\ s_2 &= .33 \\ s_3 &= .333 \\ s_4 &= .3333 \end{aligned}$$

$$\cdot \quad \cdot \quad \cdot$$

Each of the numbers[1] $s_1, s_2, s_3, \ldots$ is a "terminating decimal"; that is, we can write the entire decimal—there are no dots in any single s indicating any desire to go on like this forever. It is instructive to compute the differences $\frac{1}{3} - s_1, \frac{1}{3} - s_2, \frac{1}{3} - s_3, \ldots$. We find that

$$\frac{1}{3} - s_1 = \frac{1}{3} - \frac{3}{10} = \frac{10 - 9}{30} = \frac{1}{30} = \frac{1}{3(10)},$$

$$\frac{1}{3} - s_2 = \frac{1}{3} - \frac{33}{100} = \frac{1}{300} = \frac{1}{3(10^2)},$$

$$\frac{1}{3} - s_3 = \frac{1}{3} - \frac{333}{1000} = \frac{1}{3000} = \frac{1}{3(10^3)},$$

and, in general,

$$\frac{1}{3} - s_n = \frac{1}{3(10^n)},$$

where 10^n is 10 taken as a product n times. It thus appears that, when we take more and more terms of the decimal .33333 $\cdots$ to have successively .3, .33, .333, ..., we get numbers nearer and nearer to $\frac{1}{3}$. Furthermore, no matter how small a number you name I can carry out the division to enough places so that the terminating decimal s_n differs from $\frac{1}{3}$ by less than that number. This can be done since n can be taken large enough so that $\frac{1}{3(10^n)}$ is smaller than the number you named. This fact is conveniently expressed by the simple statement "The sequence .3, .33, .333, ... converges to $\frac{1}{3}$." We say that ".3333 $\cdots$ is the *decimal expansion* of $\frac{1}{3}$" or that ".3333 $\cdots$ *converges to* $\frac{1}{3}$."

Division of 3.0000 $\cdots$ by 7 gives .428571428571 $\cdots$, and it can be shown that

$$\tfrac{3}{7} = .428571428571 \cdots$$

in the sense that the sequence .4, .42, .428, .4285, ... converges to $\frac{3}{7}$. It can also be shown that

$$\tfrac{15052}{3300} = 4.561212121212 \cdots,$$

where the block *12* keeps repeating without end; for this reason the decimal is called a *repeating decimal* (even though there are some terms at the beginning before the repetition starts). Terminating decimals can be made into nonterminating decimals by adding zeros; for example,

$$\tfrac{1}{4} = .25 = .25000000 \cdots,$$

where the 0 keeps repeating. Thus the decimal value of $\frac{1}{4}$ can be considered to be a repeating decimal.

[1] We read s_1 "s sub-one," and so on.

We shall show later on that every repeating decimal converges to a rational number. Combining this result with the converse statement given in Exercise 21 below we see that *the rational numbers are precisely (that is, all and only) those numbers which can be expressed as repeating decimals or terminating decimals.*

The expression

$$.101001000100001000001\cdots,$$

in which each bunch of zeros contains one more than the preceding bunch, is not a repeating decimal; hence it cannot be the decimal expansion of a rational number.

EXERCISES

1. Five gross, four doza-two inches is how many feet? How would you write your answer in dozals?

2. Six hundred twenty-five eggs is how many gross in dozals?

3. In the binary scale suppose a quart is represented by the number 1. How would you represent the two quantities, a pint and a gill, in what, for the binary scale, corresponds to decimals for the decimal scale of notation?

4. Divide 1.0626 by 253 and carefully explain the reasons for your results and methods.

5. Show that the difference between $\frac{3}{7}$ and .428571 is 3/7,000,000.

6. Show that .428 is less than $\frac{3}{7}$ and that .429 is greater than $\frac{3}{7}$.

7. Find a terminating decimal whose value differs from $\frac{1}{11}$ by less than 1/1,000,000.

8. Does the addition of the decimal expansion of $\frac{1}{7}$ and $\frac{1}{3}$ give the decimal expansion of $\frac{10}{21}$?

9. The first six numbers of the decimal form of $\frac{1}{7}$ are 142857. 3(142857) = 428571; 2(142857) = 285714. Can you give any explanation of this phenomenon?[1] What multiple of 142857 is 857142? Does multiplication of 142857 by 8 fit into the scheme?

10. Find the decimal expansions of $\frac{1}{13}$, $\frac{2}{13}$, and $\frac{3}{13}$. Compare your results with those of Exercise 9.

11. Find the repeating decimal for $\frac{1}{37}$.

12. Find the repeating decimal for $\frac{1}{101}$.

13. Find a rational number whose value is between $\frac{1}{6}$ and $\frac{5}{31}$.

14. Find a rational number whose value is between two rational numbers a and b.

15. Which is greater: $\frac{4}{11}$ or $\frac{3}{8}$? $\frac{23}{11}$ or $\frac{27}{13}$?

16. Two $23\frac{1}{2}$-ounce packages of soap can be bought for 37 cents or one 4-

[1] See references **4** and **27**.

pound 5-ounce package for 55 cents. If the weights are net weights, which is the more economical purchase?

17. Consider the sequence of fractions

$$\tfrac{1}{2}, \tfrac{2}{3}, \tfrac{3}{5}, \tfrac{5}{8}, \tfrac{8}{13}, \ldots,$$

where if b/c is one fraction, the next is $c/(b + c)$. Are these increasing from left to right, decreasing from left to right, or neither? If "neither," is there a pattern? [The limiting value of the sequence is the so-called golden ratio of the ancients (see references **58** and **69**).]

18. Brand A of ginger ale sells at three bottles for 25 cents, and brand B three bottles for 20 cents. If each bottle of the former holds 16 ounces and each of the latter 13 ounces, which is the cheaper?

*19. What will be the integers b for which $1/b$ can be expressed as a terminating decimal?

*20. Obtain the decimal form of $\tfrac{1}{7}$ and find whether you can multiply it by 3 to get the decimal form of $\tfrac{3}{7}$.

*21. Show that the nonterminating decimal expansion of each rational number is a repeating decimal.

*22. Show that the part of the decimal value of $1/m$ which repeats has less than m digits if m is a positive integer; for example, $\tfrac{1}{7}$ has less than 7 digits in the repeating part of its decimal expansion.

**23. Show that if p is a prime number, the number of digits in the repeating part of the decimal value of $1/p$ is a divisor of $p - 1$.

**24. Show that the nonterminating decimal expansion of each rational number converges to the number.

12. SQUARES

The squares of the positive integers $1, 2, 3, 4, \ldots$ are $1, 4, 9, 16, \ldots$; the square of 0 is 0; and the squares of the negative integers $-1, -2, -3, -4, \ldots$ are $1, 4, 9, 16, \ldots$. Thus it is apparent that there is no integer whose square is 2.

We are now going to prove that *there is no rational number whose square is 2*. We do this by assuming that r is a rational number whose square is 2 and then showing that the assumption is false. Let $r = p/q$, where p and q are integers having no common integral factor greater than 1. (Each rational number, and in particular r, can be represented in this way.) The assumption that $r^2 = 2$ means that $(p/q)^2 = 2$ and hence that $p^2/q^2 = 2$ and

$$p^2 = 2q^2.$$

Since 2 is a factor of $2q^2$ and $p^2 = 2q^2$, 2 must be a factor[1] of p^2. Therefore, p is divisible by 2, for if p were an odd number, then p^2 would also be odd.

---

[1] This follows from the fact noted in Chapter 3 that any positive integer, in particular p^2, has only one factorization into prime factors aside from the order of the factors.

Since p is divisible by 2, we can write $p = 2t$, where t is an integer. Then $p^2 = 2q^2$ implies that $4t^2 = 2q^2$ and hence $2t^2 = q^2$. Thus q^2 must be divisible by 2. Hence q is divisible by 2. We have shown that 2 is a common factor of p and q; this contradicts the fact that p and q have no common positive integral factor greater than 1. Therefore, the assumption that there is a rational number whose square is 2 is false.

EXERCISES

1. Prove that there is no rational number whose square is 3.

2. Prove that there is no rational number whose cube is 2.

3. We know that if a number c is a factor of the product de and if c has no factors greater than 1 in common with e, then c is a factor of d. Why, then, does the fact that 1 is the greatest common divisor of p and q and $p^2 = 2q^2$ imply that p^2 is a factor of 2? Notice that 2 is a factor of p^2. By this means give a proof different from that in the text that there is no rational number whose square is 2.

4. Using the method of Exercise 3, show that there is no rational number whose square is 5.

5. Prove that there is no rational number whose square is a prime number.

*6. Prove that if n is a whole number which is not the square of an integer, then it is not the square of a rational number.

13. REAL NUMBERS

The mere fact that there is no rational number whose square is 2 indicates that the set of rational numbers is not adequate to give a solution of the equation $x^2 = 2$ and that a larger class of numbers should be created. This larger class should include the solution of $x^3 = 5$, the number π, and many others. All these numbers share the property that we can approximate them as closely as we please by means of decimals. For $\sqrt{2}$ we have the sequence

$$1., 1.4, 1.41, 1.414, 1.4142, \ldots.$$

For $\sqrt[3]{5}$ we have

$$1., 1.7, 1.70, 1.709, 1.7099, 1.70998, \ldots.$$

For π we have

$$3., 3.1, 3.14, 3.141, 3.1415, 3.14159, \ldots.$$

All these, and many more, may be associated with sequences of decimals which extend as far as we wish to compute them. For instance, to thirty places, the value of π is

$$3.14159\ 26535\ 89793\ 23946\ 26433\ 83279.$$

In fact, using modern computers, π has been computed to 100,000 decimal places. All of these have the property that by computing more and more terms we come closer and closer to the number to be represented. That is, if we square the successive decimals in the sequence for $\sqrt{2}$, we come closer and closer to 2; if we cube the successive decimals in the sequence of $\sqrt[3]{5}$, we come closer and closer to 5. If we multiply the diameter of a circle by the decimals in the sequence for π, we come closer and closer to the value of the circumference. Furthermore, in all cases, by using enough terms in the sequence of decimals we can come as close as we please to the desired number.

While we are about the process of creation, we should create a real number to "go with" or be represented by any decimal. For instance, the following decimal has a pattern which would permit us to write any number of terms:

$$.10100100010000100000 1 \cdots .$$

We want this also to represent a real number. Or, from another point of view, we could take as a postulate that there is a real number represented by this decimal.

We thus *create* enough numbers so that every decimal, whether finite or infinite in extent, "represents" a number and call the set of such numbers the real numbers. The word "represents" in this connection is very vague, and we use it only in the hope that it will convey an intuitive meaning of the situation. To give a precise definition of the real numbers is far beyond the scope of this book. Certainly the set of real numbers includes the sets of integers and rational numbers and contains, also, other numbers which are not rational. If a real number is not rational it is called irrational.

In understanding real numbers, the imagery of geometry is useful. We have seen that on a line we can set up a one-to-one correspondence between the integers and points on the line; in fact, such a correspondence is possible for

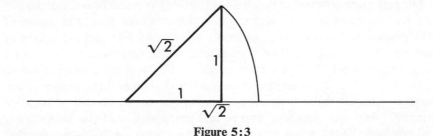

Figure 5:3

rational numbers as well. There is (Fig. 5:3) a line segment whose length is $\sqrt{2}$, since this is the length of the diagonal of a square of side 1. So there is a place on the line corresponding to $\sqrt{2}$ somewhere between the points corresponding to 1.414 and 1.415. One may set up a one-to-one correspondence between the set of real numbers and points on the line. In fact, it follows from a complete definition that a one-to-one correspondence can be set up between all points of a line and the complete set of real numbers.

On the basis of a precise definition of the real numbers it may be shown that they form a field. Also one can define inequalities of real numbers with the properties mentioned in Section 5. We shall not attempt here to establish any of these properties but shall assume them to be true.

EXERCISES

1. Show how to draw a line segment $\sqrt{3}$ inches long.
2. Prove that 2 lies between the squares of the numbers 1.4142 and 1.4143.
3. Find a rational number which is closer to $\sqrt[3]{3}$ than .1, that is, such that the difference between $\sqrt[3]{3}$ and the number you find is to be less than .1.
4. Prove that the sum of a rational number and an irrational number is an irrational number. Is the result the same if "sum" is replaced by "product"?
5. Can the sum of two irrational numbers ever be rational?
6. If $a + bx = 0$, where a and b are rational but x is irrational, prove that a and b are both zero.
*7. Draw a line $\sqrt[4]{2}$ inches long.
*8. Show that between any two irrational numbers there is always a rational number.
*9. Show that between any two rational numbers there is always an irrational number.

14. COMPLEX NUMBERS

The real numbers may be divided (as the integers were) into three classes: positive real numbers, zero, and negative real numbers. Since the square of each positive real number is positive, the square of 0 is 0, and the square of each negative real number is positive, it follows that there is no real number x such that $x^2 = -1$. It is now appropriate to ask "Is there a number whose square is -1?" If by number one means "real number," the answer is no. Anyone who has really understood the steps in the discussion of negative integers, of rational numbers, and of real numbers should rise immediately and proclaim: "Let there be a number whose square is -1." This new number is called i, the initial letter of "imaginary." The number i is called *imaginary*. Of course, this does not mean that i is imaginary in the sense that there really is no such number; we "imagine" a number i whose square is -1 in exactly the same way that we "imagine" a number whose square is 2. Anyone skilled in mathematics or in one of the numerous sciences which require use of mathematics knows that the number i plays an important role in the world of affairs. As an actual matter of fact, the usefulness of the number i is not impaired by its unfortunate name, which naturally tempts the young and gullible to think that it is somehow far more mysterious than the "real numbers."

Numbers of the form

$$z = x + iy,$$

where x and y are real numbers and $i^2 = -1$, are called *complex numbers*. We add, subtract, and multiply complex numbers $a + bi$ just as we would perform these operations on $a + bx$. That is,

$$(a + bi) + (c + di) = (a + c) + (b + d)i,$$
$$(a + bi) - (c + di) = (a - c) + (b - d)i,$$
$$(a + bi)(c + di) = ac + bci + adi + bdi^2 = (ac - bd) + (ad + bc)i,$$

since $i^2 = -1$. If we assume that the real numbers have the properties of a field, these definitions assure us that the set of complex numbers is closed under addition, subtraction, and multiplication. It is not hard to show that the associative and commutative properties hold for addition and multiplication and that 0 and 1 are the additive and multiplicative identities. The distributive property also may be shown (see Exercise 15 in Section 15).

The additive inverse of $a + bi$ is $(-a) + (-b)i$. The only other property of a field which needs to be verified is the existence of a multiplicative inverse, that is, divisibility except by zero. To show this, we assume the properties of fractions for complex numbers and have

$$\frac{a + bi}{c + di} = \frac{(a + bi)(c - di)}{(c + di)(c - di)} = \frac{(ac + bd) + (bc - ad)i}{c^2 + d^2}.$$

Thus the quotient on the left is equal to the complex number $x + yi$, where

$$x = \frac{ac + bd}{c^2 + d^2} \quad \text{and} \quad y = \frac{bc - ad}{c^2 + d^2}$$

are real numbers. We should notice here that, since c and d are real numbers, $c^2 + d^2$ can be 0 only if both c and d are zero, that is, only if $c + di = 0$. Thus the set of complex numbers forms a field.

The complex numbers, however, are not ordered. The number i, for instance, is neither greater, less than, nor equal to zero; for suppose i were positive—then its square $i^2 = -1$ would be positive, which would imply that $(-1)i = -i$ would also be positive. This is impossible, since a number and its opposite cannot both be positive (or negative).

However, it is possible to represent a complex number graphically as in Fig. 5:4 (p. 138). This representation is the key to an important use for complex numbers, for each complex number so represented will determine a vector connecting it with the point 0.

Just as for the real numbers, we define the absolute value of a complex number to be the distance between 0 and the point which represents it; that is, the absolute value of $a + bi$ is $\sqrt{a^2 + b^2}$. We write $|a + bi| = \sqrt{a^2 + b^2}$. This is a positive real number if $a + bi \neq 0$. In case the complex number is real, its absolute value is then $\sqrt{a^2}$, which is consistent with our former definition.

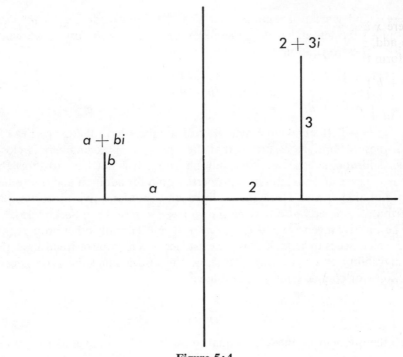

Figure 5:4

At this point we should call attention to the distinction between "complex numbers" and "imaginary numbers": $z = x + iy$ is a complex number no matter what real numbers x and y are, but it is *imaginary* only if $y \neq 0$. In other words, all real numbers are complex numbers but none are imaginary numbers. Each complex number is either real or imaginary but not both.

In more advanced mathematics we sometimes speak of things not complex numbers which we may call *numbers* (some of them are sometimes called *hypercomplex* numbers and might be called *supercomplex* numbers in accordance with the advertising of the day), but in one or more respects they misbehave. About the most widely used misbehaving "numbers" are some which act like this:

$$ab = -ba.$$

Physicists like them and so do many mathematicians, but we, being good little boys, will not associate with them.

15. SUMMARY OF NUMBER SYSTEMS

We now have four examples of fields in the sense of Section 4 of Chapter 4: the complex numbers, the real numbers, the rational numbers, and the numbers on a circle with a prime number of divisions. The last is quite different from the others in that any such field has only finitely many numbers in it. We can indicate relationships among the others by the following diagram:

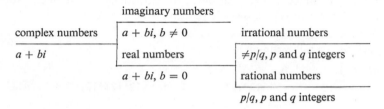

Notice that every complex number is either imaginary or real but not both and that the imaginary numbers do not form a field. (Why?) Every real number is either rational or irrational but not both, and the irrational numbers do not form a field. (Why?)

We have inequalities in the fields of real numbers and irrational numbers but not in the field of complex numbers or the field of numbers on a circle.

EXERCISES

1. Using distributive and associative laws and the fact that $i^2 = -1$, prove that
$$(1 + i)(2 + i)(3 + i) = 10i.$$

2. Express the following in the form $a + bi$, where a and b are real numbers:
$$(1 + 2i)(3 - 5i)(4 - i).$$

3. Express in the form $a + bi$ the following: $\dfrac{3 + i}{2 - i}$.

4. Express in the form $a + bi$ the following: $\dfrac{5 + 6i}{3 + 4i}$.

5. Find the absolute values of $3 + i$ and $2 - i$ and of the answer to Exercise 3. What is a relationship among the three absolute values?

6. Prove that the absolute value of the product of two complex numbers is equal to the product of their absolute values.

7. Represent graphically the numbers $1 + 2i$, $3 + i$, and their sum. What is the relationship among the three corresponding vectors?

8. Do what is asked for in Exercise 7 for $1 - 2i$ and $3 - i$.

9.   Show (by substituting $2 + 3i$ for x in the left side) that the equation

$$x^2 - 4x + 13 = 0$$

is satisfied when $x = 2 + 3i$.

10. Show that

$$x^2 - 4x + 13 = (x - 2)^2 + 9$$

and then explain why there is no real value of x for which

$$x^2 - 4x + 13 = 0.$$

11. Are any complex numbers rational? Are any imaginary numbers rational? If in either case your answer is yes, give an example.

12. Ascribe all possible names (such as "complex," "real," and so forth) to each of the following numbers:

   a. $\sqrt{2}$.   **b.** $-3\sqrt{2}/\sqrt{2}$.   **c.** $\sqrt[3]{-2}$.
   d. $\sqrt{-5}$.   **e.** $.123123\ldots$.   **f.** $.121121112\ldots$.
   g. π.

13. In the table put X's in the appropriate blanks. For instance, in the upper left hand corner there is an X because $\frac{3}{2}$ is a rational number; there will not be an X in the first row under "imaginary" because $\frac{3}{2}$ is not an imaginary number, and so on.

	Rational	Imaginary	Real	Complex	Irrational
$\frac{3}{2}$	X				
8(mod 9)					
$3i$					
$\sqrt{2}$					
$\sqrt{-3}$					

14. The same as Exercise 13, replacing the numbers on the left by 3, 0, 7(mod 83), $2i/3$, $\sqrt{5}$, $\sqrt{-5}$.

15. Assuming the field properties of the real numbers, prove that the distributive property holds for the complex numbers.

16. Why do the irrational numbers not form a field?

17. Why do the imaginary numbers not form a field?

18. Following the pattern of Chapter 1, draw circles indicating the relationships among the types of numbers we have considered so far.

19. Prove that if $a + bi = c + di$, where a, b, c, and d are real numbers, then $a = c$ and $b = d$.

*20. One might define a complex number $a + bi$ to be "greater than" $c + di$ if $a > c$, or $a = c$ and $b > d$. Which of the following properties hold for *every* set of complex numbers p, q, r, and s?

    a. $p > q$ and $q > r$ implies $p > r$.

    b. $p > q$ and $r > s$ implies $p + r > q + s$.

    c. $p > 0$ and $r > s$ implies $pr > ps$.

In each case either prove your answer to be yes or give an example in which the implication does not hold.

16. TOPICS FOR FURTHER STUDY

1. Development of the rational numbers: references **2** and **13** (Chap. 2 and supplement).

2. Irrational numbers: reference **49.**

3. Approximations to π: references **38** (pp. 72–79) and **49.**

4. Repeating decimals and revolving or cyclic numbers: references **4, 27,** and **52** (pp. 147–160).

5. Real numbers: references **17** and **37.**

6

Algebra

Myself when young did eagerly frequent
Doctor and Saint, and heard great argument
  About it and about: but evermore
Came out by the same door wherein I went.

.   .   .

Waste not your Hour, nor in the vain pursuit
Of This and That endeavour and dispute;
  Better be jocund with the fruitful Grape
Than sadden after none, or bitter, Fruit.

.   .   .

Ah, but my Computation, People say,
Reduced the Year to better reckoning?—Nay,
  'T was only striking from the Calendar
Unborn To-morrow, and dead Yesterday.[1]

1. INTRODUCTION

Omar Khayyām, so the story goes, was one of the three favorite pupils of Imám Mowaffak of Naishapur, one of the greatest of the wise men of Khorassan. These three pupils vowed that the one of them to whom the greatest fortune fell would share it equally with the rest. When one of them became

---

[1] From the *Rubáiyāt of Omar Khayyām*, translated by Edward Fitzgerald.

Vizier to the Sultan he made Omar Khayyām official astronomer in the court of the Sultan. And this man, of whom we are apt to think only as the author of the *Rubáiyát*, became one of the greatest mathematicians of his time. Among other things, he revised the Arabian calendar and wrote a book on algebra which David Eugene Smith declares "was the best that the Persian writers produced." In this book are the first recorded instances of certain results that we use today. It is rather interesting that in the fourteenth edition of the *Encyclopædia Brittanica* the only reference in the index to this Persian poet is to a quarter-page discussion of some of his remarkable results in algebra. This fact is probably in line with Omar Khayyām's opinion of the relative values of his achievements in life, all his talk notwithstanding.

But Omar Khayyām was not by any means the first algebraist. Ahmes, an Egyptian, used what might be called *algebra* about 1550 B.C., and Diophantus, a Greek, wrote the first treatise on algebra about 275 A.D. But who the early algebraists were cannot be determined until we agree on what we mean by algebra, and on this point there seems to be considerable difference of opinion. However, even though the advanced mathematician might not agree, most of those who read these pages would probably be willling to say that algebra consists fundamentally in allowing letters to stand for numbers, and either considering such a representation as a kind of formula for a general result or the basis for a general process which one could carry through with any set of numbers. Of course, symbolism of this kind came historically some time after the solution of problems that we should today solve by means of algebra.

Leaving this very brief discussion of the history of algebra to glance back over the first chapters of this book, we see that this is not the first time this symbolism, which we may call *algebra*, occurs. We have already used letters in place of numbers to gain the fundamental advantage of not saying exactly what we are talking about. To say that all men have noses (a statement which is true in the vast majority of cases) is a much stronger statement than saying that John has a nose. From the general statement I can conclude that John has a nose or that Jack has a nose or that Edward has a nose—in general, "all God's children have a nose." But to know merely that John has a nose is no help whatever in showing that Edward has one also, unless perhaps one uses John to see what a thing called a nose looks like. Furthermore, to say that a man traveling 20 miles per hour for 5 hours will have gone $5 \cdot 20 = 100$ miles expresses an isolated result, but to say that $T \cdot R = D$, when R is the constant rate in miles per hour and T is the number of hours, gives the distance, D, in miles, is to state a result for every rate and every period of time. It is the essence of algebra, as we know it, that by writing letters in place of numbers we can obtain a result that applies to a large set of numbers. The usual algebra taught in high schools tends to leave the impression that its chief use is in solving equations. It is true that later on in this chapter we shall consider this topic, but such a use of algebra is minor compared to its use in proving certain general results and in deriving useful formulas.

2. SQUARE NUMBERS

By way of illustrating the use of algebra in arriving at certain general conclusions we first consider an interesting property of the list of squares of numbers. Recall first that 3 squared,[1] which is written 3^2, is another way of writing $3 \cdot 3$. The list of the squares of the integers then, will begin like this:

1   4   9   16   25   36   49   64   81   100   121   144   169   196   225.

(We stop when we become tired.) Compiling the list becomes more and more laborious as we proceed. Unless we are very good at mental arithmetic, we cannot, without using paper and pencil, continue the list beyond what we have memorized; that is, we cannot unless we notice something about this list. There is, on investigation, a certain property of the differences of the successive terms of the sequence. We notice that $4 - 1 = 3$, $9 - 4 = 5$, $16 - 9 = 7, \ldots$, and continuing we see that the successive differences are these:

3   5   7   9   11   13   15   17   19   21   23   25   27   29.

That looks as if the next square beyond 225 is $225 + 31 = 256$. And, sure enough, multiplication shows that $16^2 = 256$! We can continue, to get $17^2 = 256 + 33 = 289$, since by now we are rather confident that the system will work.

We may consider ourselves clever to have discovered the above system, which *seems* to hold. If so, the wind will be taken out of our sails by the following question: Given $35^2 = 1225$, what is 36^2? Now, it would be altogether too much labor to compute all the squares up to 35^2 even by our clever system, and to compute 34^2 in order to apply our system would be as much labor as finding 36^2 by straight multiplication. We need to pursue our investigation further. We had to add 5 to 2^2 in order to get 3^2, 15 to 7^2 in order to get 8^2, 31 to 15^2 in order to get 16^2. Notice that $5 = 2 \cdot 2 + 1$, $15 = 2 \cdot 7 + 1$, and $31 = 2 \cdot 15 + 1$. The system thus seems to be: Add one more than twice an integer to get the square of the next higher integer. We now have what seems to be a system for getting any square from its predecessor. Will it always hold and, if so, why? No amount of carrying out special cases in the form which we just considered would give us any clew as to the reason for its working in general. But we can begin to see what the general situation is if we make use of the relationship between a number and its successor as follows: $(15 + 1)^2 = (15 + 1)(15 + 1) = 15(15 + 1) + 1(15 + 1) = 15^2 + 15 + 15 + 1 = 15^2 + 2(15 + 1)$. This is useful for it leads us to begin to see that *whatever* the number used instead of 15 in $(15 + 1)^2$, the result would be that obtained by replacing 15 by that number

---

[1] You might be interested to know that Sir Isaac Newton used a different notation for $a \cdot a$. For instance, instead of $b^2 + c^2 = h^2$ he wrote $bq + cq = hq$ (see *The Scientific American*, January 1968, p. 140).

in $15^2 + 2 \cdot 15 + 1$. The easiest way to see this is to let a stand for any number. Then $(a + 1)^2 = (a + 1)(a + 1) = a(a + 1) + 1(a + 1) = a^2 + a + a + 1 = a^2 + 2 \cdot a + 1$.

Using the letter a (or any other letter) has the advantage that it points out that whatever number we had we could deal with it exactly as we did with the a. It is a kind of formula for the process of working it out. No matter what numbers were put in for a at the beginning, the result would be what one would get by replacing a by that number in the conclusion. This, then, is an *algebraic proof* that to get the square of 1 more than a certain number, one adds the square of the certain number to 1 more than twice the number. Moreover, the algebraic proof tells us more than we started out to find; for any rational number, any real number, even any complex number, would give the same result; that is,

$$(a + 1)^2 = a^2 + 2a + 1.$$

Although geometry and algebra seem for the most part very distinct in high school mathematics, they go hand in hand many times in proving certain results. It is enlightening to see that the above result can also be proved geometrically, that is, with the help of a picture. Consider the arrays of dots shown in Fig. 6:1. In the first array a little inspection shows that to get the 4

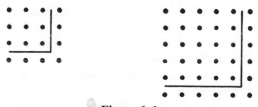

Figure 6:1

by 4 array from the 3 by 3 array we adjoin a column of three dots on the right, a row of three dots below, and a dot in the lower right-hand corner. In other words, we add $2 \cdot 3 + 1$ dots. A similar situation prevails in the second array and can be seen to prevail in any square array of dots, by the following argument. Suppose we have an n by n array of dots; if we adjoin at the bottom another row of n dots, at the right side another column of n dots, and in the lower right-hand corner one dot, we will have an $n + 1$ by $n + 1$ array of dots which has $n^2 + 2n + 1$ dots, whether n is 3 or 4 or whatever number of dots. And now we pause to ask whether this was a geometric proof because we used a picture or an algebraic one because we used a letter n for any number in describing how the process would go in general. Of course, the answer is that it is both. But it certainly is not purely algebraic, for we talked about dots. The proof is without doubt primarily geometric (the function of the algebra is merely to make our talk a little simpler) and we shall therefore call it a *geometric proof* even though we have used a little algebra.

   Notice that by using the dots we have proved our result only for positive integers, for we could not speak of half a dot or −2 dots. There is another geometric proof which will somewhat enlarge the set of numbers for which our result is true. Consider the square in Fig. 6:2. The area of the large square

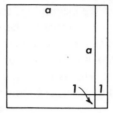

Figure 6:2

will be the area of the small one plus two rectangles a by 1 plus a square 1 by 1, which gives us the desired results.

EXERCISES

1.   For what kinds of numbers a does the last proof apply?

2.   What must be added to 3^2 to get 6^2, to 4^2 to get 7^2, in general to a^2 to get $(a + 3)^2$? Prove your result algebraically and geometrically, stating carefully for what kinds of numbers your results apply.

3.   What must be added to a^2 to get $(a + b)^2$? Give a geometric and an algebraic proof of your results.

4.   Given the two numbers 8 and 3. Computation 1 is the following: $8 + 3 = 11$, $8 − 3 = 5$, $11 \cdot 5 = 55$. Computation 2 is the following: $8^2 = 64$, $3^2 = 9$, $64 − 9 = 55$. The two computations yield the same result. Prove algebraically and geometrically that this is true for any two positive integers. Will it hold for other kinds of numbers?

*5.   Prove that the sum of the digits of any perfect square (that is, a square of an integer) has one of the following remainders when divided by 9: 0, 1, 4, 7. Can you say anything about the order in which such remainders occur as one goes from one square to the next?

*6.   Consider Exercise 5 for cubes instead of squares.

*7.   Suppose we consider the cubes and, being wise, we write the differences of successive cubes with them.

1	8	27	64	125	216	343	512	729	1000
	7	19	37	61	91	127	169	217	271

These differences do not behave as nicely as did the differences for the squares. How can one get from one cube to the next? Prove the formula that you get. Is there a geometric proof which you can find? Notice that, if you take the differences of the successive numbers of the second row of the table above, you get

$$12 \quad 18 \quad 24 \quad 30 \quad 36 \quad 42 \quad 48 \quad 54,$$

which do behave well. Why do they behave as they do?

3. TRIANGULAR NUMBERS AND ARITHMETIC PROGRESSIONS

Suppose that instead of arranging our dots in squares we arranged them in tenpin fashion as shown in Fig. 6:3. The numbers of dots in the successive

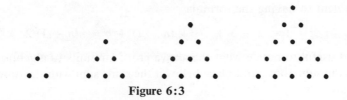

Figure 6:3

triangles, assuming that a lone dot is triangular in shape, are

$$1 \quad 3 \quad 6 \quad 10 \quad 15 \quad \cdots.$$

In a fashion analogous to our calling numbers in square array "square numbers," we call these *triangular numbers.* How can we get one triangular number from the preceding? The answer to this can easily be found, for we add 2 to the first triangular number to get the second, 3 to the second to get the third, ..., n to the $(n-1)$st to get the nth. This is, however, not really much help in finding the twentieth triangular number without finding the sum of

$$1 + 2 + 3 + 4 + 5 + \cdots + 17 + 18 + 19 + 20$$

which involves some labor. Although it is possible to find an algebraic derivation of a formula for the sum of the first n positive integers, it is easier to find it geometrically. Furthermore, we have here to do with integers alone, and there is no greater generality which an algebraic proof could give. Hence we give a geometric proof. Two sets of dots arranged in tenpin fashion can be made into Fig. 6:4, in which one array is turned upside down. Observe that

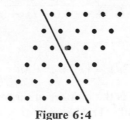

Figure 6:4

in the total array there are six rows of dots and five slanting columns. Thus, there are $6 \cdot 5 = 30$ dots in all and hence $30/2 = 15$ dots in the triangular array. Similarly, if we represented by dots the twentieth triangular number we should have a triangle with 20 dots on a side. Forming two such triangles inverting one, and placing them together to form an array like that above, we should have 21 rows and 20 slanting columns, hence 420 dots in all, which shows that in the triangular array there would be 210 dots. Therefore, 210 is the twentieth triangular number. In general, if we represent the *n*th triangular number by a triangular array of dots and adjoin another such array inverted to form a figure like a parallelogram, we should have $n + 1$ rows and n slanting columns. Thus, *the nth triangular number* is $n(n + 1)/2$. This is equivalent to having the formula

(1) $1 + 2 + 3 + \cdots + (n - 1) + n = n(n + 1)/2.$

It is useful to notice what the above proof amounts to algebraically. If we write the sum above and place below it the same sum with the order reversed, we have

(2)
$$
\begin{array}{l}
1 \ + \ 2 \ + \ 3 \ + \cdots + (n-2) + (n-1) + \ n \\
\underline{n \ + (n-1) + (n-2) + \cdots + \ 3 \ + \ 2 \ + \ 1} \\
(1+n) + (1+n) + (1+n) + \cdots + (1+n) + (1+n) + (1+n) = n(1+n).
\end{array}
$$

Each $(1 + n)$ is the number of dots in one column of the double triangular array. As in the geometric proof, the sum of the numbers in the first line of (2) is thus $n(n + 1)/2$.

This algebraic trick (and, in fact, the geometric one) can be applied to give a more general result. Suppose we have any sequence of numbers beginning with the number a and such that the difference between two successive numbers is a fixed number d. Then they will be

$$a, a + d, a + 2d, \ldots, a + (n - 1)d = l,$$

where there are n numbers in the sequence and l stands for the last one. This sequence is called an *arithmetic progression*. If we write the sequence in the reverse order and add as in (2), we have $a + l$ instead of $1 + n$ as the sum of each column and there are n columns; that is,

$$
\begin{array}{l}
a \ \ + (a+d) + (a+2d) + \cdots + (l-2d) + (l-d) + \ \ l = S \\
\underline{l \ \ + (l-d) \ + (l-2d) \ + \cdots + (a+2d) + (a+d) + \ \ a = S} \\
(a+l) + (a+l) \ + (a+l) \ + \cdots + (a+l) \ + (a+l) + (a+l) = 2S.
\end{array}
$$

Hence

(3) $a + (a + d) + (a + 2d) + \cdots + a + (n - 1)d = \dfrac{(a + l)n}{2} = S,$

where $l = a + (n - 1)d$, n is the number of terms in the sum, and l is the last one. This reduces to formula (1) when $a = d = 1$. Probably the easiest way to remember this formula is that the sum is equal to the number of terms multiplied by half the sum of the first and last terms.

In this connection, we quote a certain story about a famous mathematician, Gauss, which appears in E. T. Bell's *Men of Mathematics*[1]:

Shortly after his seventh birthday Gauss entered his first school, a squalid relic of the Middle Ages run by a virile brute, one Büttner, whose idea of teaching the hundred or so boys in his charge was to thrash them into such a state of terrified stupidity that they forgot their own names. More of the good old days for which sentimental reactionaries long. It was in this hell-hole that Gauss found his fortune.

Nothing extraordinary happened during the first two years. Then, in his tenth year, Gauss was admitted to the class in arithmetic. As it was the beginning class none of the boys had ever heard of an arithmetical progression. It was easy then for the heroic Büttner to give out a long problem in addition whose answer he could find by a formula in a few seconds. The problem was of the following sort, $81297 + 81495 + 81693 + \cdots + 100899$, where the step from one number to the next is the same all along (here 198), and a given number of terms (here 100) are to be added.

It was the custom of the school for the boy who first got the answer to lay his slate on the table: the next laid his slate on top of the first, and so on. Büttner had barely finished stating the problem when Gauss flung his slate on the table: "There it lies," he said—"Ligget se" in his peasant dialect. Then, for the ensuing hour, while the other boys toiled, he sat with his hands folded, favored now and then by a sarcastic glance from Büttner, who imagined the youngest pupil in the class was just another blockhead. At the end of the period Büttner looked over the slates. On Gauss' slate there appeared but a single number. To the end of his days Gauss loved to tell how the one number he had written was the correct answer and how all the others were wrong. Gauss had not been shown the trick for doing such problems rapidly. It is very ordinary once it is known, but for a boy of ten to find it instantaneously by himself is not so ordinary.

This opened the door through which Gauss passed on to immortality. Büttner was so astonished at what the boy of ten had done without instruction that he promptly redeemed himself and to at least one of his pupils became a humane teacher. Out of his own pocket he paid for the best textbook on arithmetic obtainable and presented it to Gauss. The boy flashed through the book. "He is beyond me," Büttner said, "I can teach him nothing more."

EXERCISES

1. Show that formula (1) holds for $n = 8$.
2. Find the sums of the following arithmetic progressions:
   - **a.** $3 + 5 + 7 + \cdots$ for 20 terms.
   - **b.** $4 + \cdots + 39$ (having 13 terms).
   - **c.** $5 + 2 + \cdots - 19$.
   - **d.** $5 + 3 + \cdots$ for 19 terms.
   - **e.** $9 + 13 + \cdots + 41$.
   - **f.** $5 + \cdots + 53$ (having 15 terms).
   - **g.** $\frac{1}{2} - \frac{1}{3} - \cdots$ for 10 terms

[1] Reference **7**.

3. Find the sums of the following arithmetic progressions:

   a. $1 + 5 + 9 + \cdots$ for 22 terms.   **b.** $5 + \cdots + 56$ for 12 terms.
   c. $7 + 4 + 1 + \cdots + (-17)$.   **d.** $\frac{1}{4} - \frac{1}{8} - \cdots$ for 24 terms.

4. When 70 boards are piled vertically in order of length, each one is 3 inches longer than the one above it. The top one is 1 *foot* long. How long is the bottom one? If they were placed end to end in a straight line, what would be the combined length?

5. Show that the sum of the twelfth triangular number and the eleventh is equal to $12^2 = 144$.

6. Are the figures given in the story about Gauss correct?

7. Prove geometrically and algebraically that the sum of the $(a - 1)$st triangular number and the ath is equal to a^2.

8. Prove that the sum of the first n odd numbers is n^2 no matter what positive integer n is. What connection does this have with Section 2?

9. Show that the differences of the squares of successive integers form an arithmetic progression. Can the same be said about the differences of the cubes of successive integers?

10. Give a geometric proof of formula (3) for $a = 2$, $d = 3$, $n = 5$.

11. An arithmetic progression has first term a and common difference d. Show that the formula for the sum of the first n terms is

$$\frac{d}{2} n^2 + \frac{2a - d}{2} n.$$

12. Suppose the formula for the sum of the first n terms of an arithmetic progression is $5n^2 + 21n$. What is the initial term and the common difference?

13. The first three terms of a sequence of numbers are 5, 8, and 21. Suppose you know that the difference of successive terms form an arithmetic progression. What will be the fourth and fifth terms of the sequence? What will be a formula for the nth term?

14. Answer the same questions as in Exercise 13 for the sequence beginning 3, 7, 23.

15. A man starts in a certain position at a salary of $300 a month. Each month his salary is increased by $10. What is the amount of his thirty-sixth salary check? What total amount of money does he receive during the 3 years?

16. Another man starts in a certain position at a salary of $300 a month. His monthly salary is increased by X dollars every 6 months. What should X be so that he will receive the same total salary during 3 years as the man of Exercise 15?

17. What is the system for winning the following game? B mentions a positive integer not greater than 5, A adds to it a positive integer not greater than 5, and so on. He who reaches 37 first, wins. If both B and A know the system and play it, who will win?

18. What would be the system for winning the above game if 5 and 37 were replaced by any other numbers?

19. Give the rule for working the following trick and explain your result. You choose a number and multiply it by 3. Tell me if the result is odd or even. If it is even, divide it by 2; if it is odd, add 1 to make it even and then divide by 2. Multiply your result by 3. Tell me what is the largest multiple of 9 less than your result. I will then tell you what number you originally chose.

20. We have formulas for the nth triangular number and the the nth square number. We might consider pentagonal numbers as illustrated by the sequence of pentagons in Fig. 6:5. The first four pentagonal numbers are 1, 5, 12, and 22. Find a formula for the nth pentagonal number.

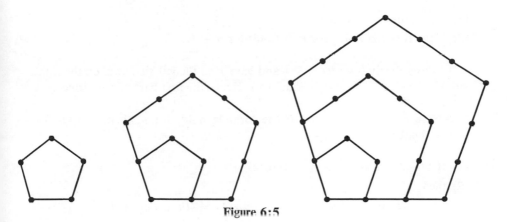

Figure 6:5

*21. In a fashion similar to that above, define hexagonal numbers and find a formula for the nth hexagonal number. Generalize this if you can to polygonal numbers, that is, for regular polygons of any number of sides.

4. MATHEMATICAL INDUCTION

There is a property of positive integers which we have not yet mentioned but which, on occasion, is very useful. It is called the property of *mathematical induction* and can be phrased as follows: Let S be a set of positive integers with the following properties:

(1) The number 1 is in S.

(2) If any integer, n, is in S, then $n + 1$ is also in S.

Then S must consist of all the positive integers.

This should not be hard to accept, for we know first that the integer 1 is in the set by hypothesis number (1). Then we apply hypothesis (2) with $n = 1$, which affirms that since 1 is in the set, $1 + 1 = 2$ is in the set. Then apply hypothesis (2) with $n = 2$, which affirms that $2 + 1 = 3$ is in the set. And so we can continue, to establish the fact that all integers are in S. But even though this property is apparent, we cannot prove it from the other properties we have already dealt with—hence we merely accept it as true without proof.

Let us illustrate this property to prove something we have already shown:

$$1 + 2 + 3 + \cdots + n = \tfrac{1}{2}n(n + 1).$$

Let S be the set of integers for which this formula is true. We want to show that S contains all the positive integers, that is, that the formula is true for all positive integers. In accordance with the property we must show two things:

(1) The formula holds for $n = 1$.
(2) If the formula holds for n, it holds for $n + 1$.

To show the first, we let $n = 1$ and have 1 on the left side and, on the right side, $\tfrac{1}{2} \cdot 1 \cdot (1 + 1)$, which is equal to 1. Hence the first part of the property is established.

Next we should show that *if* the formula holds for n, then it holds for $n + 1$. That is,

(1) If $1 + 2 + 3 + \cdots + n = \tfrac{1}{2}n(n + 1)$,
(2) then

$$1 + 2 + 3 + \cdots + n + (n + 1) = \tfrac{1}{2}(n + 1)(n + 1 + 1).$$

We can replace the sum of the first n terms on the left side by their sum, which is $\tfrac{1}{2}n(n + 1)$ by our assumption, and have

$$[1 + 2 + 3 + 4 + \cdots + n] + (n + 1) = \tfrac{1}{2}n(n + 1) + (n + 1).$$

Then, using the distributive property, we have

$$\tfrac{1}{2}n(n + 1) + (n + 1) = (\tfrac{1}{2}n + 1)(n + 1) = \tfrac{1}{2}(n + 2)(n + 1).$$

The right-hand side of this equation is equal to the right-hand side of (2). Hence our principle of mathematical induction shows us that the set S contains all the positive integers; that is, the formula (1) holds for all positive integers. This completes the proof.

Of course we had a simpler proof earlier without using this principle, but we shall see that this property is useful where our previous tricks do not apply.

Now consider another application of this property. Suppose we wish to show that $13^n - 5^n$ is divisible by 8 for all positive integers n. Let S be the set of integers n for which this is so. First, S contains 1, since for $n = 1$, $13^n - 5^n = 13 - 5 = 8$, which is divisible by 8. Now we must show that

(1) If $13^n - 5^n$ is divisible by 8,
(2) then $13^{n+1} - 5^{n+1}$ is divisible by 8.

Then (1) affirms that for some integer k,

$$13^n - 5^n = 8k.$$

By the well-defined property,

$$13^n - 5^n + 5^n = 8k + 5^n,$$
$$13^n = 8k + 5^n.$$

Hence

$$13^{n+1} - 5^{n+1} = 13 \cdot 13^n - 5^{n+1} = 13(8k + 5^n) - 5^{n+1}$$
$$= 13 \cdot 8k + 13 \cdot 5^n - 5^{n+1}$$
$$= 13 \cdot 8k + 13 \cdot 5^n - 5 \cdot 5^n$$
$$= 13 \cdot 8k + (13 - 5) \cdot 5^n$$
$$= 13 \cdot 8k + 8 \cdot 5^n,$$

which is a multiple of 8. This result can be shown much more simply by use of congruences.

There are a number of drawbacks to the method of mathematical induction. One is that one must guess by computation or other means what the result is before using such a proof. Another is that the algebra is often more troublesome than the method itself. It is also true that most of the meaningful uses of this method are in more advanced mathematics, where it is a much used tool. However, most of the exercises below are best solved by mathematical induction.

EXERCISES

1. Prove by induction that if $0 < a < b$, then $0 < a^n < b^n$ for all positive integers n.

2. Show by use of congruences that $13^n - 5^n$ is divisible by 8 for all positive integers n.

3. Show that if x and y are two fixed integers, then $x^n - y^n$ is divisible by $x - y$ for all positive integers n.

4. Show that if x and y are two fixed integers, then $x^{2n+1} + y^{2n+1}$ is divisible by $(x + y)$ for all positive integers n.

5. Show by induction or otherwise that $k(k + 1)$ is an even integer if k is an integer.

6. Show by induction or otherwise that $(2n + 1)^2 - 1$ is always divisible by 8, for n a positive integer.

7. Show that if n is an integer, $n^3 - n$ is divisible by 6.

8. Compute the first five sums of the following sequence:

$$1^3, 1^3 + 2^3, 1^3 + 2^3 + 3^3, \ldots, 1^3 + 2^3 + \cdots + n^3, \ldots.$$

What connection do these sums have with the first five triangular numbers?

Then guess a formula for the sum of the first n terms of the given sequence and prove it by mathematical induction.

9. Prove by induction or otherwise that, if n is an integer, then $n(n - 1)(2n - 1)$ is divisible by 6.

10. Consider the sequence

$$\frac{1}{2}, \frac{1}{2} + \frac{1}{2 \cdot 3}, \frac{1}{2} + \frac{1}{2 \cdot 3} + \frac{1}{3 \cdot 4}, \cdots, \frac{1}{2} + \frac{1}{2 \cdot 3} + \cdots + \frac{1}{n(n + 1)}, \cdots.$$

Compute the first four sums and guess the formula for the sum of n terms. Then prove it. (There is a clever way to prove the result without mathematical induction.)

11. Show that $1^2 + 2^2 + 3^3 + \cdots + n^2 = \frac{1}{6}n(2n + 1)(n + 1)$.

12. Consider the sequence

$$\frac{3}{2}, \frac{5}{4}, \frac{9}{8}, \frac{17}{16}, \cdots, \left(1 + \frac{1}{2^n}\right).$$

By computing the sums of the first few terms or by other means, guess a formula for the sum of the first n terms. Then by mathematical induction or otherwise prove the formula.

13. Do what is asked in Exercise 12 for the sequence

$$\frac{1}{2}, \frac{3}{4}, \frac{7}{8}, \frac{15}{16}, \cdots, \qquad \text{where the } n\text{th term is } \frac{2^n - 1}{2^n}.$$

5. COMPOUND INTEREST

Although the results and formulas we have derived have some uses outside mathematics, we dealt with them largely as mathematical curiosities. There are, however, certain formulas which very obviously affect our daily lives. One who knows no algebra must take such formulas on faith—he must rely on the authority of those who know algebra.

In ancient times the taking of interest, "usury," was considered the lowest of practices, but nowadays we all expect to receive something for the use of our money. In order that the reward shall be proportional to the amount loaned we charge a *rate of interest*. If the interest rate is 5% per year the interest on $100 is $5, and at the end of 1 year our $100 becomes $105; $1000 would accumulate interest of $50 in 1 year; and so on. In general, at an interest rate of i, P dollars will accumulate interest of Pi and thus, at the end of the year, P dollars becomes

$$A = P + Pi = P(1 + i).$$

This shows that, whatever the principal at the beginning of the year, the amount at the end of the year may be obtained by multiplying the principal by $(1 + i)$. If, now, the interest is *compounded annually*, the total amount, $P(1 + i)$, at the end of the first year draws interest the second year (just as if it were drawn out and deposited again), and, by the above reasoning, we multiply this by $(1 + i)$ to get the amount at the end of the second year; that is,

$$P(1 + i)(1 + i) = P(1 + i)^2.$$

Similarly, if interest were compounded annually for 3 years, at the end of that time it would amount to $P(1 + i)^3$. Thus, we have shown that if interest is at the rate of i compounded annually, at the end of n years P dollars will amount to

(4) $$A = P(1 + i)^n.$$

This we call the *compound interest formula*. We call A the *amount* of P dollars at the interest rate of i compounded annually for n years. Thus, for example, $100 at a rate of 3% compounded annually for 9 years would amount to

$$A = 100(1.03)^9 = 130.48.$$

One useful point of view is expressed by saying that $100 now will be worth $130.48 9 years from now if interest is 3% compounded annually.

Sometimes it is the amount rather than the principal that is known; as, for example, if I want to pay now a certain sum that at the end of 9 years will amount to $100. Then the expression is

$$100 = P(1.03)^9.$$

To find P, divide both sides of the equation by $(1.03)^9$ and have

$$P = \frac{100}{(1.03)^9} = \frac{100}{1.3048} = 76.64.$$

The value of money is not only the amount but when it is paid. At 3% compounded annually, $100 now is worth $130.48 9 years from now. However, $100 9 years from now is worth only $76.64 now. More generally, P dollars now is worth $P(1.03)^9$ 9 years from now. And the present value of R dollars 9 years from now is $R/(1.03)^9$.

The compound interest formula may also be used when interest is compounded more often that once a year. For instance, if the interest rate is 6% compounded semiannually, the interest rate for each half year is 3%. Thus the amount of P dollars at 3% compounded annually for 10 years is the same as the amount of P dollars at 6% compounded semiannually for 5 years, because the rate per interest period is the same in both cases and the number of interest periods is the same. **Hence, formula (4) may be taken to be the amount of P dollars at the rate of i per interest period, for n interest periods.** For example, $100 compounded quarterly at 8% for 10 years would be

$$A = P(1.02)^{40} = 220.80.$$

For the convenience of those computing interest, tables have been computed. We here give a small one which will be sufficient for our purposes, when $P = \$1$:

$$A = (1 + i)^n.$$

n	$\frac{1}{2}\%$	1%	2%	3%	4%	5%	6%	8%
1	1.005000	1.010000	1.020000	1.030000	1.040000	1.050000	1.060000	1.080000
2	1.010025	1.020100	1.040400	1.060900	1.081600	1.102500	1.123600	1.166400
3	1.015075	1.030301	1.061208	1.092727	1.124864	1.157625	1.191016	1.259712
4	1.020150	1.040604	1.082432	1.125509	1.169859	1.215506	1.262477	1.360489
5	1.025251	1.051010	1.104081	1.159274	1.216653	1.276282	1.338226	1.469328
6	1.030377	1.061520	1.126162	1.194052	1.265319	1.340096	1.418519	1.586874
7	1.035529	1.072135	1.148686	1.229874	1.315932	1.407100	1.503630	1.713824
8	1.040707	1.082857	1.171659	1.266770	1.368569	1.477455	1.593848	1.850930
9	1.045911	1.093685	1.195093	1.304773	1.423312	1.551328	1.698479	1.999005
10	1.051140	1.104622	1.218994	1.343916	1.480244	1.628895	1.790848	2.158925

There is another form of the compound-interest formula which has interesting consequences, as we shall see. If we let t be the annual interest rate, q the number of times it is compounded in a year, and n the number of years, then the amount at the end of n years will be given by the formula

$$A = P\left(1 + \frac{t}{q}\right)^{qn},$$

since t/q is the rate per interest period and qn is the number of interest periods.

Now let us see what happens in a particular case as q, the number of interest periods per year, increases while the annual rate remains the same. Let t, the annual rate, be 6%, or .06, and let q be the number of interest periods per year. Thus if the interest is compounded annually, $q = 1$; while if it is compounded quarterly, $q = 4$; and so on. In the table below, A denotes the amount at the end of 1 year for a sum of \$100 deposited at the beginning of the year. So the first line gives values of q, the second line the corresponding values of $A/100$ according to the formula $A/100 = (1 + .06/q)^q$, and the third line gives A to the nearest cent.

q	1	2	4	6	12	24
$A/100$	(1.06)	$(1.03)^2$	$(1.015)^4$	$(1.01)^6$	$(1.005)^{12}$	$(1.0025)^{24}$
A	\$106.00	\$106.09	\$106.14	\$106.15	\$106.17	\$106.18

Notice that there is only 18 cents difference between the first and last amounts, and as the number of periods increases the difference in amounts becomes less.

As a matter of fact, in this particular case the amount would be $106.18 for any greater number of interest periods, although if the principal had been $1000 there would be a little difference. This then naturally leads one to wonder what would happen if the number of interest periods were to increase without limit. The formula for the limiting amount turns out to be

$$Pe^{nt}$$

where e is an important mathematical constant whose value is approximately 2.718, n is the number of years, and t is the rate per year. It may be seen from a table that

$$e^{.06} = 1.0618,$$

correct to four decimal places. This verifies the statement above that if the number of interest periods per year is more than 12, the amount at the end of the year for $100 would be $106.18.

This is a formula for what is called "continuous interest," which closely approximates compounding interest daily, a practice in a number of banks today. It has a number of advantages: The formula holds when n is not a whole number, and by reference to tables one can find the amount at each instant; also it has "good customer appeal," since money draws interest for the exact time it is in the bank.

EXERCISES

1. I deposit $100 in a bank which pays interest of 2% compounded annually. What should I draw out at the end of 10 years?

2. I lend a friend $50. He promises to pay it back at the end of 3 years with interest at 6% compounded semiannually. How much should I receive?

3. What amount deposited in the bank of Exercise 1 will amount to $100 at the end of 10 years?

4. What sum am I entitled to receive now in return for a promise to pay $100 3 years from now if interest is 6% compounded semiannually?

5. If the annual interest rate is 12% and $100 is deposited in the bank at the beginning of a year, what are the amounts for interest compounded 2, 3, 4, 6, and 12 times during the year?

6. An insurance agent makes a commission of 3% on all insurance sold. How much insurance must he sell to have an income of $10,000?

7. A town has a population of 1000 persons. Each year for a period of 5 years, the population increases by 60 persons. What is the percentage of increase each year?

8. Suppose that in Exercise 7 the population decreased by 60 persons each year. What would be the percentage of decrease each year?

9. Comparing Exercises 7 and 8, is the percentage of decrease in any year for Exercise 8 equal to the percentage of increase in Exercise 7? Give reasons

why you should have expected the answer to be what it is before you carried out the computation.

10. State a problem involving the same calculations as Exercise 7 but starting with "A man has $1000."

11. In a certain country, half the population is under 16 years of age at present. During the next 16 years, each of these will have an average of two children (that is, each couple will have four children). If the net increase of the rest of the population is 20% over the next 16 years, what will be the percentage increase in total population at the end of that time?

6. ANNUITIES AND GEOMETRIC PROGRESSIONS

There is another problem connected with interest whose solution involves a little more mathematics. If you join a Christmas Club you deposit a set amount of money at regular intervals throughout the year and just before Christmas receive from the bank an amount which, as a result of the interest, is more than the sum of the amounts which you deposited. This procedure is related to the answer for the following question: For how much money paid to you at present will you agree to pay to me or my heirs $100 a year for 10 years? Such a series of payments constitute an *annuity certain*; an "annuity" because payments are annual (the term has come to include cases where payments are at any regular stated intervals), and "certain" because the number of payments does not depend on how long I live. Problems in other kinds of annuities are more complex and we shall not consider them at this point.

To simplify the discussion consider the following modification of the Christmas Club plan: P dollars is deposited at the beginning of each year for n years at an interest rate of i compounded annually. What will this amount to at the end of n years? To answer this question, notice that the first deposit will bear interest for n years, the second for $n - 1$ years, ..., the last for 1 year. Hence the amount will be

(5) $P(1 + i)^n + P(1 + i)^{n-1} + \cdots + P(1 + i)^2 + P(1 + i).$

It is possible to simplify this sum. Notice that reading from the right, each term is obtained from the preceding by multiplying by $(1 + i)$. In the arithmetic progression we obtained each term by *adding* a fixed number to the preceding. A sequence of numbers such that each is obtained from the preceding by *multiplying* by a fixed number is called a *geometric progression*. (The adjective "geometric" is used because the *ratio* of any number of the progression to the next one is the same as the ratio of the next to the one following that—ratio being an important geometric concept.) In this case we thus need to find the sum of a sequence of numbers in geometric progression. Since the finding of such a sum is useful in many situations, we shall temporarily postpone our consideration of the problem of the Christmas Club and proceed to find a formula for the sum of any geometric progression.

To approach the problem gradually, let us first consider the progression

$$S = 3 + 6 + 12 + 24 + 48 + 96 + 192,$$

which has the initial term 3 and the constant multiplier 2. It is not too laborious to add such a sum, but neither is it very enlightening. A trick is much more fun and instructive. Since 2 is the multiplier we multiply S by 2 and write it and the original sum carefully in the form

$$2S = \qquad 6 + 12 + 24 + 48 + 96 + 192 + 384$$
$$\underline{S = 3 + 6 + 12 + 24 + 48 + 96 + 192\qquad}$$

The arrangement is an invitation to subtract and we have

$$2S - S = S = -3 + 384 = 381.$$

While this trick will give us the sum of any geometric progression, it is useful on occasion to have a formula for it. Thus, we now find the sum

$$S = a + ar + ar^2 + ar^3 + \cdots + ar^{n-1}.$$

Notice that this sum has n terms. Using the above trick we have

$$rS = \qquad ar + ar^2 + ar^3 + \cdots + ar^{n-1} + ar^n$$
$$\underline{S = a + ar + ar^2 + ar^3 + \cdots + ar^{n-1}\qquad}$$

Subtraction gives

$$rS - S = -a + ar^n,$$
$$S(r - 1) = a(r^n - 1).$$

Divide both sides by $(r - 1)$ and have the formula for the sum of a geometric progression whose first term is a, whose ratio is r, and which has n terms:

$$(6) \qquad\qquad S = a\left(\frac{r^n - 1}{r - 1}\right), \qquad \text{if } r \neq 1.$$

Another formula for the sum which is sometimes more convenient may be obtained by noticing that the least term in the progression is $ar^{n-1} = l$. Then $ar^n - a = r(ar^{n-1}) - a = rl - a$. And the formula may be written

$$(7) \qquad\qquad S = \frac{rl - a}{r - 1}.$$

We apply these formulas in the following examples.

Example 1

Find the sum of the geometric progression

$$2, 10, 50, 250, \ldots, 6250.$$

Solution. Since we know the last term but not the number of terms, we use formula (7) for the sum and have

$$S = \frac{5 \cdot 6250 - 2}{5 - 1} = \frac{31248}{4} = 7812.$$

It should be noticed that with our present knowledge it would be a little difficult to find the number of terms in this progression. Since the last term is ar^{n-1}, we would have to solve for n the equation

$$2 \cdot 5^{n-1} = 6250$$

or

$$5^{n-1} = 3125.$$

We can find n by trying the various powers of 5. In fact, $n = 6$.

Example 2

Find the sum of the first 15 terms of the geometric progression below. What is the last term?

$$2, 6, 18, \ldots.$$

Solution. Since we here know the number of terms we use formula (6) and have

$$S = 2 \cdot \frac{3^{15} - 1}{3 - 1} = 3^{15} - 1 = 14{,}348{,}906.$$

The last term is $2 \cdot 3^{14} = 9{,}565{,}938$.

Example 3

What should I receive 4 years from now in return for four annual payments of $100 beginning now if the interest rate is 3% compounded annually?

Solution. The first payment bears interest for 4 years and hence, using the compound interest formula with $P = \$100$, $i = .03$, and $n = 4$, we see that at the end of the 4 years I should get $100(1.03)^4$ in return. Similarly, for the second payment I should get $100(1.03)^3$, since it bears interest for 3 years. Continuing in this manner we have

$$A = 100(1.03)^4 + 100(1.03)^3 + 100(1.03)^2 + 100(1.03)$$

as the amount which I should receive at the end of 4 years. This is a geometric progression, for each term may be obtained from the preceding by multiplying $1/1.03$. There is some gain in simplicity if we consider the order reversed and take the term $100(1.03)$ as the first term. Then the multiplier is 1.03, formula

(6) for the sum of a geometric progression applies, and putting $a = 100(1.03)$, $r = 1.03$, and $n = 4$ we have

$$A = 100(1.03) \cdot \frac{(1.03)^4 - 1}{1.03 - 1} = 103 \frac{(1.03)^4 - 1}{.03}.$$

From the interest table we see that $(1.03)^4 = 1.125509$. Hence

$$A = \frac{103(.125509)}{.03} = 430.91,$$

where the answer is given to the nearest cent.

Example 4

Knowing that his daughter should be entering college 5 years from now, a man wishes to have on hand at that time $1000 to pay part of her expenses. To this end he proposes to make five equal annual payments beginning now. If he can get interest at 5% compounded annually, what should each payment be?

Solution. Here we know the final amount but not the annual payment. Call the annual payment P. We see that the first payment would be worth $P(1.05)^5$ at the end of the 5 years, the second payment $P(1.05)^4$, and so on. They must in all amount to $1000, which gives us

$$1000 = P(1.05)^5 + P(1.05)^4 + P(1.05)^3 + P(1.05)^2 + P(1.05).$$

Our geometric progression now has $P(1.05)$ as the first term and $r = 1.05$. Thus

$$1000 = P(1.05) \frac{(1.05)^5 - 1}{1.05 - 1} = P(1.05) \frac{1.276282 - 1}{.05}$$
$$= P(5.801922),$$
$$\$172.34 = P.$$

Example 5

I wish to pay off a present loan of $1000 in five annual installments beginning 1 year from now. The rate of interest is 5%. There are two plans by which it can be done. Plan A: At the end of each year I pay one fifth of the principal (that is, $200) plus the interest on that portion of the principal unpaid at the beginning of that year (that is, the *outstanding principal*). Plan B: The annual payments are to be equal. Find the annual payments under each plan.

Solution. For plan A we can best show the solution by making a table:

	Principal un-paid at begin-ning of year	Payment on principal	Interest paid	Principal unpaid at end of year
First year	1000	200	50	800
Second year	800	200	40	600
Third year	600	200	30	400
Fourth year	400	200	20	200
Fifth year	200	200	10	0

Notice that the interest paid over the 5 years is $150. The payments under this plan are easy to compute, but it has the disadvantage that the first year I have to pay $250 in all, which is $40 more than the last payment.

Plan B: Why is it that $230 is not the annual payment under this plan? You should be able to answer that question before carrying through the computation. Let R be the annual payment. Since the first payment is made 1 year from now, its value now is $R/(1.05)$ (see Section 5). The second payment is now worth $R/(1.05)^2$, and so on. Their total present values must be $1000. Thus we have

$$1000 = \frac{R}{1.05} + \frac{R}{(1.05)^2} + \frac{R}{(1.05)^3} + \frac{R}{(1.05)^4} + \frac{R}{(1.05)^5}.$$

This is a geometric progression in which the first term is $R/(1.05)^5$ and the multiplier is 1.05. Hence

$$1000 = \frac{R}{(1.05)^5} \frac{(1.05)^5 - 1}{1.05 - 1} = \frac{R(1.276282 - 1)}{1.05^5(.05)}$$
$$= R(4.32948),$$
$$230.97 = R.$$

Notice that the total amount of interest paid in this case is almost $5 more than in plan A. It is instructive to form a table for this plan by way of comparison. Plan B is often referred to as "amortization of a debt." The word means literally "killing off." The computation proceeds in this fashion: In the first column, the interest is 5% of the $1000. The difference between this and the annual payment ($180.97 in the first column) goes toward payment of the principal. This latter amount subtracted from the top entry in the column gives the principal unpaid at the end of the year. Notice that 5% of the principal outstanding at the end of any year is the interest entry for the next year. The fact that 2 cents is still outstanding at the end of the 5 years results from the necessity to make each entry in the table to the nearest cent.

Amortization Table for a Debt of $1000 Paid in Five Years at 5%

	First year	Second year	Third year	Fourth year	Fifth year
Principal unpaid beginning of year	1000.00	819.03	629.01	429.49	219.99
Annual payment	230.97	230.97	230.97	230.97	230.97
Interest	50.00	40.95	31.45	21.47	11.00
Payment on principal	180.97	190.02	199.52	209.50	219.97
Principal outstanding at end of year	819.03	629.01	429.49	219.99	.02

In this table the entries for a year are in one column to make the computation easier. If the payments were made over a longer period of years it would be necessary to have a wider sheet of paper or change the arrangement.

EXERCISES

1. Find the sum of the following geometric progressions:
   a. $6 + 18 + 54 + \cdots + 486$.
   b. $2 + \cdots + 1024$ (having 10 terms)
   c. $3 + 6 + 12 + \cdots$ (having 10 terms).
   d. $1 - \frac{1}{2} + \frac{1}{4} - \cdots$ (having 10 terms).

2. Find the sum of the following geometric progressions:
   a. $3 + 12 + 48 + \cdots + 12288$.
   b. $2 + \cdots + 486$ (having 6 terms).
   c. $1 - \frac{1}{3} + \frac{1}{9} - \cdots$ (having 7 terms).
   d. $3 + \frac{3}{2} + \frac{3}{4} + \frac{3}{8} + \cdots$ (having 10 terms).

3. In 1 second a certain living cell divides to form two cells, in the next second each of these two divides to form two more, and so on. Estimate the number of cells at the end of 1 minute. If at the end of 1 minute the cells exactly fill a thimble, how full will the thimble have been at the end of 59 seconds? Assume that all cells are the same size.

4. There is an old story one of whose variations appears in D. E. Smith's *History of Mathematics*.[1] A certain Arab king was so pleased with the invention of the game of chess that he told its inventor that he would grant any request. Whereupon the inventor asked for one grain of wheat for the first square, two for the second, four for the third, eight for the fourth, and so on in geometric progression until the sixty-fourth square was reached. Assuming that there are 15 million grains of wheat to the ton, show that the inventor asked for more than 100 billion tons (the number "billion" being used in the U.S. sense.) Compare the number of grains for the twentieth square with the sum of the numbers of grains for the first nineteen squares.

[1] Reference **60**, pp. 549 ff.

5. In the case of each of the following state whether it is an arithmetic progression, a geometric progression, or neither. Find the sum of the first ten terms if it is either of the two named progressions:

   a. $\frac{1}{3} + \frac{1}{2} + \frac{2}{3} + \frac{5}{6} \cdots$.   **b.** $\frac{1}{3} + \frac{1}{6} + \frac{1}{12} + \frac{1}{24} + \cdots$.

   c. $\frac{1}{3} + \frac{1}{4} + \frac{1}{5} + \frac{1}{6} + \cdots$.   **d.** $\frac{1}{2} - \frac{1}{3} + \frac{2}{9} - \frac{4}{27} + \cdots$.

   e. $\frac{2}{5} + \frac{1}{15} - \frac{4}{15} - \frac{9}{15} - \cdots$.

6. In the case of each of the following, state whether it is an arithmetic progression, a geometric progression, or neither. Find the sum of the first ten terms if it is either of the two named progressions:

   a. $\frac{1}{2} + \frac{3}{4} + 1 + \frac{5}{4} + \cdots$.

   b. $(1 + \frac{1}{2}) + (1 + \frac{1}{4}) + (1 + \frac{1}{8}) + (1 + \frac{1}{16}) + \cdots$.

   c. $\frac{1}{2} + \frac{1}{4} + \frac{1}{6} + \frac{1}{8} + \cdots$.

   d. $1 - 1 + 1 - 1 + \cdots$.

7. Can three successive numbers a, b, and c be in both an arithmetic progression and a geometric progression? If so, give an example; if not, show why not.

8. A gambler plays a number of games. If he wins the first he get 1 cent, if he loses he pays 1 cent. The gain or loss in the second game is 2 cents, in the third 4 cents, and so on, the gain or loss each time being double that for the previous time. If his total capital is 2^{20} cents (about $10,000) and he loses every time, how many games can he play and still pay his losses? If he loses the first thirteen games and wins the fourteenth what will be his net gain or loss?

9. A person desires to borrow $400 from a bank and pay it back in quarterly installments over the period of a year. Since the interest rate for the bank is 6%, the year's interest on $400 is $24, and hence the bank requires that principal and interest be paid off in quarterly installments of $106 each (one fourth of $400 + 24). Is the person really paying interest at the rate of 6%?

10. If the loan of Exercise 9 were paid off as in plan A of Example 5, assuming interest is compounded quarterly, what would the quarterly payments be? If plan B of Example 5 were used, what would the payments be? How do these compare with the answer to Exercise 9?

11. The sum of $10 is deposited at the beginning of each year for 5 years. If interest is paid at the rate of 4% per year compounded annually, what should be the amount at the end of 5 years? What should be the amount at the beginning of the fifth year immediately after the fifth deposit has been made?

12. Find the annual payment and construct a table similar to that for Example 5 for the amortization of a debt of $1000 at 4% over a period of 6 years.

13. I give to a bank a government bond which will mature to $1000 5 years from now (that is, its value 5 years from now will be $1000). In return I expect the bank to pay me five equal yearly payments beginning at present. If interest is 3% compounded annually, how much should each of the payments be?

14. In return for what sum deposited now should a bank give ten semiannual

payments of $100 beginning 6 months from now? Interest is 4% compounded semiannually.

15. A man purchases a farm for $15,000, agreeing to pay $5000 cash and the balance (principal and interest) in ten equal annual payments beginning 1 year from the date of purchase. What should each payment be if interest is 5% compounded annually? Compare your result with that of Example 5, plan B.

16. Mr. Rounds wishes to accumulate a fund which will amount to $1000 at the end of 10 years. To do this he makes ten equal annual payments. Find the amount of each payment, first if it is made at the beginning of each year, second if at the end of each year. Interest is 4% compounded annually.

17. What amount deposited now in a bank paying 2% interest compounded semiannually would entitle the depositor to ten semiannual withdrawals of $1000 each beginning 6 months from now?

18. Mr. Wood bought an automobile, paying $1000 cash and $1000 at the end of each year for 3 years. What would be the cash price of the car if money is worth 6% compounded annually?

19. On January 1, 1922, a streetcar company issued $10,000 in 5% bonds to mature January 1, 1932. (That is, the bonds pay 5% of the face value each year and the total face value at the time of maturity.) If the bonds are to be redeemed by sinking funds at 4% interest, how much must be set aside from the company's earnings at the end of each year to provide for the interest and the retirement of the debt?

20. There are two plans under which a man can pay off a $10,000 mortgage in ten equal annual payments with interest at 5% compounded annually. Plan A: He pays off principal and interest in ten equal annual installments. Plan B: At the end of each year he pays the interest for that year on $10,000:$500. At the same time he deposits an amount in a savings bank giving 5% interest, the amount being so determined that at the end of 10 years he will have $10,000 in the savings bank to pay off the principal. Is the total annual payment under the two plans the same? If so, or if not, why?

21. In return for a loan of $1000 at the beginning of a certain year, a man agrees to pay at the end of each year for 5 years $200 plus the interest at the rate of 6% on the balance remaining due at the beginning of the year. What is the total amount which he pays? Does this problem have to do with an arithmetic or geometric progression or neither?

22. Show that the value of the sum $1 + 1/2 + \cdots + 1/2^n$ is
$$2 - (1/2)^n.$$

23. For the geometric progression 6, 18, 54, 162, ... the differences of successive numbers, 12, 36, 108, ... also form a geometric progression. Show that this will be true for every geometric progression. Will the progression of the differences ever be the same as the original progression? If so, when?

24. Is there a result for arithmetic progressions analogous to that in Exercise 23?

25. If a, b, and c are three numbers in arithmetic progression, what is b in terms of a and c?

26. If a, b, and c are three numbers in geometric progression, what is b in terms of a and c?

[*Note:* The b in Exercise 25 is called the *arithmetic mean* (or average) of a and c while in Exercise 26 it is called the *geometric mean* of a and c, or mean proportional of a and c.]

*27. Prove that the arithmetic mean of two positive numbers is never less than the geometric mean of those numbers.

*28. Given a rectangle. Prove that the side of a square of equal perimeter is never less than the side of a square of equal area.

*29. Porky, a guinea pig, has a litter of seven female pigs on each birthday. Each of her offspring does likewise. Assuming that none die, how many pigs will start life on the day when Porky is 3 years old? How many descendants will she have immediately after she is 5 years old? Formulas for the answers may be given and the answers estimated.

7. THE BINOMIAL THEOREM

There is a close connection between calculation of such an interest table as we have been using and a certain theorem which turns out to be useful in numerous situations. The so-called *binomial theorem* is in reality a formula for $(x + y)^n$. (The expression $x + y$ is called a *binomial* because it has *two* terms in it.) At this stage we are not concerned with getting a formula but merely in developing a quick method of calculating the result for small values of n. To this end, let us see what $(x + y)^n$ looks like for $n = 2$, 3, and 4.

$$
\begin{aligned}
(x + y)^2 &= x(x + y) + y(x + y) = x^2 + xy \\
& \underline{+ xy + y^2} \\
& x^2 + 2xy + y^2, \\
(x + y)^3 &= x(x^2 + 2xy + y^2) + y(x^2 + 2xy + y^2) \\
&= x^3 + 2x^2y + xy^2 \\
& \underline{+ x^2y + 2xy^2 + y^3} \\
& x^3 + 3x^2y + 3xy^2 + y^3, \\
(x + y)^4 &= x^4 + 3x^3y + 3x^2y^2 + xy^3 \\
& \underline{+ x^3y + 3x^2y^2 + 3xy^3 + y^4} \\
& x^4 + 4x^3y + 6x^2y^2 + 4xy^3 + y^4.
\end{aligned}
$$

We can shorten this process if we use merely the coefficients of the various expressions, that is, the numbers which multiply the various terms. For $(x + y)$ we write 1 1 and get the coefficients of $(x + y)^2$ by adding as follows:

$$
\begin{array}{ccc}
1 & 1 & \\
 & 1 & 1 \\
\hline
1 & 2 & 1
\end{array}
$$

Similarly, to get the coefficients of $(x + y)^3$ we write

$$
\begin{array}{cccc}
1 & 2 & 1 & \\
  & 1 & 2 & 1 \\
\hline
1 & 3 & 3 & 1
\end{array}
$$

Furthermore, if we knew the coefficients of $(x + y)^7$ to be

$$1 \quad 7 \quad 21 \quad 35 \quad 35 \quad 21 \quad 7 \quad 1$$

we could get the coefficients of $(x + y)^8$ as follows:

$$
\begin{array}{ccccccccc}
1 & 7 & 21 & 35 & 35 & 21 & 7 & 1 & \\
  & 1 & 7 & 21 & 35 & 35 & 21 & 7 & 1 \\
\hline
1 & 8 & 28 & 56 & 70 & 56 & 28 & 8 & 1
\end{array}
$$

We can avoid writing the coefficients of $(x + y)^7$ a second time by using the scheme

$$
\begin{array}{ccccccccc}
1 & 7 & 21 & 35 & 35 & 21 & 7 & 1 \\
1 & 8 & 28 & 56 & 70 & 56 & 28 & 8 & 1
\end{array}
$$

where now each number in the second line except the first and last is the sum of the two nearest numbers in the line above it. By this method we can compute the coefficients of $(x + y)^n$ for successive values of n.

If, then, to complete the picture we call $(x + y)^0 = 1$ we have the following table of coefficients for $n = 0, 1, 2, \ldots, 9, 10$:

n																					
0										1											
1									1		1										
2								1		2		1									
3							1		3		3		1								
4						1		4		6		4		1							
5					1		5		10		10		5		1						
6				1		6		15		20		15		6		1					
7			1		7		21		35		35		21		7		1				
8		1		8		28		56		70		56		28		8		1			
9	1		9		36		84		126		126		84		36		9		1		
10	1	10		45		120		210		252		210		120		45		10		1	

This array goes by the name the *Pascal triangle*, although it was known to a Chinese algebraist in 1300, over three centuries before Pascal's day. We shall in a later chapter find formulas for all the terms in the expression for $(x + y)^n$.

Now let us apply our newly found knowledge to the computation of an entry in the interest table. The coefficients of the expression for $(x + y)^9$ are found in the next-to-last line of the Pascal triangle, and putting $x = 1$ and $y = .03$ we have

$$(1.03)^9 = 1 + 9(.03) + 36(.03)^2 + 84(.03)^3$$
$$+ 126(.03)^4 + 126(.03)^5 + \cdots.$$

Notice that $(.03)^4 = .00000081$ and $126(.03)^4 = .0010206$. The sixth term is .0000030618. Since after the sixth term the coefficients (84, 36, 9, 1) become smaller and smaller, we can see that each term after the sixth is less than .03 times the previous term. Hence the sum of the first five terms will give us a value of $(1.03)^9$, which is accurate to *five* decimal places, and the sum of the six terms gives accuracy to six places. If we are to find the amount for a principal of $1000, five-place accuracy (that is, five terms) will give us accuracy to cents in the value of the amount; and the first six terms would suffice for a principal of $10,000. Taking the sum, then, of the first six terms we have the result that $(1.03)^9 = 1.304773$, accurate to six decimal places.

It should be pointed out that the fact that the binomial theorem is of some assistance in this connection depends vitally on the fact that y in $(x + y)^9$ is small. Fortunately (at least for purposes of this calculation) interest rates are small. If y were 1, the successive terms would, in the beginning, increase.

Considered on its own merits, the Pascal triangle has many interesting properties. For purposes of description let us speak of the slanting lines parallel to the sides as being the *diagonals* of the triangle and number them from the outside inward. Thus 1 1 1 1 $\cdots$ is the first diagonal, 1 2 3 4 $\cdots$ is the second, and so forth. Notice that the numbers of the third diagonal are apparently the triangular numbers (see Exercise 6 below). The numbers in the fourth diagonal are what are called *pyramidal numbers*, for if cannonballs, for instance, are piled in a triangular pyramid, the smallest such pyramid will contain one ball, the next will contain 4, the next 10, and so on (see Exercise 7 below).

We have seen for a number of examples why it is that the Pascal triangle gives the coefficients in the expansion of $(x + y)^n$. Let us see how we can show this for all integer values of n. Call

$$(x + y)^n = x^n + ax^{n-1}y + bx^{n-2}y^2 + cx^{n-3}y^3 + \cdots.$$

Then

$$(x + y)^{n+1} = (x + y)(x + y)^n = x(x + y)^n + y(x + y)^n,$$

by the distributive property. Then

$$
\begin{aligned}
x(x + y)^n &= x^{n+1} + ax^n y + && bx^{n-1}y^2 + && cx^{n-2}y^3 + \cdots \\
y(x + y)^n &= && x^n y + && ax^{n-1}y^2 + && bx^{n-2}y^3 + cx^{n-3}y^4 \\
& && && && + \cdots
\end{aligned}
$$

$$(x + y)^{n+1} = x^{n+1} + (a + 1)x^n y + (b + a)x^{n-1}y^2 + (c + b)x^{n-2}y^3 + \cdots$$

Comparing this with the expansion of $(x + y)^n$ it is apparent that the second coefficient in the expansion of $(x + y)^{n+1}$ is the sum of the first two coefficients in the expansion of $(x + y)^n$; the third coefficient of $(x + y)^{n+1}$ is the sum of the second and third coefficients in the expansion of $(x + y)^n$; and so on. We have thus shown that the Pascal triangle constructed as indicated gives the coefficients in the expansion of $(x + y)^n$. In a later chapter we shall have occasion to look at these coefficients again.

EXERCISES

1. Use the binomial theorem to find the amount of $100 at 12% interest compounded monthly for 20 months. Assume that the first five coefficients of $(x + y)^{20}$ are 1, 20, 190, 1140, and 4845.

2. Use the binomial theorem to find the amount of $100 at 2% interest compounded annually for 10 years; for 6 years.

3. Find the coefficients of the powers of x and y in the expansion of $(x + y)^{12}$. Use this result to answer the following: Find the amount of $100 at the end of 1 year if interest is 6% compounded monthly.
   (Notice that this is little larger than would be the amount if interest were compounded annually.)

4. Prove that the differences of successive numbers in any diagonal of the Pascal triangle are the numbers after 1 in the previous diagonal. (For instance, in the third diagonal $3 - 1 = 2$, $6 - 3 = 3$, $10 - 6 = 4, \ldots$ and $2, 3, 4, \ldots$ are the numbers after 1 in the second diagonal.)

5. Show intuitively that any entry of the Pascal triangle is equal to the sum of the numbers above it in the previous diagonal; for example, $15 = 5 + 4 + 3 + 2 + 1$.

*6. Show that the numbers in the third diagonal of the Pascal triangle are triangular numbers and use your result to find a formula for the third term in the expression for $(x + y)^n$. (*Hint:* Notice, for instance, that $10 = 4 + 3 + 2 + 1$ and show that each number in the third diagonal is the sum of the numbers in the previous rows of the second diagonal.)

*7. Show that the numbers in the fourth diagonal of the Pascal triangle are pyramidal numbers.

*8. Show that the formula for the fourth term in the expression for $(x + y)^n$ is $n(n - 1)(n - 2)/6$.

9. Show that if in any row of the Pascal triangle we form the sum of the first term, minus the second, plus the third, minus the fourth, and so on, the final sum is zero; for instance $1 - 6 + 15 - 20 + 15 - 6 + 1 = 0$ and $1 - 5 + 10 - 10 + 5 - 1 = 0$.

**10. Prove that the sum of the numbers in the nth row of the Pascal triangle is equal to 2^n.

8. NONTERMINATING PROGRESSIONS

There is another application of geometric progressions which can well be illustrated by an ancient paradox which is very well known: the story of Achilles and the tortoise. The argument, you recall, was that Achilles could never catch the tortoise because before reaching the tortoise he would have to traverse half the distance between him and the tortoise and by that time the tortoise would have moved farther along; the process continues without end "showing" that Achilles could never catch the tortoise. As a matter of fact, the same argument would apply if the tortoise remained stationary. For simplicity's sake we, therefore, assume that the tortoise was asleep and that he did not walk in his sleep. Further, suppose that when the story begins Achilles is 4 miles from the tortoise. He must traverse half the distance, that is, 2 miles, then half the remaining distance, 1 mile, and, continuing, we have a sum like this:

$$2 + 1 + 1/2 + 1/4 + \cdots,$$

each term being obtained from the preceding by taking half of it. Notice that after having gone 2 miles he had 2 left to go, after having gone $2 + 1$ he had 1 left to go, after $2 + 1 + 1/2$ he had $1/2$ left to go, and so on, after having gone $2 + 1 + 1/2 + \cdots + 1/2^n$ he had $1/2^n$ left to go. The further one goes in the sequence the less there is left to go. More precisely, $2 + 1 + 1/2 + \cdots + 1/2^n$ differs from 4 by $1/2^n$. In fact, the progress of Achilles may be seen from Fig. 6:6. From this we see that, by taking n large enough, our sum will differ

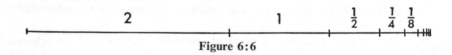

Figure 6:6

from 4 by as little as we please; not only that, but for all values of n larger than "large enough" it will differ from 4 by the same or less than "as little as we please." For example, if you select the number $1/1000$, I take $n = 10$ and not only will the sum then differ from 4 by less than $1/1000$ but for all values of n greater than 10 the sum will differ from 4 by less than $1/1000$; if you select the number $1/1,000,000$, I take $n = 20$ and not only will the sum then differ from 4 by less than $1/1,000,000$ for this value of n but also for all greater values of n. In fact, no matter how small a number you name, I can take n so large that for that value of n and for all greater values of n the sum will differ from 4 by less than that amount. We express this by saying that *the limit of* $2 + 1 + \cdots + 1/2^n$ *as n increases without bound* is 4. And we write it, for short,

$$\lim_{n \to \infty} 2 + 1 + 1/2 + \cdots + 1/2^n = 4.$$

We could express the same result in the language used for repeating decimals by saying that

$$2 + 1 + 1/2 + 1/4 + \cdots$$

converges to 4. Notice that 5 is not the limit because, although it is true that as n increases the sum comes closer and closer to 5, yet, if you gave me the number 1/2, I could not choose n so large that the sum would differ from 5 by less than 1/2. As a matter of fact, the sum differs from 5 by more than 1 for all values of n. On the other hand, $3\frac{7}{8}$ is not the limit, for although it is true that if you select, for instance, the number 1/1,000,000, I can choose n so large, $n = 3$, that the sum will differ from $3\frac{7}{8}$ by less than 1/1,000,000; but it will not be true that for all values of n greater than 3 the sum will differ from $3\frac{7}{8}$ by less than 1/1,000,000. As a matter of fact, for all values of n greater than 3 the sum will differ from $3\frac{7}{8}$ by at least 1/16. Notice, further, that the sum cannot have any limit other than 4, for suppose that it had a limit N. Then if you name the number 1/1,000,000 I must be able to find a number n greater than 20 such that the sum differs from N by less than 1/1,000,000. We know from the above that the sum differs from 4 by less than 1/1,000,000. Hence N differs from 4 by less than 2/1,000,000. We can use this process to show that N differs from 4 by as small an amount as you please and the only number N which has this property is 4 itself.

One can use the formula for the sum of a geometric progression to see that the limit of the sum is 4. By Exercise 22 at the close of Section 6 we have $1 + 1/2 + 1/4 + \cdots + 1/2^n = 2 - (1/2)^n$ and hence $2 + 1 + 1/2 + \cdots + 1/2^n = 4 - (1/2)^n$, which shows what we previously have noticed—that the sum differs from 4 by $1/2^n$.

The misconception in the "paradox" of Achilles and the tortoise is, of course, that since the distance which Achilles travels in catching the tortoise is indeed $2 + 1 + 1/2 + \cdots$ without end, it is supposed, without justification, that the time required is also without end. We can actually compute the time. Suppose that Achilles was walking at the rate of 4 miles per hour and that, as before, the tortoise was fast asleep and 4 miles away. It would take Achilles 1/2 hour to walk the 2 miles, 1/4 hour to walk the 1 mile, 1/8 to walk the 1/2 mile, and so forth. Thus the time which he would take to walk $2 + 1 + 1/2 + 1/4 + \cdots + 1/2^n$ miles would be $1/2 + 1/4 + \cdots + (1/4)(1/2^n)$ of an hour. As n becomes larger and larger the time comes closer and closer to 1. As a matter of fact,

$$\lim_{n \to \infty} [1/2 + 1/4 + \cdots + (1/4)(1/2^n)] = 1.$$

This shows that, as we knew all the time, the time taken to catch the tortoise is 1 hour.

EXERCISES

1. Show that $\lim_{n \to \infty} (1 + 1/3 + 1/9 + \cdots + 1/3^n) = 3/2$. For what values of n will the sum differ from 3/2 by less than 1/1,000,000?

2.   Show that $\lim\limits_{n \to \infty} (1 + x + x^2 + \cdots + x^n) = \dfrac{1}{1 - x}$ for all real numbers x
     such that $-1 < x < 1$.

3.   Find the $\lim\limits_{n \to \infty} (1/10 + 1/100 + \cdots + 1/10^n)$. What does this show about
     the unending decimal .1111111 ... ?

4.   What fraction has the decimal value .12121212... ?
     Solution. The result of Exercise 2 is here helpful, although we could go
     back to the formula for the sum of a geometric progression. Write

     $$.121212\cdots = 12(.01 + .01^2 + .01^3 + \cdots)$$
     $$= 12(.01)(1 + .01 + .01^2 + \cdots)$$
     $$= .12 \lim_{n \to \infty} (1 + .01 + .01^2 + \cdots + .01^n).$$

     Putting $x = .01$ in the result of Exercise 2 we have

     $$.121212\cdots = .12 \,\frac{1}{1 - .01} = \frac{12}{100 - 1} = \frac{12}{99} = \frac{4}{33}.$$

     A second method of solution is the following, which uses the same trick
     we used in finding the formula for the sum of a geometric progression. Let
     $x = .121212\cdots$ (we know from our statement in Chapter 5, Section 13,
     that there is such a number). Then

     $$100x = 12.121212\cdots$$
     $$x = .121212\cdots$$

     and subtraction gives us

     $$99x = 12$$

     and hence

     $$x = 12/99 = 4/33.$$

5.   What fraction has the decimal value .123123123$\cdots$?

6.   What fraction has the decimal value 56.101101101$\cdots$?

7.   Show that $\lim\limits_{n \to \infty} [1/2 - 1/4 + 1/8 - 1/16 + \cdots - (-1/2)^n] = 1/3$.

8.   Does the sum $1 - 1 + 1 - 1 + \cdots - (-1)^n$ have a limit as n increases
     without bound?

9.   Does the sum $1 + 2 + 4 + 8 + \cdots + 2^n$ have a limit as n increases with-
     out bound?

10. Prove $\lim\limits_{n \to \infty} (n - 1)/n = 1$.

11. Find $\lim\limits_{n \to \infty} (2n + 1)/n$.

12. A ball is dropped from a height of 16 feet. After first striking the ground it
     rises to the height of $(3/4) \cdot 16$ feet, on the second bounce it rises to three
     fourths the height it rose on the first bounce, and so it continues to rise on
     each bounce three fourths the height it rose on the previous one. Assume
     that it bounces vertically. Find how far the ball travels before coming to
     rest. How long will it take? (Assume the distance traveled by a falling body
     in time t to be $s = 16t^2$.)

13. A boy in a swing moves from north to south in an arc of 12 feet, then from

south to north in an arc of 9 feet, and so on, each time swinging through an arc three fourths the length of the previous swing. How far will he travel? If he swings strictly in accordance with the above statement and if each swing takes just as long as the next, will he ever come to rest? Give your reasons.

*14. Show that every repeating unending decimal has a rational number as a limit, that is, is the expansion of a rational number.

15. Let S_n be the sum of n numbers; suppose S_n increases as n increases (that is, each S_n is greater than its predecessor) and $\lim_{n \to \infty} S_n = A$. Show that if $A - S_B < c$, then $A - S_n < c$ for all n greater than B.

*16. One might try to use the method in Exercise 4 to find the sum of $1 + 2 + 4 + 8 + \cdots$ without end as follows.

Let $$x = 1 + 2 + 4 + 8 + 16 + \cdots.$$
Then $$2x = \quad\ 2 + 4 + 8 + 16 + \cdots.$$

Subtracting the upper from the lower we have

$$x = -1.$$

Thus apparently by the method we used successfully in Exercise 4, we have arrived at the absurd conclusion that

$$-1 = 1 + 2 + 4 + 8 + 16 + \cdots.$$

What is wrong?

9. THE WAYS OF EQUATIONS

Before we proceed to additional applications of algebra we shall do well to delve into the ways of equations, which we have already dealt with in informal fashion. An algebraic equation consists of two algebraic expressions with an equals sign between them.

Some equations involve only numbers: $\frac{5}{3} = \frac{10}{6}$ is a true equation of this kind and $3 = 5$ is a false one. Many equations contain letters as well as numbers. In the latter case, it is sometimes true that no matter what numbers we substitute for the letters, the equation is true. Examples of this type are

$$3x = 5x - 2x,$$
$$x^2 - y^2 = (x - y)(x + y).$$

These are called *identical equations* or *identities*, and the left side is said to be identically equal to the right side.

Equations which are not identities are called *conditional equations*. Examples of this type of equation are

$$3x = 3x^2/x,$$
$$x + 2 = 3,$$
$$x + y = 3,$$
$$0 \cdot x = 1.$$

The first of these examples is true for all numbers x except $x = 0$, the second only for $x = 1$, the third only for those numbers x and y whose sum is 3, and the last for no numbers x.

In the equations which we deal with here, the letters stand for numbers and we work with them as we do with numbers. The commutative property then enables us to say $xy = yx$ identically, the distributive property to say $x(y + z) = xy + xz$ identically, and so forth. Since our letters stand for numbers, we freely replace any expression by any other identically equal to it.

If we have an equation in which just one letter is involved, the *set of solutions* consists of all numbers which, when substituted for the letter, make the equation true or, as is sometimes said, "satisfy the equation." The set of solutions of $3x = 3x^2/x$ consists of all numbers with the single exception of zero. The set of solutions of $x + 2 = 3$ consists of a single number: the number 1. The set of solutions of $0 \cdot x = 1$ is the null set. An equation in a single letter is an identity if and only if the set of solutions contains all numbers.

If we have an equation in which two letters, say x and y, are involved, the set of solutions consists of all those pairs of values of x and y for which the equation holds. If the equation is $x^2 - y^2 = (x - y)(x + y)$, the set of solutions consists of all pairs of numbers. If the equation is $(x^2 - y^2)/(x - y) = x + y$, the set of solutions consists of all pairs (x, y) in which $x \neq y$. If the equation is $x + y = 3$, the set of solutions is all pairs whose sum is 3.

The usual method of finding the set of solutions of an equation, that is, solving the equation, is to derive from it a sequence of other equations until finally we have one whose solution is evident. Of course, there should be some connection between the set of solutions of the given equation and of that which forms the final step. When we can, we try to manage it so that each equation in our sequence has the same set of solutions as every other equation, that is, in which each equation is *equivalent* to every other one.

The process of getting an equivalent equation from a given one relies on the well-defined properties of equality first mentioned in Section 5 of Chapter 4. They are

(1) If $a = b$, then $a + c = b + c$.

(2) If $a = b$, then $ac = bc$.

Notice that the converse of statement (1) also holds, since, by (1) $a + c = b + c$ implies $a + c + (-c) = b + c + (-c)$; that is, $a = b$. Hence, no matter what c is, the equations $a = b$ and $a + c = b + c$ are equivalent equations. For example, $x + 3 = 5$ and $x + 3 + (-3) = 5 + (-3) = 2$ are equivalent equations. That is, $x + 3 = 5$ and $x = 2$ are equivalent equations. This is useful, since the equation $x = 2$ tells us the value of x, which is a solution of the equation $x + 3 = 5$. Of course, $x + 3 = 5$ and $x + 2 = 4$ are also equivalent equations, but the latter is not any better than the former.

We add -3 since we can look ahead to see that we will then have x by itself on the left side of the equation.

Similarly, the converse of statement (2) also holds *if $c \neq 0$*, since $ac = bc$ implies, by statement (2), $ac(1/c) = bc(1/c)$, or $a = b$. Thus $a = b$ and $ac = bc$ are equivalent equations for all values of c except zero. For example, $x/2 = 3$ and $x = 6$ are equivalent equations since we get the latter from the former by multiplying both sides of the equation by 2 and $2 \neq 0$. We chose to multiply by 2 since again we would then have x by itself on the left side of the equation.

We have shown that the following operations yield equivalent equations:

(1) Subtracting or adding the same number to both sides of an equation.
(2) Multiplying or dividing both sides of an equation by some number different from zero.

Let us apply this to the equation $3x + 7 = 4$. First, subtract 7 from both sides to get $3x = -3$ and then divide both sides by 3 to get $x = -1$. Since the second equation is equivalent to the first and the third is equivalent to the second, by the transitive property of equivalence, the third is equivalent to the first. Since the set of solutions of $x = -1$ is the single number -1, this number is also the set of solutions of the given equation $3x + 7 = 4$.

Suppose we perform one of the four operations described using an expression in x or some other letter instead of a number; will the equations then be equivalent? To answer this, let S stand for the set of values of x for which the expression is a number. Then, for all values of x in S, subtracting or adding the number described by the expression will yield an equivalent equation. For all values of x in S, multiplying or dividing both sides of the equation by the number described by the expression will yield an equivalent equation *except* for those values of x which make the expression zero.

For instance, suppose we have the equation $-x + 3 = 2x + 5$ to solve. If we add x to both sides of the equation, we get the equivalent equation $3 = 3x + 5$. Alternatively, we could add $x - 3$ to both sides of the equation to get the equivalent equation

$$0 = 2x + 5 + x - 3 = 3x + 2,$$

since for every number x, $x - 3$ is a number

Suppose we have the equation

$$\frac{1}{x - 2} = 3.$$

Although it would not help especially in solving this equation, we could add the negative of the left-hand side to both sides of this equation provided x is different from 2. (Here the set S described above consists of all numbers except 2.)

More care must be taken when we multiply or divide. Suppose we have the
equation $1/x = 2$. Here if we multiply by x, we will have an equivalent
equation except for $x = 0$. That is, if we exclude $x = 0$, the given equation
and $1 = 2x$ are equivalent. Since the latter has the solution $x = \frac{1}{2}$, which is
different from zero, the former has the same solution. Furthermore, $x = 0$ is
not a solution of the given equation.

Finally consider the equation

$$\frac{x}{x-2} = \frac{2}{x-2}.$$

If we multiply both sides of this equation by $x - 2$, we get the equation
$x = 2$, which is equivalent to the given equation except when $x - 2$ is zero,
that is, except when $x = 2$. Since $x = 2$ does not satisfy the given equation
and since $x = 2$ is the only solution of the equation we got by multiplying,
the given equation has no solutions.

The two equations $a = b$ and $a^2 = b^2$ are not equivalent, for, although
every value of a and b which satisfies the former equation satisfies the latter,
it is not true that every solution of the latter is a solution of the former. The
equation $a^2 = b^2$ merely implies that $a = b$ or $a = -b$. Thus squaring both
sides of an equation *does not* yield an equivalent equation. This has a bearing
on the usual method of solution of an equation like $\sqrt{x+1} = 3$. We square
both sides and get $1 + x = 9$ or $x = 8$. The latter equation is not equivalent
to the given one and we must substitute the final number into the given
equation to see whether or not it is a solution. In this case $x = 8$ is indeed a
solution, but if the original equation had been $\sqrt{x+1} = -3$, then $x = 8$
would not be a solution, because the radical sign means the positive square root.

The following principle is then important: *Unless, in the process of solving
an equation, each equation is equivalent to its predecessor, one cannot be sure
that all solutions of the last equation are solutions of the first.*

Of course, we usually manage it so that in the series of equations every
solution of an equation is a solution of its successor and hence that all the
solutions of the given equation are among those of the final equation. In that
case we need try in the given equation only the solutions of the final equation.
In the example above we knew that the solutions of $x = 8$ include the solutions of $\sqrt{x+1} = 3$ and also of $\sqrt{x+1} = -3$.

In dealing with equations, you should beware of artificial rules of thumb
which go by the names of "canceling" or "transposing." For example, we
may have $x + 2 = 4$ and conclude from it that $x = 4 - 2 = 2$. This process,
of course, amounts to "transposing 2 and changing the sign," but it is just as
brief, much more sensible, and much safer to say "subtract 2 from both sides."
One reason that the latter is safer than the former is that we are apt to forget
what "transpose" means, but we are not apt to forget what "subtract"
means. For instance, in $x/2 = 4$ we can "transpose" the 2 and change the

sign to get either $x = 4/(-2)$ or $x = 4 - 2$, neither of which is correct. By the time we put all the necessary restrictions on the word "transpose" we might just as well have said "subtract" and be done with it. Similarly, $3x = 12$ implies $x = 4$ because we can *divide* both sides by 3.

EXERCISES

Find all the solutions of each of the equations in Exercises 1 through 14. State exactly what you are doing in each step of the process and point out what steps lead to equivalent equations and what do not.

1. $5x - 6 = 3$.

2. $-2x + 3 = 5$.

3. $\dfrac{x}{3} - 8 = 20$.

4. $\dfrac{x}{4} + 7 = -3$.

5. $\dfrac{3}{x - 2} = 5$.

6. $\sqrt{x - 5} = 3$.

7. $x + 2 = 2x + 2 - x$.

8. $\dfrac{1}{x - 3} = 4$.

9. $\sqrt{x - 5} = -3$.

10. $\dfrac{3}{x - 2} = \dfrac{10}{2x - 4}$.

11. $1000 = 2.357P$.

12. $100 = \dfrac{2.462P}{.03}$.

13. $\dfrac{2x - 1}{x - 1} + \dfrac{1}{x + 1} - 2$.

14. $\sqrt{3 - x} + \sqrt{2x} = 0$.

15. Let f and g stand for two expressions in x and suppose we are seeking the solutions of the equation $f = g$. Suppose to do this we square both sides and solve $f^2 = g^2$. Why must all the solutions of the first equation be also solutions of the second? If a value of x is a solution of the second equation but not of the first, show that it must satisfy the equation $f = -g$.

16. Suppose that to solve an equation in x we multiply both sides of the equation by $(x - 2)$. Must every solution of the given equation be a solution of the new one? Must all solutions of the second equation be solutions of the given equation? Explain your answers.

10. THE WAYS OF INEQUALITIES

In many respects inequalities can be dealt with just as we dealt with equations, but there are important differences, as we learned in Section 5 of Chapter 5. For convenience we restate the results:

(1i) If $a < b$, then $a + c < b + c$.

(2i) If $a < b$, then $ac < bc$ if c is positive,

$$ac = bc \text{ if } c = 0,$$
$$ac > bc \text{ if } c \text{ is negative.}$$

We have exactly the same results if $<$ is replaced by $>$, $\leq$, or $\geq$. Since c is positive if $1/c$ is positive and conversely, the converse of each statement (1i), (2i), also holds and we have equivalent inequalities just as we had equivalent equations in the previous sections.

To solve an inequality in x, just as to solve an equation in x, we find the set of values of x for which the inequality is true, that is, the values of x which satisfy the inequality. For instance, to solve $-3x + 7 < 4$, we first subtract 7 from both sides of the inequality, getting $-3x < -3$. Then we divide both sides by -3 being careful to reverse the inequality as directed by the third part of property (2i). This yields $x > 1$. This inequality is equivalent to the original one. Notice that the set of values of x which satisfy the inequality is an infinite set.

EXERCISES

1. Find the set of solutions of each of the following inequalities:
   a. $3x + 7 < 4$. b. $-2x + 3 < 5$. c. $-2x + 4 \leq -2$.
   d. $x + 5 \geq 3x + 6$. *e. $\dfrac{1}{x-2} \leq 3$.

2. Find the set of solutions of each of the following inequalities:
   a. $3x + 5 > -2$. b. $-3x + 5 > 3$. c. $5x - 2 \geq 8$.
   d. $x + 5 \leq -4x + 2$. *e. $\dfrac{1}{x-1} \leq 4$.

3. Let a and b stand for two real numbers neither of which is zero, and suppose $a < b$. In which of the following cases is $1/a < 1/b$ and in which is $1/a > 1/b$?
   a. $ab > 0$. b. $ab < 0$.
   Give reasons for your answers.

4. Solve the following inequality: $|y| \leq 2$.

5. Are $a < b$ and $a^2 < b^2$ equivalent inequalities? If not, is every solution of the former inequality a solution of the latter? Is every solution of the latter a solution of the former? Discuss all possible cases.

6. Are $a < b$ and $a^3 < b^3$ equivalent inequalities? Answer the other questions of Exercise 5.

11. THE ANTIFREEZE FORMULA

This situation often arises in the winter months for those of us who own cars: I have enough alcohol in the radiator of my car for a temperature of 10°F, and the radiator is full. The newspaper says that tonight the temperature is going to drop to 0°F, the newspaper is always right, and I have no garage for my car. How much alcohol must I put in after drawing an equal amount of mixed water and alcohol from my radiator? Usually, lacking sufficient data and energy, we rely on our own guess or that of the garage man who might "play safe," that is, guess too high. But it is possible to find the exact answer without much difficulty. Of course, we are helpless unless sometime when the garage man is not looking we copy his table of the amount of alcohol required for various temperatures. It would look something like the first two columns of the following table, where the first column gives the freezing temperature when a tank of 15-quart capacity contains the amount of "alcohol' (really about 90% alcohol) listed in the second column, the amount given a little more accurately than is likely to be given in his table. We have also listed the percentage in the third column.

Temperature, degrees Fahrenheit	Quarts of "alcohol" in 15-quart tank	Percentage of "alcohol"
−20	7.7	51
−10	6.6	44
0	5.7	38
10	4.2	28
20	3.0	20

Quarts are given to the nearest tenth and the percentages to the nearest units.

Now the guessing begins. You might say, "Since I have 4.2 quarts and need 5.7 quarts, I should add 1.5 quarts." The garage man says, "Better put in two quarts." And you are both wrong. For suppose you draw off 1.5 quarts of the mixture; you then have .28(13.5) quarts of alcohol in your radiator, that is, 3.8 quarts. If you add 1.5 quarts to that, you have 5.3 quarts, which is short of the 5.7 required. On the other hand, if you follow the advice of the garage man and draw off 2 quarts you have .28(13) quarts of alcohol in your radiator, that is, 3.6 quarts. Adding 2 quarts to that gives 5.6, which is (we fooled you) again less than the 5.7 quarts required. This shows that we misjudged the garage man. You may well ask, "What is a quart of alcohol among friends?" But why avoid algebra! With a good formula we can get an accurate answer in less time than it took to go through the above computations.

Postponing the derivation of the formula, we can solve the problem at hand by letting x be the amount in quarts of the 28% mixture which must be withdrawn and replaced by alcohol. What is left after the withdrawal will be .28(15 − x) quarts of alcohol. After adding x quarts of alcohol we will have x + .28(15 − x) quarts of alcohol in our 15 quarts. Since we want a 38% mixture we require .38(15) quarts of alcohol in the resulting mixture, and hence we have

$$x + .28(15 - x) = .38(15)$$
$$x + .28(15) - .28x = .38(15)$$
$$x - .28x = .38(15) - .28(15)$$
$$(1 - .28)x = (.39 - .28)15$$
$$.72x = .1(15) = 1.5$$
$$x = 2.1.$$

This tells us that 2.1 quarts are actually needed.

It should be noted that in combining two substances in a solution there is always a certain amount of shrinkage. It turns out to be true that there is little shrinkage with alcohol and water. At the other extreme, it is well known that 1 cupful of sugar put into 1 cupful of water yields just about 1 cupful of sweet water.

As it was pointed out, the alcohol is not pure and the table is made taking this into account; the percentages are percentages of 90% alcohol. In the exercises below, unless something is said to the contrary, the alcohol is of the strength provided for in the table.

EXERCISES

1. A full 15-quart radiator has just enough alcohol for $-10°F$. How much should be withdrawn and replaced by alcohol to have just enough alcohol for $-20°F$? What would your answer be to the question if the radiator held 30 quarts?

2. A radiator whose capacity is V quarts is full of $s\%$ solution of alcohol. How much of the solution should one withdraw and replace with alcohol to have a full radiator of $S\%$ alcohol?

3. How could computation by the preceding formula be simplified by making use of the usual chart giving the number of quarts of alcohol required for various temperatures and various radiator capacities? Use this to compute the answers for Exercise 1.

    (Notice that the results of Exercises 2 and 3 with the table of solutions required for various temperatures gives us a quick means of solving all such antifreeze problems. Henceforth, to get our answer for any particular case we need merely substitute the proper values for the letters in the formula.)

4. How much of 10 gallons of a 10% solution of alcohol in water should be withdrawn and replaced by a 40% solution to yield 10 gallons of a 20% solution? How much of a 40% solution should be *added* (no withdrawal) to 10 gallons of a 10% solution to yield a 20% solution?

5. A certain company finding that it has a surplus of $2000 decides to distribute it among its workers in the form of a bonus. Seeking to give greater reward for long service it divides its workers into three classes: Class 1: those who have been employees for less than 5 years; class 2: those not in class 1 who have been employees for less than 10 years; class 3: all other employees. The bonus of each of those in the second class is to be twice that of each of the first and of each of those in the third class three times that of each in the first. If

there are 20 employees in the first class, 50 in the second, and 40 in the third, how should the bonus be divided?

6. In the situation of Exercise 5, the surplus is P dollars; the numbers in the three classes are n_1, n_2, and n_3, respectively. Find a formula for the amounts received by the workers.

7. In a certain state, the income tax is 4% of the *net income*, that is, the income after all deductions are made. One of the allowable deductions is the amount of the tax itself. If I is the income after all deductions are made except that of the tax itself, what, in terms of I, is the amount of the tax?

12. PUZZLE PROBLEMS

There is no hard and fast line between solving equations and finding formulas, but there is some difference in point of view. In Section 11, in deriving each formula we solved innumerable equations once and for all by finding a formula for the result. This has a great advantage if we are going to solve a large number of problems of the same kind but none whatever in solving one. It is chiefly in the puzzle type of problem that we are interested in a result for a single numerical problem. We do not need a general formula because if we have solved one puzzle problem we have no further interest in any other problem which differs from it only in the numbers involved. In such problems we are told something about a certain number—given various conditions about it—and wish to find what number or numbers satisfy these conditions. Hence we let some letter be the number and so manipulate the conditions that we find what number the letter must be. As a matter of fact, we did this in Section 11 when we found the antifreeze formula.

For purposes of illustration consider the following problem, which is so simple that it can be solved without any use of algebra but which we shall solve in an algebraic way to bring the method into relief: What three consecutive integers have 27 as their sum? To answer this we suppose that there are three consecutive integers whose sum is 27. The second would be 1 more than the first, the third 2 more than the first, and the sum of the three would be 3 more than 3 times the first. Since 27 is 3 more than thrice the first, $24 = 27 - 3$ must be thrice the first, and the first number is $\frac{24}{3} = 8$. In that case the others would be 9 and 10. Now $8 + 9 + 10 = 27$ and we have indeed found three consecutive integers whose sum is 27. Furthermore, they are the only three consecutive integers whose sum is 27, since we showed above that the first must be 8 if there are any.

This whole process can be expressed more concisely if, after assuming that there are three such integers, we let y be the first one. Then the others are $y + 1$ and $y + 2$ and we have $y + y + 1 + y + 2 = 3y + 3 = 27$. Thus $3y = 24$ and, dividing both sides by 3, we have $y = 8$. Each equation is equivalent to its predecessor and hence $y = 8$ is the solution and the only solution of the original equation. Thus the three consecutive numbers are 8, 9, and 10.

However, suppose we wish the answer to the following question: What number beside zero is its own double? Let x be such a number and $x \neq 0$. Then $x = 2x$ and we might proceed to its solution as follows: $x^2 = 4x^2$ and hence $0 = 4x^2 - x^2 = 2x - x$. That is, $3x^2 = x$. We are assuming that x is not zero and hence can divide both sides by x having $3x = 1$ and $x = \frac{1}{3}$. We have thus shown that if there is any number different from zero which is its own double, it is $\frac{1}{3}$, for in our sequence of equations the solution of each equation is included among the solutions of its successor. But since our last equation is not equivalent to the first one it may have, and in fact does have, a solution which does not satisfy the original equation. The last equation does show, however, that since $x = \frac{1}{3}$ does not satisfy the given equation, there is no value of x different from zero which will. It is enlightening to attempt to retrace the sequence of equations. Now $3x = 1$ implies $3x^2 = x$, which in turn implies that $4x^2 - x^2 = 2x - x$, but this does not imply that $2x = x$.

EXERCISES

1. Is there any solution to the following? If so, find it. Find four consecutive integers whose sum is 28. State very carefully the steps in the argument.

2. In attempting to solve the following problem the method given below might be used which leads to a value which does not satisfy the original condition. At what point does the retracing of the argument first break down?

    Problem: Find a number whose positive square root is 1 more than the positive square root of 1 more than the number. Let x be the number and have

    $$\sqrt{x} = \sqrt{x + 1} + 1.$$

    Square both sides to get

    $$x = x + 1 + 2\sqrt{x + 1} + 1,$$
    $$0 = 2 + 2\sqrt{x + 1},$$
    $$-2 = 2\sqrt{x + 1},$$
    $$-1 = \sqrt{x + 1}.$$

    Square both sides again and have

    $$1 = x + 1. \qquad \text{Thus } x = 0.$$

3. To find what numbers are equal to their cubes, we set up the equation $x^3 = x$. Dividing by x we get $x^2 = 1$ and $x = 1$ or -1. Both the numbers 1 and -1 are equal to their cubes, but 0 is also equal to its own cube and does not appear in the final result. Why?

4. Is there an integer such that if it is subtracted from its cube the result is 1 less than the integer? If so, find it.

5. Christopher is now twice as old as Phyllis. Two years ago he was three times as old as Phyllis. How old are they now?

Solution 1: Let p be Phyllis' age. Then $2p$ is Christopher's age. Two years ago Phyllis was $p - 2$ years old and Christopher was $2p - 2$. Since Christopher was then three times as old as Phyllis, we have

$$2p - 2 = 3(p - 2),$$
$$2p - 2 = 3p - 6.$$

Add 6 to both sides and subtract $2p$ from both sides to get $4 = p$. Each equation is equivalent to its predecessor, which shows that Phyllis is 4 years old and Christopher 8.

Solution 2: Let p be Phyllis' age and c Christopher's age. The first sentence of the statement of the problem expressed algebraically is then

$$c = 2p.$$

The children's ages 2 years ago were $c - 2$ and $p - 2$. Then the second sentence of the problem may be written

$$c - 2 = 3(p - 2).$$

Add 2 to both sides and get

$$c = 2 + 3(p - 2) = 2 + 3p - 6 = 3p - 4.$$

This, taken with $c = 2p$, gives

$$2p = 3p - 4.$$

Add 4 to both sides and subtract $2p$ to get

$$4 = p.$$

Logically, we have shown that if there are any ages satisfying the conditions imposed, we have found them. Although it is possible here to consider equivalent equations, the simplest way to show that we have a solution is to try the numbers in the problem as given. Christopher, now being 8, is twice as old as Phyllis. Two years ago they were 6 and 2, respectively, which satisfies the other condition.

6. A bottle and a cork together cost $1.05. The bottle costs exactly $1.00 more than the cork. How much does the cork cost?

7.[1] Weary Willy went up a certain hill at the rate of $1\frac{1}{2}$ miles per hour and came down at the rate of $4\frac{1}{2}$ miles per hour, so it took him just 6 hours to make the double journey. Now how far was it to the top of the hill?

8. The sum of the digits of a two-digit number is 11. If the order of the digits is reversed, the number represented is increased by 45. What is the given number?

9. A rectangular field and a square field have the same perimeter. The difference between the areas is 10,000 square feet. What is the difference in feet between the side of the square field and the width of the rectangular field?

10. A rectangular field is twice as long as it is wide. A square of equal perimeter is 10,000 square feet greater in area. What are the dimensions of the rectangular field?

11. A jet plane flies twice as fast as a propeller plane. For a 1200-mile trip the propeller plane takes 2 hours longer. What are the speeds of the planes?

[1] Puzzle 28, reference **21**.

12.[1] A man bought two cars but, owing to an unforeseen circumstance, found that he had to dispose of them. He sold them for $1200 each, taking a loss of 20% on one and making a profit of 20% on the other. Did he make a profit on the whole transaction or a loss and to what extent?

13.[2] In a recent motor ride it was found that we had gone at the rate of 10 miles per hour, but we made the return journey over the same route at 15 miles per hour, because the roads were more clear of traffic. What was our average speed?

14.[3] A man persuaded Weary Willie, with some difficulty, to try to work on a job for 30 days at 8 shillings a day, on the condition that he would forfeit 10 shillings a day for every day that he idled (that is, the net loss for each idle day was to be 2 shillings). At the end of the month neither owed the other anything, which entirely convinced Willie of the folly of labor. Now, can you tell just how many days of work he put in and on how many days he idled?

15.[4] "When I got to the station this morning," said Harold Tompkins, at his club, "I found I was short of cash. I spent just one half of what I had on my railway ticket, and then bought a penny newspaper. When I got to the terminus I spent half of what I had left and twopence more on a telegram. Then I spent half of the remainder on a bus and gave threepence to that old matchseller outside the club. Consequently, I arrive here with this single penny. Now, how much did I start with?"

16.[5] A meeting of the Amalgamated Society of Itinerant Askers . . . was held to decide whether the members should strike for reduced hours and larger donations. It was arranged that during the count those in favor of the motion should remain standing, and those who voted against should sit down.

"Gentlemen," said the chairman in due course, "I have the pleasure to announce that the motion is carried by a majority equal to exactly a quarter of the opposition." (Loud cheers.)

"Excuse me, guv'nor," shouted a man at the back, "but some of us over here couldn't sit down."

"Why not?"

"'Cause there ain't enough chairs."

"Then perhaps those who wanted to sit down but couldn't will hold up their hands . . . I find there are a dozen of you, so the motion is lost by a majority of one." (Hisses and disorder.)

Now how many members voted at that meeting?

SOLUTION: We can let N be the total number in the meeting and S the number standing. Then

(i) $$S = 5(N - S)/4.$$

---

[1] Adapted from puzzle 9 in reference **19**.   [2] Puzzle 67 in reference **19**.
[3] Puzzle 10 in reference **21**.   [4] Puzzle 9 in reference **22**.
[5] Puzzle 83 in reference **21**.

The number really wanting to vote for the motion is $S - 12$ and those wanting to vote against would be $N - S + 12$. Thus

(ii) $$S - 12 + 1 = N - S + 12.$$

The first equation gives $4S = 5N - 5S$ or $9S = 5N$ and the second gives $2S - 23 = N$. Thus $9S = 5(2S - 23) = 10S - 115$. Hence $S = 115$ and $230 - 23 = 207 = N$. Thus we have shown that if there is any solution to the problem $N = 207$ and $S = 115$. Substitution shows that these values satisfy equations (i) and (ii).

17.   The combined ages of Mary and Ann are 44 years, and Mary is twice as old as Ann was/when Mary was half as old as Ann will be/when Ann is three times as old as Mary was/when Mary was three times as old as Ann. How old is Mary?—Sam Lloyd.

  SOLUTION: Divide the statement of the problem by the / as shown and let m and a be the ages of Mary and Ann, respectively. Then we have

(i) $$m + a = 44.$$

Suppose before the first/Ann was $a - y$ years old. Then

(ii) $$m = 2(a - y).$$

At that time Mary was $m - y$ years old and let $a + z$ be the age Ann will be before the second/. We then have

(iii) $$m - y = (a + z)/2.$$

Let $m - t$ be Mary's age before the third/and have

(iv) $$a + z = 3(m - t).$$

Ann was then $a - t$ years old and we have

(v) $$m - t = 3(a - t).$$

At this point you are expected to bring out your paper and pencil and do what you are told. Solve the last equation for t and get $t = (3a - m)/2$. Substitute this in equation (iv) and have $a + z = 9(m - a)/2$. Substituting this in equation (iii) gives $m - y = 9(m - a)/4$. Solve this for y to get $y = (9a - 5m)/4$. Substitute this in equation (ii) and have $m + 2a - (9a - 5m)/2$, which implies that $5a = 3m$ and $a = 3m/5$. Substitute this value of a in equation (i) and have $m + 3m/5 = 44$. Hence $8m = 220$, $m = \frac{55}{2} = 27\frac{1}{2}$ and $a = 16\frac{1}{2}$. Notice that there was a definite system to the process. We put in new variables as we needed them. Then, working from the end, we systematically eliminated them.

  We have thus shown that if there is a solution, the values are what we have found. Now the easiest way to show that there is a solution in this case is to show that the values we have found satisfy the conditions of the problem. To do this we work from the end of the problem toward the front. Equation (v) expresses the fact that t years ago Mary was three times as old as Ann. We found that $t = (3a - m)/2$ and $a = \frac{33}{2}$ and $m = \frac{55}{2}$ gives $t = 11$ and we see that $\frac{55}{2} - 11 = 3(\frac{33}{2} - 11)$. Thus, when Mary was three times as old as Ann she was $\frac{33}{2}$ years old. Ann's age after the second/was then $3 \cdot 33/2 = \frac{99}{2}$ years. After the first/, then, Mary was $\frac{99}{4}$ years old. How old was Ann then? The answer to this question is that, since that was $\frac{5.5}{2} - \frac{9.9}{4} = \frac{11}{4}$ years ago, Ann must have been $\frac{3.3}{2} - \frac{11}{4} = \frac{5.5}{4}$ years old. And, sure enough, Mary is twice as old as that.

We could, of course, have demonstrated that we have indeed a solution by showing that every step in its derivation is reversible. But the process of the last paragraph not only proves that we have found a solution but checks the derivation itself.

18. A fish's head is 9 inches long. His tail is as big as his head plus one half his body. His body is as big as his head and tail together. How long is the fish?

19. A ship is twice as old as the boiler was when the ship was as old as the boiler is. The total age of the ship and boiler is 49 years. How old is the ship?

20. Louisa is 2 years older than Christopher. Six years ago she was twice as old as he. What are their ages now?

21. Two years ago Louisa's age was the sum of the ages of Christopher and Phyllis. Two years from now Louisa will be twice as old as Phyllis and the ages of all three will total 28 years. What are their ages?

22. Mary is 24 years old. She is twice as old as Ann was when she was as old as Ann is now. How old is Ann?

23. If Christopher were a year younger he would be twelve times as old as his son David. Eight years from now David will be one year short of one third his father's age. How old is David now? When will he be one half his father's age?

24. Ten years ago a man married a widow; they each already had children. Today there was a pitched battle engaging the present family of twelve children. The mother ran to the father and cried, "Come at once! Your children and my children are fighting our children!" As the parents now had each nine children of their own, how many were born during the 10 years?

*25. The following is a considerable simplification of a type of problem which the tax departments of certain corporations in Pennsylvania must solve every year. Let I be the given income and S be the amount of the combined capital stock, surplus, undivided profits, and so on, before all income taxes are computed. Let X be the amount of the Pennsylvania income tax, Y the federal income tax, and Z the Pennsylvania capital stock tax. We have the following:

$$X = .07(I - X - Y - Z),$$
$$Y = .19(I - X - Z),$$
$$Z = .005(S - X - Y - Z).$$

Find a formula for X, Y, and Z in terms of I and S.

The following three problems differ from those above in that not enough data are given to determine all the unknowns in the statement, although enough is given to answer the question proposed. A clever person can discover short cuts which may, in some cases, eliminate the use of algebra altogether.

*26.[1] A man rowing upstream passed a bottle floating down with the current and noticed that, when he passed it, it was opposite a certain stump. After he had rowed steadily upstream for 45 minutes it occurred to him that the bottle might have something in it and he immediately turned around and rowed downstream after the bottle, catching up with it 3 miles below the stump. Assuming that the man rowed at a constant rate relative to the stream, that the stream flowed at a constant rate, and that the bottle did not catch on any snags, find how fast the stream was flowing.

*27. Mr. X, a suburbanite, catches the 5 o'clock train for home daily and is met at his station by his chauffeur. One day he took the 4 o'clock train. On arriving at his station he decided to start walking for home without phoning for his chauffeur. His chauffeur started for the station at the customary time, met Mr. X on the way, and returned home with him. They arrived home 20 minutes before their usual time. How long did Mr. X walk?

*28.[2] If an army 40 miles long advances 40 miles while a dispatch rider gallops from the rear to the front, delivers a dispatch to the commanding general, and returns to the rear, how far has he to travel?

*29. (Lewis Carroll) In a very hotly fought battle, at least 70% of the combatants lost an eye, at least 75% an ear, at least 80% an arm, at least 85% a leg. How many lost all four members?

13. DIOPHANTINE EQUATIONS

So far in the problems we have been considering, it was sometimes true that an answer which was not a positive integer would have been senseless, but the problem was so set up that the result turned out to be a positive integer. There are certain types of puzzle problems in which there would be an unlimited number of answers were it not for the restriction that they must be positive integers. Such problems give rise to equations whose solutions are restricted to be integers; such equations are called *Diophantine equations*, from Diophantus, the Greek, who was one of the first to study them.

Consider the following such problem: A girl goes to a candy store and spends $1.00 on candy. Gumdrops are ten for a cent, caramels are 2 cents apiece, and chocolate bars are 5 cents each. She confines herself to these three kinds of candy and buys in all 100 pieces of candy. How many of each kind does she get? To answer this let g be the number of gumdrops, c the number of caramels, and b the number of chocolate bars. Since there are 100 pieces,

$$(8) \qquad\qquad g + c + b = 100.$$

Since they cost $1.00 we have

$$(9) \qquad\qquad g/10 + 2c + 5b = 100.$$

---

[1] Puzzle 47 in reference **22**.

[2] Puzzle 38 of reference **21**.

Solving for g in equation (8) gives $g = 100 - c - b$, and substituting in equation (9) we have

$$\frac{100 - c - b}{10} + 2c + 5b = 100.$$

Multiply by 10 to get

(10) $100 - c - b + 20c + 50b = 1000.$

This gives

(11) $19c + 49b = 900.$

We shall consider three systematic methods for the solution of equations like (11). The first two begin with the observation that if integers b and c can be found satisfying equation (11) then $19c = 900 - 49b$, which is 0 on a circle of 19 divisions; that is, we wish to determine an integer b so that $900 - 49b$ is 0 on this circle. Now, on such a circle 900 is 7, since $900 = 19 \cdot 47 + 7$ and 49 is 11. Hence, we wish to choose b so that $7 - 11b$ is 0 on this circle. At this point the two methods diverge.

1. The *brute force method* is to try $b = 1, 2, \ldots$ until you find a value which makes $7 - 11b$ have the value 0 on the circle of 19 divisions. At worst it would be necessary to try 18 different values of b. (Why is this so?) In this case $b = 11$ is the least positive value which works. Now $b = 11$ on the circle means that b can be 11, 30, 49, ... or $-8, -27, \ldots$ off the circle. A short way of writing this is to set $b = 11 + 19k$, where k is an integer (positive, negative, or zero). However, b and c are positive numbers and equation (11) shows that $49b$ must be less than 900. This would not be the case if b were 30 or had any greater value and b is not negative. Hence $b = 11$ and $c = 19$ is the only possible solution in positive integers of equation (11). Then equation (8) shows that $g = 70$. These values satisfy equation (9) and our solution is complete.

2. The *trickery method* shortens the labor but requires a little ingenuity. As above, we have $11b = 7$ on the circle with 19 divisions. We seek to reduce the multiplier of b. The first step is then to write $-8b = 7$ on the circle. We can further reduce the multiplier if we replace 7 by -12 and have $-8b = -12$ or $8b = 12$. Divide both sides by 4 to get $2b = 3$, make the right side even again by replacing 3 by -16, and $2b = -16$ implies $b = -8$ or $b = 11$.

One remark should be made about the observation before the discussion of the two methods. We could, of course, have written $49b = 900 - 19c$ and talked about the circle of 49 divisions, but the numbers are larger and our solution would be longer.

3. Finally we give a method using *congruence notation* more systematically. First notice that equation (11) is equivalent to $49b - 900 = -19c$. This means that $49b - 900$ is a multiple of 19, that is, in congruence notation,

(12) $49b \equiv 900 \pmod{19}.$

But 49 is congruent to 11 and 900 to 7(mod 19). Hence (12) is equivalent to

(13) $$11b \equiv 7(\text{mod } 19).$$

This means that for some integer x, $11b - 7 = 19x$ or

(14) $$11b - 19x = 7.$$

This is an equation like (12) but involving smaller numbers. So we use the method again (modulo the smaller multiplier, 11) to get

(15) $$-19x \equiv 7(\text{mod } 11),$$

which is equivalent to

(16) $$3x \equiv 7(\text{mod } 11).$$

We could repeat the process here, but it is simpler to notice that 7 is congruent to 18(mod 11) and (16) is equivalent to

(17) $$3x \equiv 18(\text{mod } 11).$$

This implies that $3x - 18$ is divisible by 11; that is, $3(x - 6)$ is divisible by 11 and hence

(18) $$x \equiv 6(\text{mod } 11);$$

that is, $x = 6 + 11y$ for some integer y. Substituting this for x in (14) and get

$$11b = 7 + 19(6 + 11y) = 121 + 19 \cdot 11y.$$

Dividing both sides by 11 we have

(19) $$b = 11 + 19y.$$

If we substitute this in (11) we get

$$19c + 49(11 + 19y) = 900,$$

which is equivalent to

$$19c = 900 - 539 - 49 \cdot 19y = 361 - 49 \cdot 19y.$$

Dividing both sides by 19 we have

(20) $$c = 19 - 49y.$$

Now c must be positive and hence, from (20), $y \leq 0$. Also b must be positive, which shows from (19) that $y \geq 0$. Hence the only value of y which will make both b and c positive is $y = 0$. Thus $c = 19$, $b = 11$ is the only solution of equation (11). This could be shortened by use of the trickery in method (2).

A problem of this kind, to be "good," should be so drawn up that there is one solution and only one. A carelessly made problem is apt to have many or no solutions. For instance, a problem which led to the equation $5x + 3y = 12$

would have no solution in positive integers x and y, while $5x + 3y = 38$ has three solutions. The reader should satisfy himself that these statements are true.

EXERCISES

1. The following Chinese problem of the sixth century is given by David Eugene Smith (reference **59**, p. 585). If a cock is worth 5 sapeks, a hen 3 sapeks, and 3 chickens together 1 sapek, how many cocks, hens, and chickens, 100 in all, will together be worth 100 sapeks? Find that solution in which the number of chickens is the largest.

2. A farmer with $100 goes to market to buy 100 head of stock. Prices were as follows: calves, $10 each; pigs, $3 each; chickens, $0.50 each. He gets exactly 100 head for his $100. How many of each does he buy?

3.[1] Three chickens and one duck sold for as much as two geese; one chicken, two ducks, and three geese were sold together for 25 shillings. What was the price of each bird in an exact number of shillings?

4.[2] A man received a check for a certain amount of money, but on cashing it the cashier mistook the number of dollars for the number of cents and conversely. Not noticing this, the man then spent 68 cents and discovered to his amazement that he had twice as much money as the check was originally drawn for. Determine the amount of money for which the check must have been written.

5. Johnny has 15 cents which he decides to spend on candy. Jelly beans are seven for 1 cent, licorice strips three for 2 cents, and caramels 1 cent apiece. He gets 40 pieces of candy. How many of each kind does he buy?

6. Percival has 50 cents to spend on candy. In *his* candy store jelly beans are ten for 9 cents, licorice strips, eleven for 10 cents, and caramels 2 cents apiece. He wants to get just as many pieces of candy as he can regardless of kind. How many of each does he purchase?

7. Find all the solutions of

$$35x + 19y = 372,$$

    where x and y are positive integers.

8.[3] A man's age at death was $\frac{1}{29}$ of the year of his birth. How old was he in the year 1900?

9. Find all the solutions of $25x + 73y = 269$, where x and y are positive integers.

10. Find all positive integer values of x, y, and z which satisfy

$$x + 3y - 4z - 8 = 0,$$
$$2x + y + 3z - 39 = 0.$$

[1] Puzzle 19 in reference **22**.
[2] Reference **65**, p. 65.
[3] Puzzle 52 in reference **22**.

11. Find all the solutions in positive integers of the pair of equations

$$2x + 3y - 7z = 5,$$
$$3x + 2y + 5z = 18.$$

12.[1] A man bought an odd lot of wine in barrels and one barrel containing beer. There were, in all, six barrels containing 15, 16, 18, 19, 20, and 31 gallons, respectively. He sold a quantity of the wine to one man and twice the quantity to another but kept the beer to himself. None of the barrels was opened. Which barrel contained the beer?

13. The equation $\frac{16}{64} = \frac{1}{4}$ is a result which can be obtained by the cancellation of the 6 in numerator and denominator. Find all the cases in which $ab/bc = a/c$ for a, b, and c integers between 1 and 9 inclusive.

*14.[2] The number 3025 has the property that if the number 30 and 25, formed of the digits in order of the two halves of the number, are added (giving 55) and the result squared (55^2), the result is the original number. How many other numbers are there composed of four different digits having the same property?

**15. Five women, each accompanied by her daughter, enter a shop to buy cloth. Each of them buys as many feet of cloth as she pays cents per foot. In each case the number of feet bought and the number of cents spent is a positive integer.

   a. Each mother spends $4.05 more than her daughter.
   b. Mrs. Evans buys 23 feet less than Mary.
   c. Rose spends 9 times as much as Clara.
   d. Mrs. Brown spends most of all.
   e. Effie buys 8 feet less than 10 times as much as Mrs. White.
   f. Mrs. Connor spends $40.32 more than Mrs. Smith. The other girl's name is Margaret. What is her surname?

*16. In *Kitty Foyle*, by Christopher Morley, appears the following story[3]:

   That clock was made in imitation of Horticultural Hall, you could see all the works, which were usually slow. "It has Philadelphia blood," Pop said. The kitchen clock, a much more homely thing, was fast. Once we had an argument, how long would it be, with the two clocks getting farther apart from each other at a definite rate, before they would both again tell the same time. Mac said that was easy and went upstairs to figure it out by algebra. Mother said they never would because two wrongs never make a right. Pop went in the back yard to think it over. Mac came downstairs once to ask, did we mean till both clocks told the same time or till they both told the correct time?

   "The correct time," Pop said. "Gee," Mac said, "I bet that'll be about a thousand years." He went upstairs again to calculate and didn't come back at all. Meanwhile they forgot to wind the clock and it stopped.

---

[1] Puzzle 76 in reference **19**.

[2] Rewording of puzzle 113 in reference **19**.

[3] From *Kitty Foyle*, pp. 28–29. Copyright, 1939, by Christopher Morley. Published by J. B. Lippincott, Co., Philadelphia.

Assume that at noon both clocks told exactly the correct time and that the Horticultural Hall clock loses time at the constant rate of 3 minutes per hour and the kitchen clock gains at the rate of 4 minutes per hour. When will both clocks next tell exactly the correct time (provided they are regularly wound)? Could the rates of losing and gaining be such that both clocks would never again agree exactly—much less tell the correct time together?

14. TOPICS FOR FURTHER STUDY

1. History of algebra: reference **60** (Chap. 6). In particular, the Pascal triangle is discussed on pp. 508–511 of this reference.
2. Classify various kinds of annuity problems and derive formulas for each.
3. Triangular, square, and, more generally, pentagonal, hexagonal, and other "polygonal" numbers: references **30** (pp. 201–216), **65** (pp. 5–11), and **72**.
4. Algebra in general: references **39** and **56**.
5. The theory of numbers and diophantine equations: references **36**, **57**, and **65**.
6. Sums of series (see Exercises 8 and 11 of Section 4): reference **65** (pp. 5–11).
7. Limits: reference **53** (Chap. 11).
8. Various puzzles: references **6**, **19**, **20**, **21**, **22**, **25**, and **42**.
9. Find the complete solution in integers of the Diophantine equation $a^2 + b^2 = c^2$ (a, b, and c are integers). Notice that a and b may be considered to be the legs of a right triangle and c the hypothenuse. See references **36** (pp. 72, 73) and **65** (pp. 37–40).
10. A new branch of mathematics called "the theory of games" is developed in references **40** (see also applications to "voting power" as well) and **73**.

7

Graphs and Averages

I gotta love for Angela,
   I love Carlotta, too.
I no can marry both o' dem,
   So w'at I gona do?

Oh, Angela ese pretta girl,
She gotta hair so black, so curl,
An' teeth so white as anytheeng.
An' oh, she gotta voice to seeng,
Dat mak' your hearta feel eet must
Jomp up an' dance or eet weell bust.
An' alla time she seeng, her eyes
Dey smila like Italia's skies,
An' making flirtin' looks at you—
But dat ess all w'at she can do.

Carlotta ees no gotta song,
But she ees twice so big an' strong
As Angela, an' she no look
So beautiful—but she can cook.
You oughta see her carry wood!
I tal you w'at, eet do you good.
W'en she ees be som'body's wife

She worka hard, you bat my life!
She nevva gettin' tired, too—
But dat ees all w'at she can do.

Oh, my! I weesh dat Angela
    Was strong for carry wood,
Or else Carlotta gotta song
    An looka pretta good.
I gotta love for Angela,
    I love Carlotta, too.
I no can marry both o' dem,
    So w'at I gonna do?[1]

1. MEASUREMENTS

No doubt Carlotta had black hair, too, but it was not so black and it may
have been curly but perhaps it did not curl much. She could probably even
sing, but it might have been more enjoyable when she did not attempt it. On
the other hand, Angela could surely carry wood but not much. This all goes
to show that the answer to "How much?" is at least as important as the
answer to "What?" is.

Different persons (as well as one person) are interested in different things.
When water is the topic, the physicist is interested in its composition and the
degree to which it will combine with other substances; the thirsty traveler, in
its purity (if he is not too thirsty). Once the property in which one is interested
is isolated, the important thing to determine, if one is to have a complete
description, is to what degree this property is possessed. We know that the
answer to "How many?" is given by stating a number. This answer is so
satisfactory that we immediately want to answer the question of "How
much?" or "To what degree?" by assigning a number to it. But this is much
harder to do. The beauty of Angela's song or the savor of Carlotta's cooking
could not be measured in terms of mere numbers.

Description needs adjectives and most adjectives imply a comparison. For
instance, to say that a star is very bright merely implies that it is brighter than
most stars which we can see. Such a statement is not as definite as "It is
brighter than Sirius"; then, we begin to have a measure of its brightness.
Since comparison plays a fundamental role in measurement and consists in
listing things in order, we pause for a moment to consider the *properties of
order* in terms of the brightness of a star. They are:

(1) If A and B are any two stars, star A is just one of three things: it is either
    brighter than B, just as bright as B, or not so bright as B.

---

[1] "Between Two Loves" from *Carmina* by T. A. Daly. Reprinted by permission of
Harcourt, Brace and Company, Inc.

(2) If A is brighter than B, then B is not so bright as A and vice versa. If A is just as bright as B, then B is just as bright as A.

(3) If A is brighter than B, and B is brighter than C, then A is brighter than C. The same is true if "brighter than" is replaced by "not so bright as" or by "just as bright as."

Since brightness has these three properties, we say that the stars *can be ordered according to their brightness*. Notice that, although the stars, naturally, do not form an ordered number system as described in Section 5 of Chapter 5, they do satisfy properties (3i) and (4i) of that section.

It is clear that when we have the properties of order we can arrange our objects according to the degree to which a certain quality appears. This is a step in measurement, but it has its disadvantages, for one must have the whole list more or less in mind in order to give meaning to a certain position in that list. The next step is to fix upon a *unit of measurement*. This is done in two essentially different ways.

One way of fixing the unit is to decide arbitrarily on one thing which shall be said to have one unit of the quality. Just as soon as we have something which weighs 1 pound we can express all weights by numbers with reference to that unit of weight. Anything which weighs twice as much (that is, as much as two things which each weigh 1 pound) we say weighs 2 pounds, anything which weighs half as much weighs $\frac{1}{2}$ pound, and so forth. By comparing our weights with a single one we have the advantage of not only having arranged our objects in an order according to the given quality, but have a definite measure in terms of a number. A knowledge of the unit of weight would enable me, for instance, to tell my friend in Australia how heavy my piano is without taking it over for him to compare with his piano. It should be remarked in this connection that the accuracy of such a system of measurement depends on the care with which the unit is defined, and its usefulness depends on its availability. To the former end the International Bureau of Weights and Measures was founded in Paris in 1875 for the purpose of agreeing upon certain International units of measurement. The units of measurement in this country are defined with reference to certain units in the National Bureau of Standards. The standard pound, for example, is a cylinder of pure platinum about 1.35 inches high and 1.15 inches in diameter. There are also certain legal definitions of units. For instance, the international conference agreed that the alternative and provisional definition of a meter should be 1,553,164.13 times the wavelength of the red light emitted by a cadmium vapor lamp excited under certain specified conditions. (The legal definition is apt to stand up better under bombing than could the platinum cylinder.) The availability of units usually is promoted through the use of certain instruments of measurement graduated with reference to the standard unit or a secondary unit obtained from it.

A second way to fix upon a unit is with reference to two extremes. This is often necessary, for in many cases the first method of defining a unit fails. To say that Angela is twice as beautiful as Carlotta or that one pan of water is twice as hot as another has no meaning (at least until a unit is defined) because two Carlottas are no more beautiful than one and two pans of luke-warm water are not hotter than one. In the case of Angela and Carlotta we might decide that no one could be more beautiful than the former, and we will call her beauty 100 and that Carlotta had no beauty whatsoever and her beauty would be 0. Our unit of beauty would then be one hundredth of the difference between, and we should at least have the basis for its numerical measurement. In the case of water, no water is hotter than steam and hence we call it 100; and none is colder than ice and hence we call it 0. This is the basis of the centigrade scale and is used, of course, not only for measuring the temperature of water but for all temperatures. To be accurate we must further specify that the basis of measurement shall be for pure (that is, dis-tilled) water at sea level. Notice that "twice as hot" in the centigrade scale is very different from "twice as hot" in the Fahrenheit scale, but "twice as heavy" is the same no matter what units are used.

The first kind of unit described is called an *absolute unit*, for it is determined with reference to an absolute zero, that is, a condition in which the property concerned is completely lacking. The pound is determined relative to absence of weight. Similarly, if one measures temperature in absolute units, that is, with reference to absolute zero, then "twice as hot" does have a meaning and this is independent of the unit chosen. Once one could agree on what was absolute lack of beauty, one could then, independent of the unit chosen, give meaning to "twice as beautiful."

These units are quite arbitrary, as may be judged from the wide variety of units in this world. But a unit, to be usable, must be comparable in magni-tude with the quality which it measures. Vegetables are measured in pounds but aspirin in grains.

After determining units of measure one devises instruments for measuring: scales for weight, the thermometer for heat, and goodness knows what for beauty. The more accurate the instrument, the more accurately is our com-parison established.

EXERCISES

1. According to which of the following properties can objects be ordered: weight, size, temperature, color, distance, value, loudness, place of manu-facture? If, in any cases, your answers are debatable, give a short debate. In the cases in which the answer is unequivocably yes, by what must one replace "brighter than" in the three properties of order above?

2. Name at least one unit for each of the following qualities: length, area, strength of a person, specific gravity, amount of electric current, brightness,

excellence of work in a course, hardness of jewels, strength of a salt solution. (It is expected that students will, if necessary, supplement their knowledge by the use of dictionaries and encyclopedias to answer these questions.) In which cases is the unit an absolute unit?

3. Describe briefly instruments used to measure at least three of the qualities listed above.

2. GRAPHS OF SETS OF MEASUREMENTS

It was pointed out in Chapter 5 that different numbers could, with advantage, be thought of as corresponding to points on a line. But graphical representation is chiefly useful in presenting the relationship between two sets of values. It depicts such relationships in the large more vividly than does a mere table of numbers.

We are all more or less familiar with the way in which a graph is drawn, but it may be worthwhile to consider one case in detail. Suppose our table of values is as shown in Table 1, where t is the time in seconds and s is the dis-

Table 1

t	0	1	2	3
s	0	16	64	144

tance in feet traversed by a falling body, the distance being measured from the position of the body when $t = 0$. We draw a pair of perpendicular lines and choose the horizontal line on which to represent the values of t and the vertical line those of s. We call the former line the *t-axis* and the latter the *s-axis*, and their point of intersection the *origin*. Then on each of the lines we choose the values of t and s at the origin. In this case and in many others it is convenient to take $t = 0$ and $s = 0$ at this point, but it is not always advisable to do so. We then choose a scale on each axis. In this case a convenient scale on the s-axis would be $\frac{1}{4}$ inch to 20 feet and on the t-axis, $\frac{1}{4}$ inch to a second. (It is here necessary to make the scale of s small in order to keep the graph on the paper.) Each value of t will then correspond to some value on the t-axis and similarly for s. Then we represent the pair of values $t = 1, s = 16$ by the point which is the intersection of the vertical line through $t = 1$ on the t-axis and the horizontal line through $s = 16$ on the s-axis. Our table of values then will be represented by a set of four points as in Fig. 7:1, and we call these points the *graph* of the table of values.

The number pair (2, 64), for example, is often written beside the point which is its graph, and 2 and 64 are called the *coordinates* of the point. In order to avoid having to refer at length to "the point with $t = 2$ and $s = 64$," it is agreed that the first number of a pair of coordinates shall be measured horizontally and the second vertically. It is often convenient to use

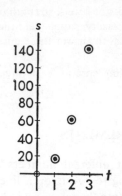

Figure 7:1   Figure 7:2

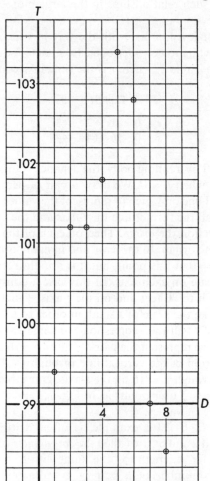

Figure 7:3

"coordinate paper,"that is, paper which has rulings of horizontal and vertical lines. If we do that our graph above looks like Fig. 7:2.

As has been indicated, it sometimes is not convenient to have both variables take on the value 0 where the axes of the graph cross. Such a situation occurs with the temperature shown in Table 2, T, on various days, D, of a patient with the measles. Here we can best show the graph by taking $T = 99$ at the origin and letting the vertical side of each square be .2°. We can have $D = 0$ at the origin and let the horizontal side of each square be 1 (see Fig. 7:3).

Another less stringent agreement has to do with which set of data is to be laid out on the horizontal axis and which on the vertical. It is often not too important a matter, for the reasons for doing one thing or another may be largely esthetic or intuitive, although nonetheless definite. Let us approach the matter with reference to our particular example. Here s and t vary and hence are called *variables*. The relationship we call a *function*. Whether or not there is a definite formula establishing the correspondence between one variable and the other, the idea of a correspondence itself rather implies that when we know the value of one, the value of the other is thereby determined. In that case we are thinking of the one as the so-called *independent variable* and the other as the *dependent variable*. It is customary to assign the horizontal axis to the independent variable and the vertical axis to the dependent variable.

In many cases the choice of the independent variable is largely a matter of instinct with a little logic behind it. In Table 1 we are apt to think of the dis-

Table 2

D	1	2	3	4	5	6	7	8
T	99.4	101.2	101.2	101.8	103.4	102.8	99	98.4

tance as depending on the time and hence measure t along the horizontal axis. We might have trouble in persuading someone that the distance depended on the time rather than the time on the distance, but most persons would agree with our choice especially since the time is recorded at regular intervals in the table. In Table 2 the temperature depends on the day. The situation is somewhat different in Table 3, where v is velocity in revolutions per minute and p

Table 3

v	1200	1500	1800	2100	2400	2700	3000	3300	3600
p	.94	1.06	1.15	1.15	1.07	.94	.75	.51	.31

is horsepower. Here a priori it might be difficult to decide whether the horsepower or revolutions per minute were first determined. But the table itself gives us a clue. In the first place, the values of v are given at regular intervals, indicating that in the experiment from which the table arose, the values of v

were first decided upon and then the horsepower calculated for these values. In the second place, any value of v in the table determines one accompanying value of p, whereas in one case there are two values of v for the same p. Either one of these considerations would lead us to take v as the independent variable and graph it along the horizontal axis.

More formally, we call y a *function* of x if there is at most one value of y for each x. Thus, for Table 3, p is a function of v but v is not a function of p.

EXERCISES

1. In each of the following, state which variable you would consider the independent and which the dependent variable:
   a. The rate of interest on $1.00 compounded annually for 10 years and the amount accumulated.
   b. The height of a triangle of given base and its area.
   c. Variables a and b, where the relationship is that a is 1 when b is a rational number and a is 0 when b is an irrational number.
   d. The number of students registered in Cornell and the price of bread in New Zealand.
2. Draw the graph of each of the following tables of values.

Table 4

E	0	6000	12,000	18,000	24,000	30,000	36,000
P	30.0	23.8	19.0	15.0	11.8	9.5	7.5

   a. Table 4, where P is pressure in pounds and E is elevation in feet above sea level.

Table 5

t	0	10	15	35	45	60
s	0	$6\frac{2}{3}$	10	$23\frac{1}{3}$	30	40

   b. Table 5, where s is the distance in feet a body travels in t seconds.

Table 6

T	−20	−15	−10	−5	0	5	10	15	20	25	30	35	40
W	1	1.5	2.3	3.4	4.9	6.8	9.3	12.8	17.2	22.9	30.1	39.3	50.9

   c. Table 6, where 1 cubic meter of air at temperature $T°C$ can hold W grams of water vapor.

Table 7

T	0	10	20	30	40	50	60
V	7500	7000	5500	3000	1300	600	300

d. Table 7, where it is estimated that T years hence a certain piece of property will be worth $\$V$.

Table 8

F	0	26	40	37	30	21	11	4	1
t	0	1	2	3	4	5	6	7	8

e. Table 8, where t seconds after starting a train, a locomotive exerts a force F tons more than the resisting forces.

Table 9

F	5′	6′	8′	10′	12′	15′	25′	50′	100′
W	2′4″	2′6″	2′9″	3′0″	3′2″	3′4″	3′8″	4′0″	4′2″

f. Table 9: When working with a portrait attachment on a camera, focus at F feet to work at W feet and inches.

3. Draw the graph of each of the following tables of values.
   a. The following table, where E is the elevation above sea level in feet and P is the pressure measured by inches in a column of mercury:

E	0	1000	2000	3000	4000	5000	6000
P	29.9	28.9	27.8	26.8	25.8	24.9	24.0

   b. The following table, where E is the approximate elevation in feet above sea level for which the boiling point of water is $T°F$:

T	212	211	210	209	208	207	206
E	0	511	1024	1539	2056	2575	3096

   c. The areas of squares with sides 1, 2, 3, 4, and 5 inches, respectively.
   d. The areas of cubes with sides 1, 2, 3, 4, 5, and 6 feet, respectively.

3. INTRODUCTION TO THE GRAPHS OF CERTAIN SIMPLE FORMULAS

If we know a formula connecting two variables we can construct our own table of values and find the graph of the relationship, that is, the graph of the formula or function. The formula and the picture augment each other in describing the relationship. Together they have the advantages of a talking picture over a photograph and a phonograph record.

Since various lines and curves have formula or equations associated with them, it is possible to prove certain geometric theorems by translating geometry into algebra, working with the algebra, and translating the results back into geometry. This interplay of the two subjects is called *analytic geometry* and is much more recent than the mere drawing of graphs, which dates back to three centuries before Christ. René Descartes is generally given the credit for inventing analytic geometry, which he discussed in a book published in 1637, but the same ideas seemed to have occurred to others at slightly earlier dates. In this section, however, we shall content ourselves with merely drawing a few simple graphs of equations.

In this section and, in fact, usually, when we are considering the graphs of equations or formulas, we choose the same scale on both axes and take as the origin the point whose coordinates are $(0, 0)$. If, then a scale is fixed, we know that to every pair of positive real numbers a and b there corresponds a point a units to the right of the vertical axis and b units above the horizontal one. To the pair $(-a, b)$ corresponds a point a units to the left of the vertical axis and b units above the horizontal axis. The points corresponding to $(a, -b)$ and to $(-a, -b)$ are also shown in Fig. 7:4, always assuming that a and b are

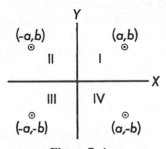

Figure 7:4

positive numbers. To every pair of real numbers there belongs just one point in the plane and to every point in the plane belongs just one pair. [Each pair is called an *ordered pair* because the order is important; that is, the point (a, b) is different from the point (b, a) unless $a = b$.] Notice that the axes divide the plane into four parts or *quadrants*, which are customarily numbered as in the figure.

The way we have chosen to set up a correspondence between the points of the plane and pairs of real numbers is only the most common of many ways to do this. The two axes need not be perpendicular nor even straight lines. Sometimes a point is located other than by its distance from two lines. But we shall remain loyal to our first love and shall not even cast a roving eye toward any other coordinate systems, however beautiful they may seem to be. Unless something is said to the contrary, we shall abjectly take the x-axis to be horizontal and the y-axis vertical.

It should be pointed out that when we are dealing with an equation or formula we have more than a mere table of values. No matter how extensive a table of values we made, we should not be able to find all the pairs of values of the variables which satisfied the equation. So that when we have a formula (and as we shall see later under some other circumstances), it is desirable to connect the points found from the table of values by a smooth curve indicating that there are other points whose coordinates satisfy the equation but whose values we have not computed. We assume that all such points will lie in a smooth curve. Thus we consider the graph of an equation to have two important properties: *The coordinates of every point on the graph satisfy the equation, and every pair of values of the variables satisfying the equation are the coordinates of a point on the graph.*

EXERCISES

1. Draw the graphs of the following equations:
   a. $x = 3$.   b. $y = -5$.   c. $x = -7$.
   d. $y = x$.   e. $y = -x$.   f. $y = 2x$.
   g. $y = -3x$.   h. $-2y = x$.
2. Draw on a single set of axes the graphs of the following equations:
   a. $y = 2x$.   b. $y = 2x + 3$.
   c. $y = 2x - 1$.   d. $y = 2x + 4$.
   If b is known merely to be a definite constant, what can you say about the graph of $y = 2x + b$?
3. Draw on a single set of axes the graphs of the following equations:
   a. $x = 3y$.   b. $x = 3y + 2$.
   c. $x = -3y + 2$.   d. $x + 3y + 2 = 0$.
4. Draw the graph of $y = -x + b$ for three different values of b.
5. Draw on one pair of axes the graphs of the equations
   $$y = x + 3, \quad y = 3x + 3, \quad y = -5x + 3, \quad y = -2x + 3.$$
   If a is known merely to be some constant, what can you say about the graph of $y = ax + 3$?

4. THE EQUATION OF A STRAIGHT LINE

When we describe numerically the steepness of a road we specify the *grade*, which is the distance it rises per unit horizontal distance. We use the same ratio in describing the steepness of a line except that in this connection we call it the slope. That is, the *slope* of a line is the distance it rises from left to right per unit distance measured horizontally. In Fig. 7:5 the grades of the lines AB are the ratio r/s and the slopes are the ratio r/t. Since any road not on a level may be uphill or downhill, depending on which way you are going, we customarily consider ourselves going from left to right in dealing with a straight line on a graph or, for that matter, with a curve. It is natural to consider the distance a line "rises" as *negative* when it is *descending* from left to right. Hence the first line in the figure has a positive grade and positive slope,

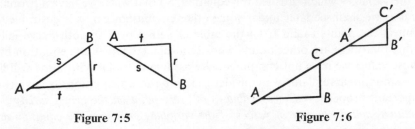

Figure 7:5   Figure 7:6

while the second line has a negative grade and a negative slope. It is permissible to speak of *the* slope of a line because the ratio does not depend on the points of observation. This may be seen to be true from Fig. 7:6, because ABC and $A'B'C'$ have corresponding sides parallel and hence are similar. It follows that $CB/AB = C'B'/A'B'$ and $CB/AC = C'B'/A'C'$. Furthermore, suppose P is some fixed point on a curve and it is true that the slope of the straight line PA is the same as that of PC no matter what points on the curve A and C are; then the curve is a straight line. This is true because we can consider A fixed for the moment; then no matter where C is on the curve it must be on the straight line PA, since there is only one line through P with the given slope. Our results then may be summarized in the following statement: **To say a curve has a fixed slope is equivalent to saying it is a straight line.**

We are now ready to deal with equations of straight lines. Suppose we have $y = 2x$. On this curve the value of y is always twice that of x. This means that if we take P to be the origin, the slope of PA is 2 no matter what point A is, provided only that its coordinates satisfy the equation. Hence the graph of $y = 2x$ is a straight line through the origin with the slope 2. Similarly, **the graph of $y = mx$ is a straight line through the origin with the slope m.**

Suppose we have the equation $y = 2x + 3$. Here then, for each value of x, y is 3 units more than it is on the line $y = 2x$. We therefore have Fig. 7:7, and no matter what point A' is, $P'A'AP$ is a parallelogram and the slope of $P'A'$ is the same as that of PA, which is 2. Thus, $y = 2x + 3$ is the equation of the line through P' and A'. Similarly, **the graph of $y = mx + b$ is a straight line through the point $(0, b)$ with the slope m.**

Again, if we were given the equation $3x + 2y = 5$, we could solve it for y and have $y = -(\frac{3}{2})x + \frac{5}{2}$, which, by what we have shown above, is the equation of a straight line with slope $-\frac{3}{2}$ and passing through the point $\frac{5}{2}$. In fact, if $b \neq 0$, we can solve for y in the equation $ax + by + c = 0$ and have $y = -(a/b)x - c/b$, whose graph is a straight line of slope $-(a/b)$ and which passes through the point $(0, -c/b)$. If $b = 0$, the given equation reduces to

$ax = -c$. Here, if $a \neq 0$, we have $x = -c/a$, which is a straight line parallel to the y-axis and $-c/a$ units from it. If $a = 0$, our equation reduces to $c = 0$, which is satisfied by no values of x and y unless $c = 0$, in which case we really do not have much of an equation. Hence: **The graph of** $ax + by + c = 0$ **is a straight line provided** a **and** b **are not both zero. Also every straight line has such an equation, which is therefore called a linear equation.**

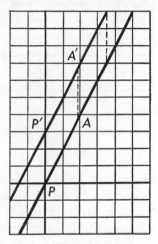

Figure 7:7

It is worthwhile comparing two special cases of the above equation. If $b \neq 0$ and $a = 0$, it reduces to $y = -c/b$, which may be written $y = 0 \cdot x - c/b$ and is a line of 0 slope, that is, a horizontal line. However, if $b = 0$ and $a \neq 0$, we have $x = -c/a$, which is a vertical line. Its slope is not a number, for if we refer to Fig. 7:5 we see that for such a line $t = 0$ and the ratio r/t would not be a number. Here we have a little conflict between mathematical terminology and common usage. In mathematical terms, a vertical line has no slope because we cannot express the slope by means of a number, but in common usage a vertical line has a greater "slope" or steepness than any other line. On the other hand, in mathematical terms a horizontal line has a slope, the number zero, while in common usage a horizontal line has no "slope" or steepness at all. We shall, of course, here conform to mathematical usage.

Example

Draw the graph of $3x + 4y = 8$.

Solution. Solve for y and have $y = -(\tfrac{3}{4})x + 2$. Determine the point P with the coordinates $(0, 2)$. The slope is $-\tfrac{3}{4}$ and hence determine the point A, which is 4 units to the right of P and 3 units down. PA is the line.

Solution 2. We know the graph is a straight line. Hence any two points will determine the graph. When $y = 0$, $x = \frac{8}{3}$, and when $x = 0$, $y = 2$. We find these two points on the graph and draw the line through them (see Fig. 7:8).

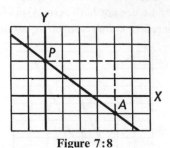

Figure 7:8

5. GRAPHS OF LINEAR INEQUALITIES

There are occasions when graphs of inequalities are useful. Suppose Mrs. Murphy does not want to spend more than $3.00 on fruit and the only fruits that look good to her are melons at 25 cents each or oranges at 50 cents a dozen. What are the various possibilities for her purchases? (Assume that no melons are cut and oranges are sold only in dozens.) If we let m denote the number of melons and d the number of dozens of oranges, the resulting inequality is

$$25m + 50d \leq 300.$$

We can simplify this by dividing both sides of the inequality by 25 to get

$$(1) \qquad\qquad\qquad m + 2d \leq 12.$$

We could form a table of values, by taking first $d = 0$ when the inequality affirms $m \leq 12$; then $d = 1$, when the inequality becomes $m \leq 10$; and so on. But it is simpler to represent this graphically.

Suppose we select m to be measured along the horizontal axis and d along the vertical axis. Then draw the graph of the line

$$m + 2d = 12.$$

This is shown in Fig. 7:9. What is the connection with the inequality (1)? Subtract m from both sides to get $2d \leq 12 - m$; divide both sides by 2 to get

$$(2) \qquad\qquad\qquad d \leq \tfrac{1}{2}(12 - m).$$

Now, on the line, $d = \frac{1}{2}(12 - m)$. Hence, for (2) to hold, the value of d for each m must be less than or equal to its value on the line. In other words, the values of m and d which satisfy inequality (1), or its equivalent inequality (2), are given by the points below and on the line $m + 2d = 12$. Of course, neither *m* nor *d* can be negative, which shows that all the points (m, d) satisfying (1) must lie inside the triangle OAB of the figure, including line segments $\overline{OB}$, $\overline{OA}$, and $\overline{AB}$, where A and B are the points where the line intersects the axes. Furthermore, *m* and *d* must be integers, and hence we have as our final con-

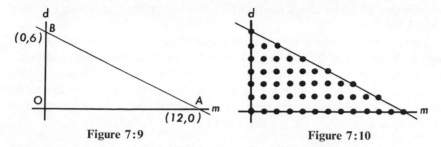

Figure 7:9   Figure 7:10

clusion: The pairs of values of *m* and *d* which satisfy the inequality (1) are coordinates of those points inside of or on the triangle bounded by the axes and the line $m + 2d = 12$ and whose coordinates are both integers. These are designated by dots in Fig. 7:10.

If *m* and *d* were not restricted to be integers but could be any nonnegative real numbers, the graph of the inequality (1) would consist of all the points within or on the triangle OAB.

If *m* and *d* were real numbers without restriction, the graph of (1) would be all the points below or on the line $m + 2d = 12$.

Sometimes more than one inequality may be involved. Suppose that, for some unknown reason, Mrs. Murphy also required that the number of melons plus the number of dozens of oranges must be greater than 7. This requires the inequality

(3) $$m + d > 7.$$

By the same methods as before, we can see that the points which correspond

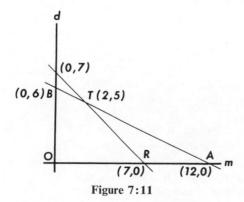

Figure 7:11

to this inequality lie above the line $m + d = 7$. Then we can draw the graph of $m + d = 7$ on the same axes as of $m + 2d = 12$, as shown in Fig. 7:11 (p. 207). Let T denote the point of intersection $(2, 5)$ and R the point where the line $m + d = 7$ cuts the m-axis. Then the values of m and d which satisfy both (1) and (3) will correspond to the coordinates of those points with integral coordinates which lie above the line TR, above or on the line m, and below or on the line TA.

There is another way of determining the set of points satisfying an inequality like $2x + 3y - 7 < 0$. This depends on a key intuitive idea which is difficult to prove. Since the graph of $2x + 3y - 7 = 0$ is a straight line, we would expect intuitively that all the points satisfying the inequality would lie on one side of the line. We may state this assumption more precisely as follows: Let A and B be two points not on the line $2x + 3y - 7 = 0$; then

(1) The line segment $\overline{AB}$ does not intersect the line $2x + 3y - 7 = 0$ if and only if $2x + 3y - 7 < 0$ is true for both A and B or false for both A and B.

(2) The line segment $\overline{AB}$ intersects the line $2x + 3y - 7 = 0$ if and only if $2x + 3y - 7 < 0$ is true for one of A and B and false for the other.

Notice that statements 1 and 2 are equivalent because each part of statement 1 is the contrapositive of a part of statement 2. Another way to look at statement 2 is the following: If $2x + 3y - 7$ is positive at point A and negative at point B, then for some point of the segment $\overline{AB}$, $2x + 3y - 7$ is equal to zero; that is, some point of the segment lies on the line $2x + 3y - 7 = 0$. On the basis of this assumption we see that since the point $(0, 0)$ has coordinates for which the inequality is true, the solution of the inequality consists of all points on the same side of the line as the origin, that is, all points below the line. Using this method we also can see that the points satisfying the inequality $2x + 3y > 7$ are the points on the opposite side of the line $2x + 3y - 7 = 0$ from the origin, since the point $(0, 0)$ is not one for which the inequality $2x + 3y > 7$ is true.

EXERCISES

1. Indicate by figure and description the values of x and y which satisfy the following inequalities:
   a. $3x + 2y < 6$.
   b. $3x + 2y < 6$ and $3x + 3y > 6$.
   c. $-4x + 7y < 8$ and $x + y > 6$.

2. Indicate by figure and description the solutions of the following:
   a. $2x - 3y > 7$.   b. $y - 3x \le 0$.
   c. $3x - 2y - 7 > 0$ and $5x - y - 7 \ge 0$.

3. Suppose that in addition to not wanting to spend more than $3.00, Mrs. Murphy, in the example of the text, decided to get at least twice as many melons as dozens of oranges. Indicate graphically and by description her requirements.

4. Jack Sprat could eat no fat and his wife could eat no lean. Jack's minimum daily requirement is 1.2 pounds of lean and his wife's 1 pound of fat. Now pork is 30% lean and 70% fat while beef is 60% lean and 40% fat; but beef costs $\frac{3}{2}$ as much as pork. How many pounds of each kind of meat should they buy to meet the minimum requirements and still be as economical as possible?

5. Indicate graphically the points (x, y) for which $|x| < 2$ and $|y| > 4$.

6. NONLINEAR GRAPHS

The graph of any formula connecting two letters can be drawn simply by constructing a table of pairs of values satisfying the formula, determining the corresponding points on the graph, and connecting them in order by a smooth curve. For instance, suppose $y = x^2 - x$. We compute the following table of values:

x	0	1	2	3	-1	-2	-3
y	0	0	2	6	2	6	12

We know that no matter what x is we can find a value of y. Then, assuming that the points lie on a smooth curve, we have the graph given in Fig. 7:12. Notice that the lowest point in the curve is apparently $(\frac{1}{2}, -\frac{1}{4})$.

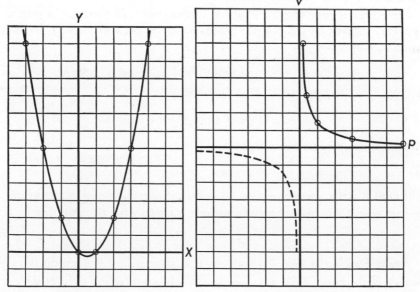

Figure 7:12   Figure 7:13

Suppose we wish the graph of $pv = 6$, where p stands for pressure in pounds and v is volume in cubic feet. Taking the point of view that pressure determines volume, we take p to be the independent variable and measure it along the horizontal axis. Our table of values is

p	$\frac{1}{2}$	1	2	6	12
v	12	6	3	1	$\frac{1}{2}$

Since we cannot have a negative value for the volume or pressure, we assign no such values to the letters. Hence, letting one square represent two units for each axis, the graph is only the solid curve in Fig. 7:13 (p. 209). However, if p and v had no physical meaning which required that they be positive, we should also have the dashed part of the graph.

As a final example, we draw the graph of $x^2 - y^2 = 9$. Solve this for y^2 and get $y^2 = x^2 - 9$; that is, $y = \sqrt{x^2 - 9}$ or $-\sqrt{x^2 - 9}$. If $x^2 < 9$ (that is, x^2 is less than 9), y is imaginary. But our graph contains only points whose coordinates are both real. Hence in our table of values no such values of x occur. Also, for each value of x there will be two values of y which are equal except for sign. The table of values is thus

x	-3	3	4	-4	5	-5	7	-7
y	0	0	$\pm\sqrt{7}$	$\pm\sqrt{7}$	± 4	± 4	$\pm\sqrt{40}$	$\pm\sqrt{40}$

where $\pm\sqrt{7}$ means $\sqrt{7}$ or $-\sqrt{7}$, and so on. The approximate value of $\sqrt{7}$ is 2.6 and of $\sqrt{40} = 2\sqrt{10}$ is 6.3 (see Fig. 7:14).

These graphs may also be used for inequalities. For instance, a glance at Fig. 7:12 shows that y is positive except when x is between 0 and 1, including 0 and 1. That is,

$$x^2 - x > 0 \qquad \text{if and only if } x > 1 \text{ or } x < 0.$$

Another way to write this is

$$x^2 - x < 0 \qquad \text{if and only if } 0 < x < 1.$$

Also, this same figure shows that

(1) $$y > x^2 - x$$

will hold for the coordinates of all points above the curve in the graph of $y = x^2 - x$. Of course, there are various ways to write (1). Another is

$$y + x > x^2.$$

Again, looking at Fig. 7:14 we see that since $x = \sqrt{9 + y^2}$ on the right branch of the curve, then $x > \sqrt{9 + y^2}$ at all points to the right of the curve of the figure. In the same fashion, we can see that $x < -\sqrt{9 + y^2}$ holds for all points to the left of the left branch of the curve. Thus we could conclude that

$$|x| > \sqrt{9 + y^2}$$

holds for all points which are not between the two branches of the curve nor on either branch. This curve and that of Fig. 7:13 are called hyperbolas.

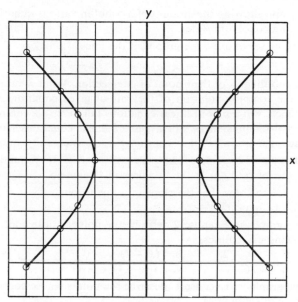

Figure 7:14

One could also solve the inequality $x^2 - y^2 > 9$ by noting that the graph of $x^2 - y^2 = 9$ divides the plane into three regions. We make the fundamental assumption for the curve which we made for the straight line: If one point in any region satisfies the inequality, then so do all points in the region. So we try a point in each of the three regions. If $x = y = 0$, the inequality is not true, and thus all points between two parts of the graph do not satisfy the inequality. But if $x = 4, y = 0$ or $x = -4, y = 0$, the inequality is true. This checks with our previous result.

EXERCISES

1. Draw the graphs of the following equations:
   a. $3x - 2y = 9$.   b. $x = 5$.   c. $y = x^2 - 3x + 2$.
   d. $y = x/(x - 1)$.   e. $x^2 - y^2 = 1$.   f. $x^2 + y^2 = 25$.
   g. $s = 16t^2$, where s is the distance in feet a body falls in t seconds.
   h. $S = (r^4 - 1)/(r - 1)$, where S is the sum of a geometric progression having four terms, whose first term is 1 and whose fixed ratio is $r, r \neq 1$.

2. Draw the graphs of each of the following equations:
    a. $y = 5$.   b. $x = y^2 - 3y + 2$.
    c. $x^2 + y^2 = 36$.   d. $y = (x - 1)/x$.
    e. $y^2 - x^2 = 4$.   f. $y = x(x - 1)(x + 1)$.

3. The length plus the girth of any parcel to be sent by parcel post must not exceed 100 inches. Draw a graph showing the maximum allowable length for various values of the girth.

4. Draw a graph showing the amount of alcohol in various quantities of a 30% solution of alcohol and water.

5. Draw on one set of axes, the graphs of the lines

    $$x + 2y = 3 \qquad \text{and} \qquad 3x - 2y = 1.$$

    Find the coordinates of their point of intersection. What relationship do they have to the common solution of the pair of given equations? Under what circumstances will a pair of linear equations fail to have a common solution? Under what circumstances will a pair of linear equations have more than one common solution? Phrase your answers geometrically and algebraically.

6. Show that $(x + 2y - 3) + (3x - 2y - 1) = 0$ is the equation of a line through the intersection of the two lines in Exercise 5.

7. Use Fig. 7:6 to show that the slope of a line through the points (x_1, y_1) and (x_2, y_2) is $(y_2 - y_1)/(x_2 - x_1)$.

8. Using the graphs in Exercise 2 describe the points whose coordinates satisfy the following inequalities:
    a. $x < y^2 - 3y + 2$.   b. $(x - 1)/x > 0$.   c. $y^2 - x^2 > 4$.
    d. $x^2 + y^2 < 36$.   e. $x(x - 1)(x + 1) > 0$.

9. Using the graphs in Exercise 1, describe the points whose coordinates satisfy the following inequalities.
    a. $x^2 - 3x + 2 > 0$.   b. $x^2 + y^2 < 25$.   c. $x^2 - y^2 > 1$.
    d. Points which satisfy both parts a and b.   e. $x/(x - 1) < 0$.

7. GRAPHS OF TABLES OF OBSERVED VALUES

We have found two methods of obtaining tables of values: (1) by means of certain physical observations and (2) by the use of certain formulas. In the case of tables found from simple formulas experience leads us to expect that if we find certain values of the variables, which satisfy the formula, graph the points corresponding to these values, and draw a smooth curve through the points, then the coordinates of any point on the curve will satisfy the formula. So also in many types of physical tables, the recorded values of the variables are only a few of the totality of such values and we expect the graph of such a table to be a smooth curve within the limits of the physical problem.

For instance, in Table 5 the body will have traveled a certain distance in 5 seconds even though such an entry is not in the table. There certainly would be a value s for every positive value of t. If we assume that the body does not move in a fitful fashion, we may consistently consider the graph to be a smooth curve and connect with a smooth curve the points corresponding to the entries in the table. Then on the above assumptions we can find from the graph the value of s when $t = 5$ and even the value of s when $t = 70$. These are guesses, of course, but they are intelligent guesses.

It is, as a matter of fact, on this assumption that nature is not too whimsical a lady that most of our orderly (and disorderly) lives are based. If I telephone a friend at 12 and 12:30 only to find that the line is busy, I am apt to conclude that it was busy for the whole half hour; if the interval were 15 minutes, I should be even more sure of it. This is what might be called *between-sight* and our faith in it increases as the betweenness narrows. Foresight is more intriguing but not so reliable. Having observed the line busy at 12 and 12:30 I might not be so sure about my guess of not being able to reach my friend at 12:45 unless I heard on the party line the voice of the neighborhood gossip.

It is, of course, highly important to keep in mind the limitations imposed on the table of values by the physical situation. In Table 9, for instance, there would be no values for W when F less than zero or, depending on the camera, for F greater than 100 feet. Furthermore, we shall soon have examples of tables of values whose graph is not a smooth curve. One such table would be a record of the number of eggs laid by a certain hen on the various days of her life.

EXERCISES

1. For which of the following tables of values of formulas should the graph be a smooth curve? Whenever such a curve is appropriate, draw it.
   a to f. Tables 4 to 9.
   g. $S = n(n + 1)/2$, the formula for the sum $1 + 2 + 3 + \cdots + n$.
   h. $A = (1.01)^n$, the amount of \$1.00 after n years at the interest rate of 1% compounded annually.
   i. The cost of sending a letter first class depending on its weight: 6 cents per ounce or fraction thereof.
   j. $s = 16t^2$, where s is the distance in feet fallen in t seconds.

2. Use the above graphs to estimate in Table 6 the value of W when $T = 12$ and the value of T which makes $W = 35$; in Table 7 the value of V when $T = 25$ and the value of T which makes $V = 6000$; in Table 8 the value of F when $t = 5.5$ and the value of t at which F has its greatest value; in Table 9 the value of W when F is 75 feet.

3. The cost (\$$C$) of publishing a certain number (N) of pamphlets is computed by the formula $C = N/80 + 140$. Draw the graph of the formula for values of N from 1000 to 10,000. Use your graph to find the cost of publishing 2500 pamphlets. How many pamphlets may be published for \$180? (Such a graph enables the printer to quote prices without computation. Graphs are used in this way in a variety of commercial enterprises.)

4. Draw the graphs for Exercises 1g and 1h, using the appropriate kind of graph.

8. FINDING A FORMULA WHICH FITS OR ALMOST FITS

We have seen how, given a formula, we can find a table of values associated with it and the resulting graph. Often in physical situations we are given a graph or table of values and want to find a formula which fits or almost fits the graph and/or the table of values. The advantage of having a formula is that we can often use it to approximate data not given in the table itself. Of course, we want the simplest formula, and it would be convenient if there were some way in which we could tell from the table of values what kind might fit.

Since in this book we do not have space to do anything more than show how one might approach this problem, let us assume, for simplicity's sake, that x is the independent variable and that we know the values of y for $x = 0, 1, 2,$ and 3. Usually the values of the independent variables are in an arithmetic progression in any case, and the methods we have will be essentially the same when the first four integers are replaced by numbers in any other arithmetic progression.

The best way to approach the subject seems to be to look at the tables of values of certain simple expressions and note what characteristics of the table could be used to tell what type of expression would best fit. We begin with the linear function $y = ax + b$ and compute the following table of values:

x	0	1	2	3	$\cdots$	n	$n + 1$
y	b	$a + b$	$2a + b$	$3a + b$		$na + b$	$a(n + 1) + b = an + a + b$

We see that the first four values of y form an arithmetic progression with the first term b and the common difference, a. This is shown conclusively by the last two values given in the table, since, no matter what n is, the difference between the values of y corresponding to $x = n$ and $x = n + 1$ is

$$(an + a + b) - (na + b) = a.$$

The difference does not depend on n, that is, is the same for all n.

Now let us use this information to find a formula for a table in which the values of y form an arithmetic progression, such as

x	0	1	2	3	4
y	2	5	8	11	14

Comparing this table with the above, we see that the initial value, b, is 2 and the common difference, 3, is a. Thus the formula which should fit the table is

$$y = 3x + 2.$$

We know that this must be the correct formula, since, from what we did above, the values of y, for the given x, form an arithmetic progression with common difference 3 and first term 2. We can check this formula by substituting the values of x in the first line of the table.

If the values of y were close to an arithmetic progression, the values given by the formula would approximate those of y. It is possible by more advanced methods to define what we mean by the "closest fit" and then determine the line which has this property. But we do not have space to go into this here.

Of course, a straight line will not approximate all sets of data. Let us now consider the next simplest equation,

$$y = ax^2 + bx + c.$$

Here we make a table of values of x, those of y, and those of D, the differences between successive values of y:

x	0	1	2	3
y	c	$a + b + c$	$4a + 2b + c$	$9a + 3b + c$
D		$a + b$	$3a + b$	$5a + b$

Here the differences, D, appear to form an arithmetic progression whose first term is $a + b$ and whose common difference is $2a$. To verify that this is true for all values of x, we make a table of values for $x = n - 1, x = n, x = n + 1$:

x	$n - 1$	n	$n + 1$
y	$an^2 - 2an + a + bn - b + c$	$an^2 + bn + c$	$an^2 + 2an + a + bn + b + c$
D		$2an - a + b$	$2an + a + b$

(Finding the values of y given requires the use of a pencil and paper.) The difference between the two values of D is

$$2an + a + b - (2an - a + b) = 2a,$$

which is the same for all values of n. Thus the differences D form an arithmetic progression.

Now we apply the above to a numerical case and find a formula for y when the differences form an arithmetic progression. Consider the following:

x	0	1	2	3	4
y	3	5	10	18	29
D		2	5	8	11

Then, looking at the table for the formula $y = ax^2 + bx + c$ with $x = 0, 1, 2, 3$ we see that $c = 3, a + b = 2$, since it is the first term of the arithmetic progression of the D's, and $2a = 3$, since 3 is the common difference of the arithmetic progression of the D's. Thus $a = \frac{3}{2}, b = 2 - a = 2 - \frac{3}{2} = \frac{1}{2}$, and the formula is

$$y = (\tfrac{3}{2})x^2 + (\tfrac{1}{2})x + 3.$$

We can check this with the last table by substituting values for x. Again, why do we know that this works? The reason is that, for the values we found, the differences of successive values of y form the arithmetic progression determined by the table. The numbers of the arithmetic progression, together with the value of y for $x = 0$, determine all the other values of y.

We shall carry this one step further, to the cubic function. Suppose

$$y = ax^3 + bx^2 + cx + d.$$

Then we will have to find the second differences. Here is our table:

x	0	1	2	3
y	d	$a+b+c+d$	$8a+4b+2c+d$	$27a+9b+3c+d$
D_1		$a+b+c$	$7a+3b+c$	$19a+5b+c$
D_2			$6a+2b$	$12a+2b$

So it appears that the second differences, D_2, form an arithmetic progression with $6a + 2b$ as the first term and $6a$ as the difference, although two terms is scarcely enough to show it. To show this we might compute the values of y for $x = n, n + 1, n + 2, n + 3$, but this would be rather tedious. In fact, we need only the values of y corresponding to $x = n$ and $x = n + 1$, which are:

If $x = n$, then $y = an^3 + bn^2 + cn + d.$

If $x = n+1$, then $y = a(n+1)^3 + b(n+1)^2 + c(n+1) + d$
$\qquad\qquad\qquad = an^3 + n^2(3a+b) + n(3a+2b+c) + a+b+c+d.$

The difference between these two values of y is

$$3an^2 + (3a + 2b)n + (a + b + c).$$

This is a quadratic expression in n. From our previous discussion, we know that the differences for a quadratic expression form an arithmetic progression; that is, the second differences D_2 in the table above form an arithmetic progression. We could have used this same argument to save trouble for the quadratic expression in the previous paragraph.

Having established the fact that for a cubic expression the second differences form an arithmetic progression given above, let us look at a numerical example:

x	0	1	2	3	4
y	-1	5	21	53	107
D_1		6	16	32	54
D_2			10	16	22

Hence $10 = 6a + 2b$, because this is the first term in the arithmetic progression for D_2, and $6a = 6$, because $6a$ is the common difference. Hence $a = 1$ and $10 = 6 + 2b$ implies that $b = 2$. Now the first value of D_1 is $a + b + c$, which in this case is equal to 6. Thus $6 = 1 + 2 + c$, or $c = 3$. The first value of $y = -1$, which is therefore equal to d in the formula. Hence the formula which should fit is

$$y = x^3 + 2x^2 + 3x - 1.$$

This may be verified by substituting values of x.

Let us summarize what we have found. Suppose the values of x are the nonnegative integers; then

(1) If the corresponding values of y are in an arithmetic progression, the formula is $y = ax + b$.

(2) If the differences between successive values of y are in an arithmetic progression, the formula is $y = ax^2 + bx + c$.

(3) If the differences of the differences (the second differences) form an arithmetic progression, the formula is $y = ax^3 + bx^2 + cx + d$.

From this it should not be hard to guess what would happen for expressions of the fourth degree or higher.

EXERCISES

1. Find the equations of the lines through
   a. $(3, -2)$ and $(5, 1)$.   b. $(6, 4)$ and $(-2, -3)$.
   c. $(-2, 1)$ and $(-2, 5)$.   d. $(0, 0)$ and $(-1, -8)$.
   e. $(1, -5)$ and $(3, -5)$.

2. Show that there is a line $y = ax + b$ passing through the points (r, s), (t, u) if $r \neq t$. What happens if $r = t$?

3. Table 10 gives the weight (W grams) of potassium bromide which will dissolve in 100 grams of water at various temperatures. Draw the graphs of the points of the table. Find and check the equation for the line on which all points of the graph lie. From the graph of the line find W when $T = 23$ and check by using the formula.

Table 10

T	0	20	40	60	80
W	54	64	74	84	94

4. Table 11 gives the volume of a certain quantity (V cubic centimeters) of gas at several temperatures ($T°$). Draw the graph of the points, and find and check the formula giving V in terms of T. Find V when $T = 50$.

Table 11

T	-33	-6	12	27	42
V	160	178	190	200	210

5. Table 12 gives the average weights (W pounds and M pounds) of women and men for various heights (h inches). Draw the graphs of W and M as functions of h. In each case find the equations of lines on which the points very nearly lie.

Table 12

h	61	63	65	67	69	71
W	118	126	134	142	150	158
M	124	132	140	148	157	166

6. Find the equation of a line which very nearly goes through all the points of the graph of the percentage of alcohol required for various temperatures as given in Section 11 of Chapter 6. What will be the percentage required for a temperature of $-5°$?

7. Table 13 gives values of S for various values of n, where $S = 1 + 2 + 3 + \cdots + n$. By the methods of this section, find a formula for S in terms of n.

Table 13

n	1	2	3	4	5
S	1	3	6	10	15

8. Given the Table 14, where r, s, and t are three distinct numbers. Show that there is a quadratic expression which exactly fits this table.

Table 14

x	0	1	2
y	r	s	t

9. Find a formula which fits or almost fits Table 15.

Table 15

t	2	5	8	11	14
s	0	15	48	99	169

10. Find the formula which fits or almost fits Table 16.

Table 16

x	0	2	4	6	8
y	0	10	45	102	184

11. A contractor agreed to build a breakwater for $250,000. After spending $66,000 as indicated in Table 17, he threw up the job without receiving any pay because he estimated that the cost would increase according to the same law and that it would require 6 months more to complete the work. Find a formula approximating the table and use it to estimate what the loss would be had he finished the breakwater.

Table 17

Time in months	0	1	2	3	4
Investment in thousands	10	12.2	22.4	40.6	66.6

12. Find a formula which fits or almost fits the Table 18 and, using the formula, find the values of y corresponding to $x = 5$, $x = -1$, and $x = \frac{1}{2}$.

Table 18

x	0	1	2	3	4
y	0	0	6	24	60

13. Why is it true that there is an expression of the form $y = ax^3 + bx^2 + cx + d$ which fits Table 19 no matter what r, s, t, and u are, provided that for some r, s, t, and u, one or more of a, b, c, and d may be zero. (It is not necessary to determine the cubic expression.)

Table 19

x	0	1	2	3
y	r	s	t	u

14. Find a formula of the form $S = an^3 + bn^2 + cn + d$ which fits Table 20, values of S in terms of n, where

$$S = 1^2 + 2^2 + 3^2 + \cdots + n^2.$$

Table 20

n	1	2	3	4	5
S	1	5	14	30	55

15. Find a formula of the form S in Exercise 14 fitting the Table 21. These are the so-called pyramidal numbers—the number of cannonballs in various

triangular pyraminds. For any *n*, the value of *S* is the sum of the first *n* values of *S* in Table 13.

Table 21

n	1	2	3	4	5
S	i	4	10	20	35

16. Suppose for $x = 0, 1, 2, 3, 4, \ldots$ the corresponding values of *y* are the numbers of the geometric progression 1, 2, 4, 8, 16, ... What is true about the successive differences of these values of *y*?

17. Show that no matter how far one continues to take differences of differences in Table 22, one will not get an arithmetic progression. Reasoning by analogy from your experience above, what conclusions would you draw from this?

Table 22

n	1	2	3	4	5	6	7	8
S	1	1	2	3	5	8	13	21

Each *S* is the sum of the two previous ones. These values of *S* form the so-called *Fibonacci series*.

9. FREQUENCY TABLES

Generally speaking, the graphs we have been considering have involved infinitely many points—usually a curve—and this has been acknowledged by joining by a curve the scattered points corresponding to the pairs in the table of values. The justification for this procedure is that were we interested mainly in tables of values which arose either from measurements of speed, temperature, time, and so on, or from formulas which gave a significant result whenever any real number (within limits) was substituted for the independent variable. The entries in our table of values were merely samples from a set of infinitely many values which we computed or observed to give us an idea of the trend.

In this section we wish to examine certain special tables and their graphs in which, either in the nature of the case or due to some definite agreement, the dependent variable does not have a value for each real value of the independent variable between any two limits. For instance, in the formula $S = n(n + 1)/2$ for the sum of the first *n* integers, *n* has only positive integral values. We can, to be sure, find a value of *S* from the formula when $n = \frac{1}{2}$ or $n = -1$, yet there would be no sense in speaking of "the sum of the first half integer" or "the first minus one integers."

In many statistical tables, there are a few entries and no more. Suppose it is required to measure the heights of 1,000,000 men, shall we say, in the army. The heights will range from, perhaps, 62 to 77 inches. The height of a given man can be any real number between these limits. The processes of measurement used in such circumstances are not accurate enough, however, to distinguish differences in height less than perhaps $\frac{1}{32}$ inch. In fact, the usual procedure is to group together all those whose heights fall between 62 and 63 inches, 63 and 64 inches, . . ., and between 76 and 77 inches, where each interval includes the lower but not the upper measurement; to these groups are assigned the heights 62.5, 63.5, . . ., 76.5 inches, respectively. With each of these groups is associated the number of men in the group. Thus we get a table of values like the following:

h	62.5	63.5	$\cdots$	76.5
n	n_1	n_2	$\cdots$	n_{15}

where $n_1 + n_2 + \cdots + n_{15} = 1,000,000$. Notice that there is certainly no value of n for values of h between 62.5 and 63.5, for instance, and hence it would be *very misleading* to connect by a curve two points of the table of values. Similar remarks could be made for the following.

Example

One hundred shells were fired at a target and the results tabulated according to their striking F feet short of (when a minus sign is used) or beyond (when a plus sign is used) the target. The results are then arranged into ten groups so that in the table, $N = 25$ for $F = -5$ means that 25 shells fell between -10 and 0 feet from the target, and so on. The table is as shown in Table 23.

Table 23

F	-35	-25	-15	-5	$+5$	$+15$	$+25$	$+35$
N	2	7	16	25	25	16	7	2

Tables such as these which record the frequency with which a certain event occurs are called *frequency tables*. We have seen that they call for a different kind of graph. Sometimes successive points are joined by straight lines forming what is called a *frequency polygon*. Here we use what is called a *histogram*, and to show what we mean by such a form of graph we construct it for Table 23. We begin by constructing the graph of the *points* of the table in the usual way. Then, since $+5$ covers the range from 0 to 10 we construct a rectangle whose base is the line segment from points 0 to 10 on the F-axis (horizontal) and whose altitude is 25, the corresponding value of N. So we proceed to form Fig. 7:15.

The histogram form of graph has the definite advantage that it shows the range which a specific number stands for; for example, 15 is the number used for the range 10 to 20. Even in such cases as Table 27, in which no range of values is involved, the histogram indicates lack of smoothness of the data.

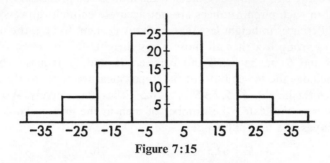

Figure 7:15

EXERCISES

1. In Table 24 N denotes the number of telephone calls, during a certain period of the day, which last T seconds. (The calls have been grouped, of course, as the heights were in the examples above.) Construct a histogram for this table.

Table 24 [1]

T	50	150	250	350	450	550	650	750	850	950
N	1	28	88	180	247	260	133	42	11	5

2. Draw the histogram for the mortality table, Table 25, where L denotes the number of survivors at age A of a group of 100,000 alive at age ten.

Table 25

A	10	15	20	25	30	35	40	45		
L	100,000	96,285	92,637	89,032	85,441	81,822	78,106	74,173		
A	50	55	60	65	70	75	80	85	90	95
L	69,804	64,563	57,917	49,341	38,569	26,237	14,474	5,485	847	3

3. In Table 25 let all people dying between their tenth and fifteenth birthdays be regarded as dying at age 12.5, and so on. For instance, we should say that the last three people die at age 97.5. Let D denote the number dying at age a, thus defined, and complete Table 26. Construct a histogram for this table. What connection would there be between it and the histogram for Table 25?

[1] J. F. Kenney, *Mathematics of Statistics*, p. 18. D. Van Nostrand Co., Princeton, N.J., 1939.

Table 26

a	12.5	17.5	22.5	$\cdots$	87.5	92.5	97.5
D							

4. A set of 10 coins was tossed 1000 times, and the number H of heads appearing each time was recorded. Letting N denote the number of times H heads appeared, the table of observations shown as Table 27 resulted. Construct the histogram for this table.

Table 27

H	0	1	2	3	4	5	6	7	8	9	10
N	2	10	41	119	206	246	204	115	44	11	2

5. The number of letters received in a certain office on the five working days of the week is given by Table 28. Use the most appropriate kind of graph to represent this.

Table 28

M	T	W	Th	F
100	50	75	60	65

6. The marks of a class of ten students are as follows:

$$60, 63, 65, 68, 72, 73, 75, 85, 86, 90.$$

Exhibit these marks graphically, first giving the marks just as they are; second, grouping them in four ranges: 60–69, 70–79, 80–89, 90–1000. In each case use the form of the graph which you consider most appropriate.

7. Name the form of graph most suitable for each of the tables of values.
   a. Table 29, where A is the number of apples having S seeds.

Table 29

S	4	5	6	7	8	9
A	9	4	14	21	24	23

   b. Table 30, where v is the velocity in feet per second of a falling body after t seconds.

Table 30

v	11	27	43	59
t	1	2	3	4

   c. The principal unpaid at the end of each year under plan A and plan B of Example 5 of Section 6 of Chapter 6.

d. Table 31, where *I* is the inaccuracy in inches allowed on a ring gauge when measuring distances of *d* inches, the lower measurement in each interval excluded.

Table 31[1]

d	.029 to .825	.825 to 1.510	1.510 to 2.510	2.510 to 4.510
I	.00020	.00024	.00032	.00040

d	4.510 to 6.510	6.510 to 9.010	9.010 to 12.010
I	.00050	.00064	.00080

e. Table 32, where if a man pays $100 a year beginning at age 40 until age *A* he can receive, beginning at age *A*, *M* dollars per months for the rest of his life.

Table 32[2]

A	50	55	60	65	70
M	4.97	9.22	15.31	24.27	37.70

f. The graph of Table 20 in Section 8.
g. The graph of Table 13 in Section 8.
h. The graph of Table 12 in Section 8.

10. AVERAGES

One of the most maltreated terms in the English language is the word "average." The "average man" thinks or does thus and so, weighs such and such, and therefore we are expected to think, do, and weigh likewise. He is set up as an example for all of us to emulate and woe be it to him who is branded as being "below average"! The average is one of the greatest comforts of mankind, for we have the feeling that if we can only take the average everything will come out all right. The average of several inaccurate measurements somehow ought to be better than any one of them, although we do hesitate to use this panacea with our results of several additions of a given column of figures.

But the general use of this idea of average, however loosely it is employed, is a sign of a very definite and legitimate need to characterize a whole set of data by means of a single one. As the eminent British statistician, R. A.

---

[1] International Business Machines Corp., *Precision Measurement in the Metal Working Industry*, Chap. IV, p. 6. Reprint for the University of the State of New York, 1941.
[2] E. C. Harwood and B. H. Francis, *Life Insurance from the Buyers' Point of View*, p. 145. American Institute for Economic Research, Cambridge, Mass., 1940.

Fisher, puts it: "A quantity of data which by its mere bulk may be incapable of entering the mind is to be replaced by relatively few quantities which shall adequately represent the whole, or which, in other words, shall contain as much as possible, ideally the whole, of the relevant information contained in the original data." Although an average does not give a very accurate idea of a trend in a set of data, it does describe the situation "in a nutshell."

We shall deal with five kinds of averages, defining them in terms of two tables:

Table 33					Table 34				
G	1	4	6	8	x	x_1	x_2	$\cdots$	x_n
N	3	2	1	4	f	f_1	f_2	$\cdots$	f_n

where Table 33 is the record of the grades of a certain class of ten students (three have grade 1, two the grade 4, and so on) and Table 34 is a general frequency table in which x_1 occurs f_1 times, x_2 occurs f_2 times, and so on. We say that f_1 is the *frequency* of the observation x_1.

The *mode* is defined to be the measurement which has the greatest frequency. In Table 33 it is 8, since more students have that grade than any other. In Table 34 it is the x corresponding to the greatest f. There may be several modes. For instance, if, in Table 33, four students had grade 1, four grade 8, and less than four each of the other grades, there would be two modes: 1 and 8.

The *median* is the middle observation or halfway between the two middle ones. This can perhaps be more easily understood by merely listing all the grades in order for Table 33 as in

(A) $\qquad\qquad\qquad$ 1 1 1 4 4 6 8 8 8 8.

There are ten grades in all, half of ten is 5, and hence the "middle" grade would be 5, which is halfway from 4 to 6. In case there were eleven grades, the median would be that one which occurs sixth from either end. It, of course, would be laborious to write every table in form (A), and hence the usual process for finding the median would be this: Find the total number of observations, that is, the sum of all the frequencies, which we may call F. If F is odd, divide $F + 1$ by 2 and the median will be the observation corresponding to the $[(F + 1)/2]$th frequency counting from either end of the table. If F is even, the median will be the observation halfway between the *two* middle observations—between the $(F/2)$th and the $(1 + F/2)$th observations.

The *arithmetic mean*, which is usually what is meant by the term "average," is the sum of all the observations divided by the total number of frequencies. Using form (A) of Table 33 we have

$$\frac{1 + 1 + 1 + 4 + 4 + 6 + 8 + 8 + 8 + 8}{10}.$$

This is obtained with less writing from Table 33 directly as

$$\frac{3 \cdot 1 + 2 \cdot 4 + 1 \cdot 6 + 4 \cdot 8}{3 + 2 + 1 + 4} = 4.9.$$

For Table 34, the arithmetic mean would be

$$\frac{f_1 x_1 + f_2 x_2 + \cdots + f_n x_n}{f_1 + f_2 + \cdots + f_n}.$$

The *geometric mean* is the *F*th root of the product of the grades or observations. Using form (A) we would have

$$\sqrt[10]{1 \cdot 1 \cdot 1 \cdot 4 \cdot 4 \cdot 6 \cdot 8 \cdot 8 \cdot 8 \cdot 8}.$$

Using Table 33 directly we may represent this with less writing as

$$\sqrt[10]{1^3 \cdot 4^2 \cdot 6 \cdot 8^4} = 3.6, \text{ approximately.}$$

For Table 34, it would be

$$\sqrt[F]{x_1^{f_1} \cdot x_2^{f_2} \cdot \ldots \cdot x_n^{f_n}},$$

where F, you recall, is the sum of the values of f. This mean cannot easily be computed except by means of logarithms, the use of which is a labor-saving device beyond the scope of this book.

The *harmonic mean* is the total number of frequencies, F, divided by the sum of the reciprocals of the observations. Using form (A) we have

$$\frac{10}{\frac{1}{1} + \frac{1}{1} + \frac{1}{1} + \frac{1}{4} + \frac{1}{4} + \frac{1}{6} + \frac{1}{8} + \frac{1}{8} + \frac{1}{8} + \frac{1}{8}}.$$

Using Table 33 directly, this amounts to

$$\frac{10}{\frac{3}{1} + \frac{2}{4} + \frac{1}{6} + \frac{4}{8}} = 2.4.$$

For Table 34 it would be

$$\frac{F}{\frac{f_1}{x_1} + \frac{f_2}{x_2} + \cdots + \frac{f_n}{x_n}}.$$

In terms of the histogram of a distribution, the mode is that x (or those x's) at the center (or centers) of the base (or bases) of the highest rectangle (or rectangles). The median is that number on the horizontal axis with the property that a vertical line through it bisects the total area of all the rectangles. (Why?) The arithmetic mean is that number on the horizontal axis with the property that a vertical wire through it will support in equilibrium a sheet of cardboard or other material of homogeneous weight cut in the form of the histogram. (That is, the histogram can be balanced on a single point of support at the point which represents the frequency for the arithmetic mean.) Notice that the above discussion would apply only to averages associated with a frequency table because for computation, except for the mode, it was necessary to find the total number of frequencies which, by necessity, would need to be a finite number. However, even for the graph of a continuously varying function, such as speed in terms of time, one can draw the graph and define the arithmetic mean and median in terms of the area under the curve. This, however, requires more advanced methods than we can use in this book.

While the first three averages are more or less catholic in their application, the last two are used for special purposes. The geometric mean is the proper average to use when ratios of values are important, because of the property that the ratio of the geometric means of two sets of corresponding measurements is equal to the geometric means of their ratios. Suppose, for instance, that one box has the edges a, b, and c and another has edges A, B, and C. The ratio of the geometric means is

$$\frac{\sqrt[3]{abc}}{\sqrt[3]{ABC}}$$

On the other hand, the ratios of corresponding sides are a/A, b/B, and c/C, whose geometric mean is

$$\sqrt[3]{\frac{a}{A} \cdot \frac{b}{B} \cdot \frac{c}{C}} = \sqrt[3]{\frac{abc}{ABC}} = \frac{\sqrt[3]{abc}}{\sqrt[3]{ABC}}.$$

In comparing the relative sizes of the boxes, it is the ratios which are important. This same average would be appropriate if a, b, and c were the respective prices of three certain commodities in a certain year, and A, B, and C their prices another year, for it is the ratios of corresponding prices that is significant in describing the price level.

The geometric mean is also appropriate in connection with rates of growth. If a town doubles its population one year and during the next year multiplies by eight, the appropriate mean for the 2-year period is the geometric mean of 2 and 8: 4. This is because quadrupling its population during each of 2 years results in a population 16 times that in the beginning and multiplying its population first by 2 and then by 8 also gives 16. If one used the arithmetic mean of 2 and 8, which is 5, then the population at the end of 2 years would be multiplied by 25, which would be too much.

The harmonic mean has some connection with harmony in music. One of the most pleasing chords is C, E, G, which correspond to the frequency ratios of 4, 5, 6. This means that the lengths of the strings which produce these notes will be in the ratios $\frac{1}{4}$, $\frac{1}{5}$, and $\frac{1}{6}$. Here $\frac{1}{5}$ is the harmonic mean of $\frac{1}{4}$ and $\frac{1}{6}$, which may be shown by using the definition or by noting that the reciprocal of $\frac{1}{5}$ is the arithmetic mean of the reciprocals of $\frac{1}{4}$ and $\frac{1}{6}$ (see below). Another good illustration of the use of this mean is afforded by Exercise 13 of Section 12 of Chapter 6. We can see what is happening better if we substitute letters for the given numbers and say that the "out journey" was at the rate of a miles per hour and the return journey at b miles per hour. Let d be the one-way distance. The total time is then $d/a + d/b$, and the distance traveled is $2d$. The average rate of the trip is the distance divided by the time and hence is

$$\frac{2d}{\dfrac{d}{a} + \dfrac{d}{b}} = \frac{2d}{d\left(\dfrac{1}{a} + \dfrac{1}{b}\right)} = \frac{2}{\dfrac{1}{a} + \dfrac{1}{b}},$$

which is the harmonic mean of these two rates.

An interesting connection between the harmonic and arithmetic means is shown by letting h be the harmonic mean of a and b. Then

$$h = \frac{2}{\dfrac{1}{a} + \dfrac{1}{b}}$$

implies that

$$\frac{1}{h} = \frac{1}{2}\left(\frac{1}{a} + \frac{1}{b}\right).$$

This tells us that the reciprocal of h is the arithmetic mean of the reciprocals of a and b. This was noted in the previous paragraph.

EXERCISES

1. Find the median and mode of Tables 24, 27, and 29.

2. After grouping, the weights of 1000 eight-year-old girls yielded the following frequency table.[1] Find the mode, median, and indicate how you would find the arithmetic mean for this table.

W (lb)	29.5	33.5	37.5	41.5	45.5	49.5	53.5	57.5	61.5	65.5
f	1	14	56	172	245	263	156	67	23	3

3. In a certain test, the 12 girls in a class made a mean grade of 72 and the 15 boys made a mean grade of 70. Find the mean grade of the entire class. (By "mean" is meant the arithmetic mean.)

4. Compare the arithmetic mean, mode, and median for Tables 35 and 36. On the basis of your results can you make any guesses as to the behavior of the three averages for various frequency tables?

Table 35

x	0	1	2	3	4	5	6
f	2	7	13	16	13	7	2

Table 36

x	0	1	2	3	4	5	6
f	1	2	19	16	1	1	20

[1] Kenney, *ibid.*, p. 18.

5. Concoct a set of grades of 10 students, 9 of whom have grades below the arithmetic mean.

6. Would it be possible for the arithmetic mean of the grades of a class of ten to be above 70 and yet seven of the class to get grades below 60?

7. In baseball circles one speaks of a player's "batting average." Is this one of the averages we have discussed above? If so, which one?

8. A certain animal quadruples its birth weight in the first month and during the second month doubles its weight. What is the appropriate mean to use in stating its "average" monthly rate of increase? Using this mean find the average.

9. Suppose the arithmetic mean of the grades of the members of a certain class is computed. What other average would you need to know to determine whether at least half the class have grades above the arithmetic mean?

10. What, in terms of averages, does the following statement mean? "He is in the upper fifty per cent of the class."

11. A person drives 30 miles an hour for 2 hours and 70 for 1 hour. What is his "average speed?" Use the appropriate mean and explain why it is appropriate.

12. Prove that if b is the arithmetic mean of a and c, then the three numbers a, b, and c form an arithmetic progression.

13. Prove that if b is the geometric mean of a and c, then the three numbers a, b, and c form a geometric progression.

14. How would you give a numerical evaluation of the "average man" and "average diet"?

15. Find the geometric mean of the four numbers 2, 6, 9, and 12 and compare it with the arithmetic and harmonic means.

16. A man drives at the average rate of 15 miles per hour through the towns. On a certain 50-mile journey he passes through three towns each 1 mile long. If he is to make the journey in 1 hour, what must be his average rate in miles per hour on the open road (that is, outside the towns)? What would be your answer if there were t towns each 1 mile long?

17. In Pennsylvania the speed limit in the towns is 15 miles per hour and outside the towns 50 miles per hour. Mr. Thims driving from city A to city B, finds that one tenth of the distance is through the towns. Can he legally average 40 miles per hour for the trip? Does your result depend on the distance between A and B? Explain your answers.

18. The population of a certain city is doubled in one decade and quadrupled in the next. What would be the most appropriate average to use in answering the following question: What was its average ratio of gain per decade? Explain the reasons for your answer.

19. Prove: To say that b is the harmonic mean of a and c is equivalent to saying that $1/a$, $1/b$, $1/c$ form an arithmetic progression.

20. A sequence of numbers is called an *harmonic progression* if each number of

the sequence is the harmonic mean of the number preceding it and the
number following it. Prove that

$$1, \tfrac{1}{2}, \tfrac{1}{3}, \tfrac{1}{4}, \ldots$$

is an harmonic progression.

21. Prove that any harmonic progression can be written in the form

$$1/a_1,\ 1/a_2,\ 1/a_3, \ldots,$$

where $a_1, a_2, a_3, \ldots$ form an arithmetic progression.

*22. Prove that the geometric mean of two unequal positive numbers is always
less than the arithmetic mean and greater than the harmonic mean. How do
the means of a pair of equal numbers compare?

*23. Fill in the frequencies in the frequency table.

x	-1	0	1
f	f_1	f_2	f_3

so that the magnitudes of median, mode, and arithmetic mean have various
possible numerical orders. For instance, if the values of f are 2, 1, and 3,
the median is $\tfrac{1}{2}$, the mode is 1, and the mean is $\tfrac{1}{6}$, giving an example of a
case in which the mean is the least, the median next, and the mode greatest.
Some orders are impossible to attain.

*24. Prove that the geometric mean of two positive numbers is equal to the
geometric mean of their arithmetic and harmonic means.

*25. Does the result of Exercise 24 hold for three positive numbers? Either prove
it or give an example in which it is not so.

11. TOPICS FOR FURTHER STUDY

1. Graphs (analytic geometry): references **2** and **53**. Of course, there are
many textbooks on this subject.

2. Curve fitting (that is, finding a curve which fits a given table): This is very
briefly treated in reference **2**. For further treatment see various books on
"difference equations" and statistics.

3. Linear programming: This is a branch of mathematics concerned with
graphs on inequalities and of this there are many industrial applications.
See references **2** and **40**.

4. Fibonacci series: references **36** and **69**.

5. Mathematics and music: references **32** and **59**.

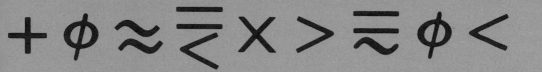

Permutations, Combinations, and Probability

1. ROUTES AND PERMUTATIONS

In Chapter 7 the frequency tables which we considered were obtained from experiments or from observing things happen. There are other frequencies of a rather different character which can be calculated by means of a kind of mental experiment where we see in our mind's eye in how many ways certain things occur or happen.

Suppose, for instance, that there are two roads connecting Ithaca, I, and Cortland, C, and three roads connecting Cortland and Syracuse, S. A driver starting from Ithaca to Syracuse may be interested to know in how many different ways he can plan his route. A look at Fig. 8:1 (p. 232), answers the question almost immediately. The possible routes to Syracuse are

$$11, 12, 13, 21, 22, 23,$$

and one sees that there are six of them. Notice that one obtains the answer by *multiplying* 2 (the number of roads connecting Ithaca and Cortland) by 3 (the number of roads connecting Cortland and Syracuse).

Another example of a problem leading to the question: "In how many ways can it be done?" is the following: Someone wants to distribute a knife, a

lighter, and a pen among three of his friends: John, Henry, and Robert. In
how many different ways can he do it?

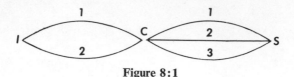

Figure 8:1

If one decides to give the knife to John, the lighter to Robert, and the pen
to Henry, we shall denote it symbolically by

$$k \quad l \quad p$$
$$J \quad R \quad H$$

The letters k, l, and p are abbreviations of "knife," "lighter," and "pen" and
J, R, and H the initials of the friends. It is easy to see now that the answer to
our problem is 6, since the following possibilities are the only ones:

$$
\begin{array}{ccc ccc ccc}
k & l & p & \quad k & l & p & \quad k & l & p \\
J & H & R & \quad J & R & H & \quad H & J & R \\[1em]
k & l & p & \quad k & l & p & \quad k & l & p \\
H & R & J & \quad R & J & H & \quad R & H & J
\end{array}
$$

If two sequences of symbols differ only in order we say that they are
permutations of each other. Thus, k, l, p is a permutation of p, k, l (and vice
versa); 3, 1, 2 is a permutation of 2, 1, 3; but a, 1, 2 is not a permutation of
1, 6, 2; and 1, 1, 3 is not a permutation of 1, 3, 3. (See Chapter 4, Section 8.)

The problem we just solved was equivalent to finding how many different
permutations of three symbols there are. It is obvious that the nature of the
symbols is immaterial and that just as there are six permutations of 3, 1, 2, so
there are six permutations of k, l, p or *, $/$, 1.

Now we shall find the number of permutations of four symbols. To
simplify the writing we shall assume that our symbols are the numbers
1, 2, 3, and 4. Let us consider first those permutations which begin with 1.
Every such permutation can be written in the form

$$1 \quad a \quad b \quad c$$

where a, b, c is a certain permutation of the three remaining symbols: 2, 3,
and 4. As we know, there are six permutations of three symbols and there are,
therefore, six permutations of four symbols which begin with 1. The same
reasoning leads to the immediate result that there are six permutations of four
symbols which begin with 2, six which begin with 3, and six which begin with
4. Hence the total number of permutations of four symbols is

$$6 + 6 + 6 + 6 = 4 \cdot 6 = 24.$$

Exactly the same kind of reasoning leads to the results that there are

$$24 + 24 + 24 + 24 + 24 = 5 \cdot 24 = 120$$

permutations of five symbols. Notice that $120 = 1 \cdot 2 \cdot 3 \cdot 4 \cdot 5$.

Even though the answers to the route problem and the knife-lighter-pen problem are both 6, the problems themselves are very different, for no permutations enter into the former. These two problems have been introduced in this section to emphasize the necessity of treating each problem on its own merits without any attempt at classification.

Essentially different is the following problem: How many different numbers are there between 100 and 1000, each of whose digits is one of 1, 2, 3, 4, 5, and 6 and in which no two digits are the same? Here it would be too much work to write all the possibilities. We could have any one of six digits in the first (left-hand) place. For the second place (the tens place) there would be five possibilities, no matter what we had in the first place. That is, there are five possibilities for each digit selected for the first place. Hence in the first two places there would be $6 \cdot 5 = 30$ different possibilities. For the units place we could not choose either digit already used and hence have the choice of four digits. We could represent this schematically as follows:

Place	First	Second	Third
Number of possibilities	6	5	4

Thus the total number of such numbers is $6 \cdot 5 \cdot 4 = 120$. If we had started enumerating the possibilities for the third place, then the second, and finally the first, we would have the same product.

Notice that if in the above problem, digits could be repeated, the number of possibilities would be $6 \cdot 6 \cdot 6 = 216$.

The chief difficulty in dealing with the problems of this chapter centers around methods of counting. Since most of the counting we shall be doing arises from two fundamental situations, let us consider them in more general terms. Suppose we have two sets of elements X and Y. The elements could be numbers, letters, or what you will. We want to count the number of pairs (x, y), where x is an element of X and y an element of Y. It does not matter if X and Y have elements in common; in fact, they could be the same set. If X has r elements and Y has s elements, then for each x there will be s possibilities for y. Thus the total number of pairs is

$$s + s + \cdots + s = rs.$$

We could check this by noting that for each y there are r possibilities for x, and thus sr possibilities in all. If we had three sets X, Y, and Z with r, s, and t elements, respectively, then to count the number of triples (x, y, z) we see that for each x, the number of pairs (y, z) is st and hence the number of triples is rst. And so one could continue for four sets, five sets—in fact, a finite number of sets.

The second fundamental situation is a kind of variation on the first. Suppose we have just one set X with r elements and want to count the number of pairs (x_1, x_2), where both x_1 and x_2 are in X. If there are no restrictions, we have the same situation as if X and Y were the same above and the number of pairs would be $r \cdot r = r^2$. The difference comes when we require that x_1 and x_2 shall be different elements of X. Then there are r possibilities for x_1 and only $r - 1$ to choose from for x_2. Again for each x_1 there are $r - 1$ possibilities for x_2 and hence the number of pairs is $r(r - 1)$. If we were concerned with triples of elements of X, no two the same, the number of possibilities would be $r(r - 1)(r - 2)$. And so we could continue for any number.

You might want to look at the previous problems in the light of this discussion.

EXERCISES

1. There are 3 roads connecting A and B and 4 roads connecting B and C. In how many different ways can one reach C by way of B starting from A?

2. There are 2 roads from A to B, 3 roads from B to C, and 3 roads from C to D. In how many different ways can one reach D from A?

3. Prove that there are 720 permutations of six symbols.

4. How many different six-digit numbers can be formed by using the digits 1, 2, 3, 4, 5, and 6, no digit being used twice?

5. How many different four-digit numbers can be formed by using the digits 1, 2, 3, 4, 5, 6, and 7, no digit being used twice? What would be the answer if repetitions were allowed?

6. How many different three-digit numbers can be formed using the digits 1, 2, 3, 4, 5, 6, and 7, no digit being used twice? What would be the answer if repetitions were allowed? Compare the answers here with those in Exercise 5.

7. Prove that there are $n!$ permutations of n symbols, where $n!$ means $1 \cdot 2 \cdot 3 \cdots n$.

8. How many different six-digit numbers can be formed by using the digits 0, 1, 2, 3, 4, and 5, no digit being used twice?

9. How many of the numbers in Exercise 8 are divisible by 5?
    SOLUTION: A number is divisible by 5 if and only if the last digit is either 5 or 0. The numbers are therefore of the form

    $$a \ b \ c \ d \ e \ 5 \quad \text{or} \quad a \ b \ c \ d \ e \ 0.$$

    There are 120 numbers of the second kind, for the number of permutations of 5 symbols is 120; and $120 - 24 = 96$ numbers of the first kind, for we must exclude the 24 permutations of a, b, c, d, e which begin with 0. Hence the answer is $120 + 96 = 216$.

10. How many of the numbers in Exercise 4 are even? Of Exercise 8?

11. How many of the numbers of Exercise 4 are divisible by 3? How many are divisible by 9? By 6?

12. How many different six-digit numbers can be formed containing only the digits 1, 2, 3, 4, 5, and 6, in which any digit may occur any number of times?

13. An organization has fifteen charter members and four officers: president, vice president, secretary, and treasurer. How many different possibilities are there for the four officers?

14. Four couples are to be seated at the head table. If no two men are to sit next to each other, how many seating arrangements would there be?

15. Three couples and one bachelor are to be seated at the head table. If no two men are to sit next to each other, how many seating arrangements would there be?

2. COMBINATIONS

Three problems somewhat different from those which we have considered above are the following:

Problem 1

An examination contains five questions, but the student is supposed to answer only two of them. In how many different ways can he make his choice? Enumerating the questions one can see that the following are the only possibilities:

$$(1, 2) \quad (2, 3) \quad (3, 4) \quad (4, 5)$$
$$(1, 3) \quad (2, 4) \quad (3, 5)$$
$$(1, 4) \quad (2, 5)$$
$$(1, 5)$$

where (2, 5), for instance, means that the student decides to answer questions 2 and 5. The answer to the problem is 10.

Problem 2

In how many different ways can a student choose three of five questions? Again the answer is 10, for choosing three questions to be answered is equivalent to choosing two questions which the student will leave unanswered; we can, therefore, use the same table as above with the understanding that (2, 5), for instance, means that the student leaves *unanswered* the questions numbered 2 and 5.

Problem 3

A student is to read two of five of Shakespeare's tragedies. In how many different ways can he make up his mind? The answer here is also 10, and the problem differs from the first only by the nature of things involved (tragedies instead of questions).

Notice that in Problems 1 and 3 we found that the number of different sets of two "elements" each which could be chosen from a set of five elements. (We use the word "element" to be noncommittal.) In the first problem, the elements were the numbers of the questions and in the third the elements were five of the tragedies of Shakespeare. In Problem 2 we noticed that the number of sets of two elements which can be chosen from five is the same as the number of sets of three elements, since every time we choose a set of two elements from five, what is left is a set of three elements.

Suppose S stands for a set of n symbols. If m is a natural number less than or equal to n, S will have subsets having m symbols. The number of different such subsets we denote by

$$_nC_m$$

which is sometimes called "the number of combinations of n things taken m at a time." In Problems 1 and 3 we found that a set of five elements has 10 subsets of two elements; that is,

$$_5C_2 = 10,$$

and in Problem 2,

$$_5C_3 = {_5C_2} = 10.$$

Above we found the number of subsets by writing them all down. But this procedure would obviously fail in case we wanted to find, for instance, the value of $_{100}C_{81}$. [By using formula (1) below let the interested reader estimate how long it would take.] There is, however, a way of evaluating the symbol $_nC_m$ for all positive integers m and n, with $m \leq n$. We shall first illustrate the method to find $_6C_3$. Let the set of symbols be S: $\{1, 2, 3, 4, 5, 6\}$. We wish to find how many subsets there are with three elements. To be more specific, suppose we have a set of six different cards and wish to find the number of different 3-card "hands" which can be chosen from the six. First, if we draw three successive cards as in the last problem of Section 1, we see that there are $6 \cdot 5 \cdot 4 = 120$ possibilities. But we have counted each *hand* several times. In fact, the following drawings would all give the same hand:

$$
\begin{array}{cccccc}
1 & 1 & 2 & 2 & 3 & 3 \\
2 & 3 & 1 & 3 & 1 & 2 \\
3 & 2 & 3 & 1 & 2 & 1
\end{array}
$$

the six permutations of the numbers 1, 2, and 3. We would have the same number if the hand were the ace, king, and queen of hearts, or indeed for any 3-card hand. Hence to get the number of different hands, we should divide by 6. The answer then is

$$_6C_3 = \frac{6 \cdot 5 \cdot 4}{1 \cdot 2 \cdot 3} = \frac{120}{6} = 20.$$

Similarly, if we wished to count the number of 3-card hands from seven cards we would have

$$_7C_3 = \frac{7 \cdot 6 \cdot 5}{1 \cdot 2 \cdot 3} = 35.$$

Another way to write the last expression is

$$_7C_3 = \frac{7 \cdot 6 \cdot 5 \cdot 4 \cdot 3 \cdot 2 \cdot 1}{(1 \cdot 2 \cdot 3)(4 \cdot 3 \cdot 2 \cdot 1)} = \frac{7!}{3!4!}.$$

(This is more compact but is no great help in the computation.) Here 7! is a short notation for the product of the first seven positive integers. (See Exercise 7 of Section 1.) We call it "seven factorial." Another notation sometimes used is $7\lfloor$.

Now let us see what happens in general. Suppose we wish a formula for the number of sets of r elements which can be chosen from a given set of n elements. There are n different possible choices for the first, $n - 1$ for the next, $n - 2$ for the next, and so on. Thus if we were to take into account the order in which the elements are chosen we would have

$$n(n - 1)(n - 2) \cdots (n - (r - 1))$$

choices for r elements. [Notice that the reason for $(r - 1)$ is that the first n can be written $n - 0$ and there are r terms.] But we have counted each set of r elements $r!$ times since there are $r!$ permutations of r things. Hence

$$_nC_r = \frac{n(n - 1)(n - 2) \cdots (n - r + 1)}{r!}, \qquad 0 < r < n,$$

where there are r terms in the product which appears in the numerator. If we multiply numerator and denominator of this expression by $(n - r)!$, we have

$$(1) \qquad _nC_r = \frac{n!}{r!(n - r)!}, \qquad 0 < r < n.$$

Why must we have $0 < r < n$? You will find in other books different notations for $_nC_r$. Three are $C_{n,r}$, $C(n, r)$, and $\binom{n}{r}$.

We now extend formula (1) to the cases in which $r = 0$ and $r = n$. From the point of view of combinations, it is reasonable to exclude $r = 0$, since to speak of n things taken zero at a time has no meaning. However, $_nC_n$ should be 1, since there is one set of n things which can be chosen from a set of n things. But if we use formula (1) for $r = n$ we have, substituting blindly,

$$\frac{n!}{(n - 0)!0!} = \frac{n!}{n!0!} = 1.$$

This would be fine if and only if 0! were equal to 1. So we *define* it to be 1. There is no inconsistency in this, because there is no sense to the expression

"The product of the first zero positive integers." So, with this understanding, we have $_nC_n = {_nC_0} = 1$ and formula (1) holds even if $r = 0$ or $r = n$. However, it still has no meaning for r a negative integer or r greater than n. There are ways of extending the meaning of the factorial for rational and other values of r, but we shall not consider them here.

There is one property of formula (1) which is useful and easy to see. It is

(2) $$_nC_r = {_nC_{n-r}},$$

because

$$_nC_{n-r} = \frac{n!}{(n-r)!(n-(n-r))!} = \frac{n!}{(n-r)!r!}.$$

One can also see this without using the formula. Every time one chooses a set of r things from n things, there remain $n - r$ things unchosen. So, in a given set of n elements, the number of subsets of r elements is the same as the number of subsets of $n - r$ elements.

EXERCISES

1. Given the set of numbers 1, 2, 3, 4, 5, and 6. Write all the subsets of four distinct numbers and all those of two different numbers. Show why the number of subsets of the first kind is equal to the number of subsets of the second kind.

2. Do what is asked in Exercise 1 for the set of five letters A, B, C, D, and E and subsets of two and three different letters.

3. Find the number of different 5-card hands one can make from a complete deck of 52 cards.

4. The number of 6-card hands which can be drawn from a deck of 52 cards is what number multiplied by the number of 5-card hands?

5. How many different bridge hands can be dealt? Estimate the magnitude of your answer.

6. Using formula (1) find $_7C_r$ for $r = 1, 2, 3, 4, 5, 6$. Which is the largest? Find $_8C_r$ for $r = 1, 2, 3, 4, 5, 6, 7$. Can you see any way the latter set of values could be computed from the former?

7. Compute $_{10}C_r$ for $r = 1, 2, 3, \ldots, 9$ and compare this with the last line of the Pascal triangle in Section 7 of Chapter 6. Can you see any reason for what you have noticed?

8. If, for a given number n, the values of $_nC_r$ are computed for $r = 1, 2, 3, \ldots, n - 1$, for what values of r will $_nC_r$ be the greatest?

9. Why in formula (1) do we first require that $0 < r < n$?

10. Suppose we have the rectangle of Fig. 8:2 whose dimensions are 3 units by 4 units, the vertical and horizontal units being of different lengths. In how many different ways can one move from A to B if one is allowed to move either one unit to the right or one unit up each time?

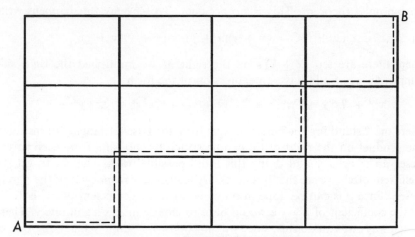

Figure 8:2

Solution. It is not hard to see first of all that there will be $3 + 4 = 7$ moves. Let us denote them by 1, 2, 3, 4, 5, 6, and 7. The way from A to B is completely determined by choosing the vertical moves. For instance, $(2, 5, 7)$ denotes the way of the dashed line on the diagram because the second, fifth, and seventh steps are vertical. [Show on the diagram the way $(3, 6, 7)$.] Now $(2, 5, 7)$ is a combination of three symbols out of seven, and since every such combination determines one and only one way of reaching B from A and since conversely each way determines one and only one combination of three symbols (out of seven), the total number of ways is $_7C_3 = 35$. The student should prove that the answer is also $_7C_4$ by considering horizontal steps.

11. A checkerboard is so placed that a black square is in the lower left-hand corner. A checker is placed on the third black square counting from the left in the first row and it is to be moved on black squares to the corresponding square in the seventh row. In how many ways can this be done assuming that no other pieces are on the board and that the checker is not a king; that is, it can move only forward? What would your answer be if it were to be moved to the second black square counting from the left in the eighth row? If to the third black square from the left in the eighth row?

*12. Chess is played on all the squares of a checkerboard. A pawn beyond the second row has at any time at most three possible moves; to the square directly in front of it and, if it can take a piece, to one of the squares immediately to the right or left of the one in front of it. A pawn is placed at the left end of the fourth row. In how many different ways can it reach the eighth row?

3. RELATIONSHIPS WITH THE BINOMIAL THEOREM

You noticed in Exercise 7 of Section 2 that the values of $_{10}C_r$ are, except for the first and last, the coefficients in the expression for $(x + y)^{10}$ by the binomial theorem. This is no accident. To see why this happens write

$$(3) \qquad (x + y)^{10} = (x + y)(x + y)\cdots(x + y)(x + y),$$

where there are ten $(x + y)$'s on the right. If we multiplied the right side completely we would have an expression of the form

$$x^{10} + ?x^9y + ?x^8y^2 + ?x^7y^3 + \cdots + ?x^2y^8 + ?x \cdot y^9 + y^{10}$$

where the ? stand for coefficients we get from the Pascal triangle, for instance. The product on the right of (3) can be gotten by choosing from each parentheses an x or a y and doing this in all possible ways. That is, to get the coefficient of x^{10} we see that it is necessary to choose x from each of the parentheses. Since this can be done in only one way, the coefficient of x^{10} is 1. To get the coefficient of x^9y we would have to choose nine x's and one y. There are ten ways we can do this, since we can choose the single y from any one of ten parentheses. In other words, the coefficient of x^9y is $_{10}C_1 = 10 = {}_{10}C_9$. Next, to get the coefficient of x^8y^2 we choose an x from eight parentheses and y's from the other two. Thus this coefficient would be the number of sets of 8 x's we can choose from 10, that is, the number of sets of two y's we can choose from ten. Hence this coefficient is

$$_{10}C_8 = {}_{10}C_2 = \frac{10 \cdot 9}{2} = 45.$$

Next, the coefficient of x^7y^3 will be the number of sets of seven things we can choose from ten, that is, the number of sets of three things we can choose from ten, which is

$$_{10}C_7 = {}_{10}C_3 = \frac{10 \cdot 9 \cdot 8}{1 \cdot 2 \cdot 3} = 120.$$

So we can continue. Thus, in general, we have: The coefficient of $x^{n-r}y^r$ in the binomial expansion of $(x + y)^n$ is $_nC_r = {}_nC_{n-r}$. That is,

$$(4) \quad {}_nC_nx^n + {}_nC_{n-1}x^{n-1}y + {}_nC_{n-2}x^{n-2}y^2 + \cdots$$
$$+ {}_nC_1xy^{n-1} + {}_nC_0y^n = (x + y)^n.$$

This, together with formula (1) in Section 2 for $_nC_r$ gives us a means of computing the coefficient of any term in the binomial expansion of $(x + y)^n$.

Now you will recall that the numbers in each row of the Pascal triangle can be obtained by adding the two nearest in the row above. This relationship, of course, must hold for the corresponding $_nC_r$. This can be shown by use of the formula (left as an exercise) or by consideration of the properties of sets. Let us consider a specific case. A little piece of the Pascal triangle is

$(x + y)^{10}$ $\qquad\qquad\qquad _{10}C_4x^6x^4 + {}_{10}C_5x^5y^5,$

$(x + y)^{11}$ $\qquad\qquad\qquad\qquad _{11}C_5x^6y^5,$

and from the property of the triangle we know that

$$_{11}C_5 = {}_{10}C_4 + {}_{10}C_5.$$

Why, in terms of sets and subsets, is this true? Suppose we consider the set of the first eleven positive integers 1, 2, 3, . . ., 10, 11. Then $_{11}C_5$ is the number of subsets of five. We could separate these into two kinds of subsets:

(1) Those containing the number 1.
(2) Those not containing the number 1.

There is no overlapping between the two kinds, and between them they include all. To count those of kind 1 we see that the other four must be chosen from the ten positive integers greater than 1. Thus there are $_{10}C_4$ of kind 1. To count those of kind 2, we note that we have to choose five from the numbers 2, 3, . . ., 11. The number of different such choices is $_{10}C_5$, the number of subsets of kind 2. Thus, for both kinds,

$$_{11}C_5 = {}_{10}C_4 + {}_{10}C_5.$$

Notice that the first subscripts on the right are one less than that on the left and the last subscript on the right is the same as that on the left. Thus, in general, we would expect the relationship

(5) $$_nC_r = {}_{n-1}C_{r-1} + {}_{n-1}C_r.$$

The general proof of this is left as an exercise.

EXERCISES

1. Without using the results of Section 2, show that $_{13}C_5$ is the coefficient of $x^8 y^5$ in the binomial expansion of $(x + y)^{13}$.

2. How many 5-card hands from a complete deck of 52 cards are there containing the king of hearts? How many not containing the king of hearts? What relation does this have with a formula of Section 2?

3. Use the method given in Section 2 to show formula (5) for $n = 13$ and $r = 7$.

4. Do what is asked in Exercise 3 but now for $n = 15$ and $r = 5$.

5. Prove formula (5) in the general form.

6. Using the formula for $_nC_r$, prove formula (5) for $n = 13$ and $r = 7$.

7. Using the formula for $_nC_r$, prove formula (5) in the general form.

8. Multiply $(x + y)(x + y)(x + y)(x + y)$ and show that before combining terms there are 16 different terms in the result. If there were n terms $(x + y)$ in the product, how many different terms would there be after multiplication and before combining terms?

9. Since in Exercise 8 the coefficient of each term is 1 before combining terms, what will be the sum of all the coefficients before combining terms? What does

this show about the sum of the coefficients of $(x + y)^n$ after like terms are combined? What does this tell about the sum

$$_nC_n + _nC_{n-1} + \cdots + _nC_1 + _nC_0,$$

that is, the sum of the numbers in the row of the Pascal triangle which is numbered n? What does this mean in terms of the subsets of a set of n elements?

4. PROBABILITY

We often hear such statements as: "His chances of being elected are small," "The odds are against his success," and the like. These are equivalent to "He is not likely to be elected" and "I do not think he is going to succeed." Such statements seem much more specific in the form: "The chances are 1 to 3 he will be elected," but the accuracy is somewhat spurious. One can make a really definite statement by saying, "I will bet three dollars to your one that he will lose." This shows that we think our chances of guessing correctly are three to one, but of course one cannot in such a situation determine exactly what his chances are.

There are, however, cases in which one has a definite, well-based reason for a statement about chances. If I threw one die I should say that my chances of throwing a 1 would be one to five. My reason for this statement would be that there are six faces of a die of which only one is a 1. If the die were honest I should expect throwing a 1 to be "equally likely" to throwing any one of the other five. Hence throwing a 1 should be one fifth as likely as throwing something else. "I feel it in my bones that in the long run" a 1 should appear one sixth of the time and might support this statement by saying that if it did not, I should suspect that the die was loaded. This circuitous "reasoning" leads us nowhere. The mathematician avoids this tangle by saying that if a thing can happen in any one of n equally likely ways and if, of these, f are favorable, then the *probability* of its happening favorably is

$$f/n.$$

Hence if in throwing a die one face is as likely to appear as any other, the probability of throwing a 1 is $\frac{1}{6}$. Notice that the odds in this case are one to five, the ratio of the number of favorable outcomes to the number of unfavorable outcomes. Similarly, odds of one to ten are equivalent to the probability of $\frac{1}{11}$.

Even though we state without qualification that the probability is $\frac{1}{6}$ or $\frac{1}{11}$ or whatnot, we must not lose sight of the fact that we do so on the assumption that one thing is *equally likely* to another. This appears in the determination of the probability of throwing two heads with a pair of coins. We might reason: One of three things may happen: I may throw two heads, two tails, or one head and one tail; hence the probability of throwing two heads is $\frac{1}{3}$.

But this is wrong, because we are twice as likely to throw one head and one tail, as may be seen by noticing that one head and one tail may appear in one of two different ways, each of which (we assume) is equally likely to throwing two heads. Hence we find that the probability of throwing two heads with a pair of coins is $\frac{1}{4}$.

From a slightly more general point of view, in probability theory we have to consider a set of happenings, S, divided into two mutually exclusive subsets: F, the subset of favorable happenings and, U, the subset of unfavorable happenings. If n denotes the number of elements in S, f the number of elements in F, and u the number of elements in U, then $n = f + u$. If any two happenings are equally likely, we call f/n the *probability* of a favorable happening.

Such probability which we determine by calculation we call *a priori probability* as opposed to *a posteriori probability*, which is determined as a result of experiment or statistical study. We shall consider the latter in Section 5.

There is one common misconception of probability that is very hard to scotch. It is the notion that the "law of averages" is nature's guardian who keeps a stern eye on her to see to it that whenever at night she misbehaves in one direction, she makes up for it the next morning by misbehaving in another direction. For example, suppose you have tossed a coin nine times and every time it has turned up heads. What is your chance of throwing tails the next time? From one point of view the chance is very small, for since the probability that an "honest" coin fall heads nine times out of nine is $\frac{1}{512}$, I should be led to suspect that the coin is not "honest" and is more likely to fall heads than tails. But if the coin is honest, that is, if any time it is as likely to fall heads as tails, then one of the times is after it has fallen heads nine times and the probability of its falling tails the tenth time is one half.

Example 1

Find the probability of throwing a 7 with a pair of dice; an 8; a 12.

Solution. One can get a total of 7 in six different ways, as shown in the following table:

Die I	1	2	3	4	5	6
Die II	6	5	4	3	2	1

One can get a total of 8 in five different ways, which the reader should list, and a 12 in just one way. To find the total number of ways a pair of dice can fall, we could list them all and perhaps the reader will prefer to do it that way. But we can save ourselves labor by seeming to change the subject and talking about three towns: A, B, and C. Number six roads from A to B and six from B to C. Being undecided which road to take, we toss the dice, and let the first die tell us what road to take from A to B and the second what road from B to C. Then, clearly, the number of different routes from A to C by way of B will be the number of different ways the dice can fall. But the number of routes is $6 \cdot 6 = 36$. Hence, on the assumption that each way in which the dice may fall is equally likely to occur, the probability of throwing a 7 is $\frac{6}{36} = \frac{1}{6}$, of an 8 is $\frac{5}{36}$, and of a 12 is $\frac{1}{36}$.

Example 2

What is the probability that if four dice are tossed at least two of them will show a 6?

Solution. Notice first of all that "at least" means that 2 sixes shall appear. In other words, we wish the probability that exactly 2 sixes, exactly 3, or exactly 4 will appear. To do this on the usual assumption that one appearance is as likely as any other, we need to compute the number of ways exactly 2 sixes may appear, add to this the number of ways exactly 3 sixes may appear and 4 sixes may appear.

First we compute the number of ways exactly 2 sixes may appear. Again we change the subject and think of five towns A, B, C, D, and E and six roads from each to the next. Here we must take exactly two roads numbered 6 in our journey from A to E. If we choose roads 6 from A to B and from B to C, then the roads numbered 6 from C to D and D to E are to be avoided and we have only five roads open for each of the last two legs of the journey. Hence there are $5 \cdot 5$, or 25, permissible routes from C to E by way of D. Thus, taking roads numbered 6 only from A to B and from B to C we have 25 routes. The story would be the same if we had elected to take roads 6 from A to B and from D to E, and so forth. Hence the total number of routes using exactly two roads 6 is 25 multiplied by the number of ways we can pick two from four routes numbered 6. Our answer for this part, then, is

$$_4C_2 \cdot 25 = 150.$$

It is easier to compute the number of ways exactly 3 sixes appear for if we choose road 6 from A to B, B to C, C to D we have five roads not numbered 6 to choose from for the journey from D to E. There are four different towns we can leave on a road not numbered 6 and our answer for this part is $4 \cdot 5 = 20$.

There is just one journey using all routes 6 and hence our total permissible routes are in number

$$150 + 20 + 1 = 171.$$

The total number of routes is $6^4 = 1296$. Hence the probability of throwing at least 2 sixes in a toss of four dice is

$$171/1296.$$

Example 3

Find the probability that in tossing ten coins exactly three are heads.

Solution. If the first three are heads, the last seven must be tails if the throw is what we wish it to be and there is only one way in which the seven can be tails. Hence the number of different possibilities for three heads is the same as the number of subsets of three heads there are in a set of ten heads; that is, $_{10}C_3 = 120$. This is the number of "favorable" happenings. On the other hand, there are $2^{10} = 1024$ ways in which the ten coins can fall, for we may think of eleven towns in a row each connected with its predecessor by two roads. There will be two routes connecting town 10 with town 11, 2^2 routes from town 9 to 11, 2^3 from 8 to 11, ..., 2^{10} routes from town 1 to town 11. Hence the probability of throwing exactly three heads out of ten is

$$\frac{_{10}C_3}{1024} = \frac{120}{1024} = \frac{15}{128}.$$

Example 4

What is the probability that if five dice are thrown exactly three will show a 2 or exactly two show a 4 or both?

Solution. Here we must first count the number of favorable happenings. Let S be the set of throws in which exactly 3 are twos and T the set in which exactly 2 are fours. Then the number of favourable outcomes is

$$n(S) + n(T) - n(S \cap T),$$

where $n(S)$ denotes the number of happenings in S. (See Chapter 3, Section 2.) Now

$$n(S) = {_5C_3} \cdot 5^2 = 250,$$

because there are $_5C_3$ ways in which the 3 twos may occur and for each of these there are $5 \cdot 5$ possibilities for the other two. Similarly,

$$n(T) = {_5C_2} \cdot 5^3 = 1250.$$

Finally, $n(S \cap T) = {_5C_3} = 10$, because once you have designated which of the dice will show a 2, there is only one possibility for the other two to show a 4. Hence the number of favorable happenings is $250 + 1250 - 10 = 1490$. The probability is $1490/6^5 = 745/3888$.

Example 6

Find the probability of drawing from a deck of playing cards a five-card hand containing two of one kind (that is, denomination) and three of another, for example, two kings and three queens. (This is called a *full house* in the game of poker.)

Solution. Since we know from the result of Exercise 3 of Section 2 that there are 2,598,960 five-card hands, this will be the denominator of the value of our probability and the numerator will be the number of different full houses in the deck. We first find the number of three-card hands containing three cards of one denomination. Since there are 13 different denominations and 4 such different three-card hands for each denomination, there will in all be $4 \cdot 13 = 52$ different three-card hands whose cards are all three of the same denomination. The denomination of the remaining two cards may be any one of the remaining twelve, and for each denomination there will be $_4C_2$, or 6, different pairs. Hence there are $12 \cdot 6 = 72$ possible pairs after the three of a kind have been selected; in other words, for each three of a kind there are 72 possible pairs to complete the full house. Thus the number of full houses is $52 \cdot 72 = 3744$, and the probability of drawing such a hand is 3744/2,598,960, which is approximately 3/2000.

Example 6

A man playing draw poker holds the following hand: the king of hearts and of diamonds, the queen of hearts and of diamonds, and the jack of hearts. He is undecided whether to discard the jack and hope to draw a queen or king, thereby having a full house, or to discard the two diamonds and hope to get a "straight" (five cards in sequence) or a "flush" (all cards of the same suit). Which is more likely: that discarding the jack he gets a full house, or discarding the diamonds one of his other hopes is realized? Assume he has the rest of the deck to draw from.

Solution. For the first alternative, there are two queens and two kings remaining in the deck and hence his probability of getting a full house is $\frac{4}{47}$. For the second, notice that to get a straight he must draw a 10 and either an ace or a 9. Hence there are $4 \cdot 8 = 32$ different ways he can complete his straight. If he is to get a flush he must draw two hearts out of the ten remaining in the deck. There are $_{10}C_2 = 45$ different such pairs. Now we might carelessly say that there are $32 + 45 = 77$ different pairs of cards which the player can draw, each one of which will yield a straight or a flush. But we have included in both the 32 and the 45 the cases in which the 10 and ace or 9 are hearts, that is, in which the resulting hand is both a straight and a flush, a so-called *straight flush*. These two hands we have counted twice. Hence $77 - 2 = 75$ is the correct number of desirable hands. There are $_{47}C_2 = 47 \cdot 23$ different possible two cards which can be drawn from 47 cards. Hence the probability of getting either a flush or a straight when two cards are discarded is

$$\frac{75}{47 \cdot 23} = \frac{1}{47}\frac{75}{23}$$

This is less than $\frac{4}{47}$. Hence the player should try for a full house. Notice that this conclusion is reached without taking into account another consideration in its favor—that even if neither a queen nor a king is drawn under the first alternative, the player still has two pairs in his hand.

Example 7

Suppose that four persons each toss independently a coin. What is the probability that exactly two of them will fall heads?

Solution. It is easy to enumerate all the possible results as follows:

```
Person  I   h h h h h h h h t t t t t t t t
Person  II  h h h h t t t t h h h h t t t t
Person III  h h t t h h t t h h t t h h t t
Person IV   h t h t h t h t h t h t h t h t
```

where h stands for heads and t for tails. The event of getting exactly two heads occurs in six cases (underlined in the table). The total number of possible situations is $16(=2^4)$, hence the probability

$$\frac{6}{16} = \frac{3}{8}.$$

Example 8

What is the probability that if five coins are tossed, exactly three will fall heads?

Solution. We could again construct a table of possibilities, but we prefer to use our knowledge of combinations. Denoting the coins by I, II, III, IV, and V, we mean by the symbol (II, III, V) the fact that the second, third, and fifth coins will fall heads and the others fall tails. There are obviously $_5C_3 = 10$ such symbols, and hence there are 10 different occurrences of the event of getting exactly three heads. Thus the total number of possibilities is

$$1 + {_5C_1} + {_5C_2} + {_5C_3} + {_5C_4} + {_5C_5} = 32 = 2^5$$

because we get either no heads (1 possibility) or 1 head ($_5C_1$ possibilities) or 2 heads ($_5C_2$ possibilities) or 3 heads ($_5C_3$ possibilities) or 4 heads ($_5C_4$ possibilities) or 5 heads ($_5C_5$ possibilities). The answer then is

$$\frac{10}{32} = \frac{5}{16}.$$

Notice that 2^5 could also be arrived at by methods like those used in the solution of Example 3.

This example has a connection with Section 3. Suppose we have the product

$$(x + y)(x + y)(x + y)(x + y)(x + y).$$

Each toss of five coins could give us a term in this product if we interpret a head to mean "choose x" and a tail to mean "choose y." For instance, if the first three coins were heads and the last two tails, we would choose x from each of the first three $(x + y)$'s and y from the last two. Each toss then would correspond to a term in the product before combining terms. Hence the number of terms before combining is equal to the number of different tosses. For instance, $_5C_3$ is the number of different tosses that will correspond to the term x^3y^2 or to x^2y^3.

EXERCISES

In all these problems the dice, coins, and cards are assumed to be "honest."

1. Find the probability of getting the totals of 2, 3, 4, 5, 6, 9, 10, 11 with two dice.

2. Find the probability of getting a total of 6 by tossing three dice.

3. Prove that the probability that exactly 4 of 8 coins will fall heads is $_8C_4/2^8$.

4. Find the probability that exactly 2 of 4 coins will fall heads. Should your answer differ from that of Example 7?

5. Generalize the above results for the case of m heads of n coins.

6. Prove that the probability that at least 2 of 4 coins will fall heads is

$$\frac{_4C_2 + {_4C_3} + {_4C_4}}{2^4} = \frac{11}{16}.$$

7. Find the probability that when four coins are tossed none or just one will fall heads. Find one connection between your result and the answer to Exercise 6.

8. In six throws of a die, what is the probability that the 2 appears exactly once?

9. Find the probability that at least 3 out of 6 coins will fall heads.

10. Which is more probable, getting a total of 7 with 2 dice or getting at least 4 heads by tossing 5 coins?

11. Without using the results of Example 2 solved above, find the probability that if 4 dice are tossed, at most one will show a 6. Why does this give another method of answering the question of Example 2?

12. Someone has remarked "The dice have no memory and no conscience." What could this mean?

13. A coin has been tossed 100 times and each time it has come up heads. What is the probability that the 101st time it will also be a head?

14. What is the probability that if a coin is tossed 101 times, it will be a head every time? Compare your answer with that of Exercise 13.

15. Find the probability that at least 3 of 6 dice will be fives.

16. Find the probability that *exactly* 2 of 5 dice will be sixes.

17. What is the probability that, tossing a coin and 2 dice, one will get a head and a total of 8 on the dice?

18. Find the probability that, tossing 4 coins and 4 dice, at least two heads and exactly one 6 will occur.

19. Find the probability that in tossing 10 coins exactly 4 are heads.

20. Little Lulu tore all the labels from 15 cans her mother brought home from the store and thoroughly mixed them. The cans were all of the same size; 4 contained corn, 6 beans, and 5 tomatoes. Her mother picked out a can and opened it regardless. What is the probability (1) that it contained corn, (2) that it contained beans or tomatoes?

21. Find the probability that from a deck of playing cards one will draw a 13-card hand
    a. In which all the cards are of the same suit.
    b. Containing four aces.
    c. Containing the A, K, Q, J, 10 of one suit.
    d. Compare the answers to parts b and c.

22. Find the probability that from a deck of playing cards one draws a 5-card hand
    a. Containing 4 of a kind, that is, 4 of the same denomination.
    b. All of whose cards are of the same suit: a flush.
    c. Whose cards are in sequence: a straight. An ace may be used at the top or bottom of the sequence.

23. One draws a 5-card hand from a deck of playing cards. Find the probability that the hand contains two pairs, for example, two kinds and two jacks.

24. In an experiment in extrasensory perception, four cards are placed face down before the subject. What is the probability that he names each card correctly, assuming that one guess is equally likely to any other?

25. In Exercise 24 what is the probability that exactly two cards are guessed correctly, under the same assumptions?

26. Four couples meet for an evening of bridge. Each man selects his partner for the first set of games by drawing from a hat the name of a woman. What is the probability that each man starts with his wife? What is the probability that each man will be at the same table with his wife? (Two couples are at each table.)

27. Six couples meet for an evening of bridge. Answer the same questions asked in Exercise 26.

28. A true-false test contains ten questions. What is the probability that marking the answers at random one will have 60% or more correct? Would the answer be the same if there were only five questions?

29. Answer the question of Example 6 solved above for the hand containing the queen of hearts and of diamonds, the jack of hearts and of diamonds, the ten of hearts.

*30. A recent president was undecided whether to run for a second term or not.

Suppose when he reached a decision, he decided to tell it to only one person: Wred Appleson. Now everyone knows that Wred Appleson tells the truth only one third of the time. Thus, if the president said yes, the probability that Mr. Appleson would communicate it as yes is one third and as no, two thirds. Similarly, if the president said no, the probability that Mr. Appleson would communicate it as yes is two thirds and as no, one third. Now, Mr. Appleson has a crony who is just as much of a liar as he is. Compute the probabilities for statements of yes or no from the crony. Suppose the crony had a crony who had a crony, and so on. What could you say about the eventual outcome? Does it depend on what the president said in the first place? Does it depend on the fraction used (that is, one third and two thirds)?

5. A POSTERIORI PROBABILITY

Using the methods of Example 3 and Exercise 19 we could complete a table of the probability of throwing 0 heads, 1 head, 2 heads, ..., 9 heads, 10 heads in one throw of 10 coins (Table A). The values of the numerators

Table A

$$\frac{1}{1024}, \frac{10}{1024}, \frac{{}_{10}C_2}{1024}, \cdots, \frac{{}_{10}C_2}{1024} \frac{10}{1024} \frac{1}{1024}.$$

are the numbers of the last line of the Pascal triangle given in Section 7 of Chapter 6.

Table B

Number of heads	0	1	2	3	4	5	6	7	8	9	10
Frequency	1	10	45	120	210	252	210	120	45	10	1

Such a table as Table B might also be the result of such an experiment as the following. Suppose 32 persons toss a set of ten coins 32 times each and record the number of times the set of ten showed no heads, one head, and so on. If they then pooled their results we should expect not a very great deviation from Table B, since the set of ten coins would have been tossed $32 \cdot 32 = 1024$ times. In fact, for any large number of tosses we should expect the ratios of the various frequencies to the total number tossed to be not far from the ratios of Table A. Notice that Table 23 in Chapter 7 is in accord with this statement, for there the results were so close to the "theoretical" values of Table A, which can be obtained without experiment (a priori probability), that it would not much matter whether we based any predictions on the experimental or theoretical table. But the points of view are somewhat different. In the case of the theoretical table we calculated in how many ways each number of heads could occur and, assuming vaguely that each way was "just as likely to happen" as any other way, divided the number of ways in which the specified number of heads could occur by the total number of ways in which the coins could fall. On the other hand, the experimental table showed us how the coins fell in a particular experiment. Then our computation of the probabilities is based on the supposition that, "in the main," things will continue to happen as they have happened in the past. Owing to the difficulty of finding a priori probability in most situations, statements of probabilities are more apt to be based on experimental or observational tables. And these considerations, vague as they are, are often sufficiently accurate for useful predictions. The mortality table, for example, is exact enough to provide the foundation for life insurance. Although the insurance company does not know which of a group of 100,000 individuals aged ten will die between his thirtieth and fortieth birthdays, it does know that about 7300 are practically certain to die within that period of years—barring a surprising advance in medicine or a disastrous epidemic or war before or during that period. To say that the probability that a boy aged ten will die between his thirtieth and fortieth years is 7300/100,000 is merely expressing in terms of the individual the fact that 7300 of 100,000 such people die. This statement is correct only if we think of it as another means of expressing what the mortality table tells us about 100,000 people. We could apply it to an individual in the sense of Section 4 only if one individual were as likely to die as another. Lacking that knowledge, we cannot say truthfully that any individual aged ten which you may select has 73 out of 1000 chances of dying between thirty and forty.

This point may be clarified by another example. A man is running a dart booth at a fair and finds from experience that one tenth of those who try succeed in hitting a balloon with a dart. He could then say quite properly that the probability of anyone winning a prize is $\frac{1}{10}$ and he can use this as a basis for the amount he will charge and the value of his prizes. But his statement does not in the least mean that whoever steps up to the board has one chance in ten of securing a prize, because not all are equally skillful. Some may win whenever they play and others never; the probabilities are 1 and 0, respectively.

The measure of both kinds of probability is a number not less than 0 and not greater than 1 associated with each of the recognized ways in which an event may turn out. In terms of the graph, this number is the ratio of the area of the appropriate rectangle in the histogram to the sum of the areas of all the rectangles.

EXERCISES

1. What is the probability that (see Table 25, page 222)
   a. A person aged 10 will be alive at age 70?

    b. A person aged 65 will be alive at age 75?

    c. A person aged 40 will be alive at age 80?

    d. What, precisely, does each of these statements mean?

2. What is the probability that
    a. A person aged 10 will die before he is 70?
    b. A person aged 65 will die before he is 75?
    c. A person aged 40 will die before he is 80?
    d. Precisely what does each of these statements mean?

3. What is the probability, in tossing 10 coins, of getting
    a. Exactly 7 heads?
    b. Exactly 5 tails?
    c. Not less than 8 heads?
    d. Either 2 or 3 heads?

4. What is the probability that, in tossing ten coins, one gets
    a. Exactly 3 heads?
    b. Exactly 5 heads?
    c. Not more than 7 heads?
    d. Either 3 or 5 heads?

5. Use Table 24 in Chapter 7 to find the probability that a call will last
    a. Just 450 seconds.
    b. Not more than 150 seconds.
    c. Not less than 750 seconds.
    d. More than 750 seconds.

6. Use either Table 25 or the table computed in Exercise 3 of Section 9 of Chapter 7 to find the probability that a person aged:
    a. 10 will die between his 30th and 35th birthdays.
    b. 30 will die before he is 35.
    c. 60 will not die before he is 70.
    d. 25 will die between his 60th and 70th birthdays.

7. Do you agree with the conclusions in the following paragraph which appeared in a popular magazine under the heading: "One Third of Your Children Doomed"? It reads: "Being of a cheerful turn of mind, we have tried to leave to insurance companies the somber task of charting life's overwhelming hazards. One particular hazard, however, has increased so steadily that we can hardly be held a spoilsport for calling the attention of 1935s parents to the fact that their children, according to the latest estimates, stand one chance in three of meeting death or serious injury from that deathlike convenience, the motor car. Take a room containing three children: one of them is destined to be killed or badly hurt by a car before he has completed his normal life span. Statistically, the motor car is life's ugliest joke; its toll makes war seem like a spring outing."

8. Is the above quotation an argument for having no more than two children?

9. The story is told that farmer Silas figured that since a recent flood a certain rickety bridge near his farm had one chance in ten of collapsing when he was on it. One day, recalling that since the flood he had crossed the bridge nine times, he realized that he had used up all his chances and hence would in the

future keep off the bridge. Assuming that the probability figure was correct, would you behave as he did after crossing the bridge nine times? Why?

6. MATHEMATICAL EXPECTATION

It is rather startling to notice that those who make most use of probability are of two classes diametrically opposite in their attitude toward money: the reckless gambler who will stake his all on the turn of a wheel, and the timid conservative who provides for every possible mishap by taking out all available insurance; in the first case all losers pay the winner and in the second all winners pay the loser.

The use which they make of probability is by way of what is called *mathematical expectation*. We illustrate it first by the description of a simple game. Suppose seven people choose one of their number as the banker. Each of the remaining six selects one of the numbers 1, 2, . . ., 6 as his number, no two having the same number. Then the banker throws one die and the number which turns up determines the winner. The question is: If the winner is to receive $6, how much should each pay the banker for the privilege of playing? The obvious and correct answer is $1, since then the banker will receive just enough money to pay the winner. (As a matter of fact, they could dispense with the banker by forming a pool.) If the banker were running the game he might charge each player $1.25 and make a tidy profit. Notice that the $1 which a player should pay if the banker breaks even is the product of the probability of winning and the amount of the prize. The *mathematical expectation* is defined to be the product of the amount to be won and the probability of winning it in problems like this when there is only one way of winning and only one amount to be won. Further modifications of this are indicated in examples below. In this particular case the mathematical expectation is $1 and at that figure each player should expect neither to lose nor gain much after many games of an evening. On a payment of $1.25 he should expect to lose 25 cents per game for the evening as a whole. As a matter of fact, the overcharge of 25 cents is small compared to the usual practice of those who run gambling games.

Another way to look at mathematical expectation in terms of a similar game is as follows: Suppose there is only one person playing; he selects a number between 1 and 6 and a die is thrown. If his number comes up he is to receive $6, otherwise nothing. Since his chance of winning is one of six, it is natural to think that he should pay one sixth of $6 to play. Notice that if he plays six times and wins just once, he will just break even, since he will pay $1.00 six times to play and will receive his prize of $6.00. Of course he cannot expect always to break even. Let us see what would be the probability of exactly breaking even if he plays six times. To do this he must win exactly once. The number of possibilities for winning the first time and none of the others is

$1 \cdot 5 \cdot 5 \cdot 5 \cdot 5 = 5^5$. There are the same number of possibilities for him to win only the second time, in fact, for any time. Hence the number of possibilities for winning is $6 \cdot 5^5$. Thus the probability is

$$\frac{6 \cdot 5^5}{6^6} = \frac{5^5}{6^5} = (5/6)^5$$

(playing six times), which is about 0.4.

Let us consider an example in insurance. If a fire insurance company found that over a period of time one of every sixty cars in a certain community burned, it would expect to charge a premium of 1/60 of the amount of the policy plus expenses and should then in the long run take in just enough money to pay all claims plus expenses, on the assumption that fires *will do* the same damage that they *have done*. Here the insurance company assumes the risk and, quite rightly, charges for it. On the other hand, companies which operate fleets of trucks usually assume the risk themselves.

It is important to note that given a set of sixty cars it is by no means certain that exactly one will be burned. Some cars are more susceptible to burning than others. Even if this were not the case and we had the strange situation that any car would be equally likely to burn to any other, the chance of exactly one of a given sixty burning would be $(59/60)^{59}$, which is about 0.37.

Example 1

A man draws one card from a deck of playing cards. If he draws a red ace he is to receive \$2, if a king \$1, otherwise nothing. How much should he pay to play?

Solution.　To clarify the situation we at first change the problem slightly and suppose that two persons, A and B, play, the former winning only with a red ace and the latter only with a king. The probability that A will win is then $\frac{2}{52}$ and his expectation is $2 \cdot \frac{2}{52} = \frac{1}{13}$ of a dollar. Similarly, the probability that B will win is $1 \cdot \frac{4}{52} = \frac{1}{13}$ of a dollar. Together they would pay $\frac{2}{13}$ dollar. Now in the problem as given, the single player wins when either of the players A or B would win and hence he should pay what they together pay. Therefore, he should pay $\frac{2}{13}$ dollar which is \$.15 to the nearest cent. This is called his *mathematical expectation*.

Example 2

A man 40 years old wishes term insurance to the amount of \$1000 for 5 years. That is, if he dies within 5 years the company pays \$1000 to his heirs; if he lives for that length of time the company pays nothing. He chooses to pay for his policy in a single premium at the beginning of the 5-year period.

What should he pay, neglecting any charge the company may make for its expenses?

Solution. Table 25 in Chapter 7 shows that out of 78,106 persons alive at 40 years of age, 74,173 are alive at age 45. Hence his probability of dying is 3933/78,106 = .050354. Thus his premium should be 1000(.050354) or $50.35.

Note: Term insurance is not the most common kind of insurance. If, at the beginning of each year, one procured term insurance for that year, he would pay each year a higher premium than the last since his probability of dying increases (this is not necessarily so in any individual case but it is true statistically for a large number of cases). In ordinary life insurance, the payments are the same one year as the next and in man's earlier years he accumulates a reserve to offset the increase in cost of insurance as he grows older. On the other hand, the usual group insurance is term insurance, in which (unless the employer contributes) the younger men pay more than they would pay for individual term insurance and the older men pay less.

Example 3

An arrow is spun on a cardboard disc divided into seven sectors as in Fig. 8:3. Players put money on any or all of the sectors. If the arrow stops on a

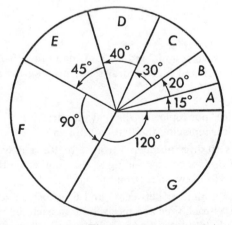

Figure 8:3

dividing line, no one loses and it is spun again. If it stops in sector A, each person having money in A is given a times as much money as he put down in that sector and the banker collects the money in all the other sectors. If the arrow stops in B, b times as much is paid to all whose money is in B, and so forth. Assuming the arrow is just as likely to stop in one spot as in another, determine the correct values of a, b, c, d, e, f, and g.

Solution. Sector A is $\frac{15}{360} = \frac{1}{24}$ of the total area. Hence anyone putting money in section A has a mathematical expectation of $\frac{1}{24}$ of what he gets if he wins. Hence if he wins he should get 24 times what he put in. Similarly, the values for b, c, d, e, f, and g are 18, 12, 9, 8, 4, and 3. We may check our results by finding that 1 is the value of the sum

$$\tfrac{1}{24} + \tfrac{1}{18} + \tfrac{1}{12} + \tfrac{1}{9} + \tfrac{1}{8} + \tfrac{1}{4} + \tfrac{1}{3}.$$

Notice that those who put money in A, for instance, actually *gain* 23 times as much as they put in, for each dollar placed in the sector is returned accompanied by 23 other dollars. Thus the *odds* for sector A would be 1:23.

Example 4

In a so-called true-false test the examinee marks a given answer true or false. How should such a test be scored?

Solution. If a large number of persons took the test, marking the answers at random, we should expect about half the answers to be right and half to be wrong. Since the mathematical expectation of each person should be zero, it follows that if the score for a correct answer is $+1$, the score for a wrong answer should be -1.

EXERCISES

1. A bag contains 8 white and 2 red balls. A man draws from the bag and is to receive $1 for each red ball he draws but nothing for a white ball. What should he pay to play if he
   a. Draws once?
   b. Draws twice, not replacing the first ball drawn?
   c. Draws twice, replacing the first ball drawn?

2. A hat contains 10 discs numbered from 1 to 10. A player drawing receives the number of dollars corresponding to the number on the disc. How much should one pay to draw once from the bag?

3. A man chooses a number between 1 and 6 inclusive and a die is tossed six times. He pays $1 each time and receives $6 for each time his number occurs. What is the probability that he will break even? What is the probability that he will lose? What is the probability that he will gain?

4. A man chooses a number between 1 and 6 inclusive and a die is tossed six times. If the number he chose appears exactly once he gets $6, otherwise nothing. What should he pay to play?

5. How much should a man aged 50 pay in a single premium for a 5-year $1000 term insurance policy? Make no allowance for the expenses of the insurance company.

6. How much should a man aged 60 pay in a single premium for a 5-year $1000 term insurance policy? Make no allowance for the expenses of the company.

7. In a certain city one out of every 5000 cars is stolen and not recovered. Allowing no expenses for the company, what should be the premium for a $1000 policy?

8. A board has squares numbered 2 to 12. Players put their money on one or several squares. Two dice are tossed and the sum of the spots showing determines the winning square. What odds should be marked on each square? (See Example 3 worked out above.)

9. A game, one of whose names is "chuck-a-luck," is played as follows. A player chooses a number from 1 to 6 inclusive and throws 3 dice. If his number appears on just 1 die he receives 10 cents, if on just 2 he gets 20 cents, if on all 3, 30 cents. Otherwise he is paid nothing. How much should he play to play?

10. Suppose the game described in Exercise 9 were altered so that if the player's number appeared just once he had his money returned and 5 cents in addition, if it appeared exactly twice he received 10 cents beyond what he paid, and if all three times, 15 cents. Otherwise he is to lose what he paid. How much should he pay to play?

11. The statistics of an insurance company show that over a period of years one of every hundred cars in a certain community burns. However, it has found that in section A in the town twice as many cars burn in any year as in the remainder of the town, even though there are just as many cars in section A as outside it. Allowing for no company expenses, what should be the fire-insurance rates in the two parts of the town?

12. If a true-false test of ten questions is scored as in Example 4, what is the probability that a person marking the answers at random should get 0 for a total score?

13. If a "multiple-choice test" has four answers listed for each question, exactly one of which is correct, how should it be scored?

7. TOPICS FOR FURTHER STUDY

For this subject, the topics are so varied and interlaced that it will be better for you to select them yourself.

See the following references: **23** (p. 247), **38** (Chap. 7), **40** (Chaps. 3 and 4), **44**, **47**, **48** (Vol. 2, part 7), and **71**.

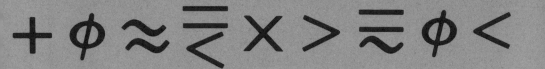

9

Mirror Geometry or
the Theory of
Enantiomorphous Figures

"It seems very pretty," she said when she had finished it; "but it's *rather* hard to understand!" (You see she didn't like to confess, even to herself, that she couldn't make it out at all.) "Somehow it seems to fill my head with ideas—only I don't exactly know what they are!"
—From *Through the Looking Glass.*

1. INTRODUCTION

The preceding chapters have dealt chiefly with numbers or with the study of simple geometric figures by means of numbers. We now turn to some topics in geometry where numerical relationships are either absent or occur only incidentally.

In the algebraic work you were expected to remember only a few of the more elementary properties of numbers and algebraic expressions. Here we ask you to remember from your study of Euclidean geometry only some of the simpler theorems. Thus we shall want to know that it is possible to draw one and only one line perpendicular to a given line and through a given point. A theorem we require is that two triangles are congruent if two sides and the included angle of one are equal in measure to the corresponding sides and the included angle of the other. A few other results from Euclidean geometry will be used in this chapter.

2. SYMMETRY AND REFLECTION

A vase shaped as in Fig. 9:1 is generally regarded as being more beautiful than one shaped as in Fig. 9:2. In many sorts of beautiful objects there is a

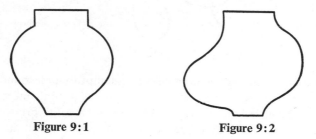

Figure 9:1   Figure 9:2

quality we somewhat loosely refer to as "symmetry." Even in a painting where the two halves of the picture certainly do not "match" there are often certain elements of "symmetry." In architecture, particularly that derived from Roman models, the symmetry of the façade is often complete.[1]

The vague notion of symmetry which we have just suggested is, of course, of no use to the mathematician until he has distilled its essence and preserved his analysis of its structure in a definition. Accordingly, we first simplify our discussion by limiting it to figures in a plane and then proceed to set up our definition.

One significant fact must not escape our attention if we are to think clearly about symmetry. Symmetry (of the kind we are here discussing) is *with respect to some particular straight line*. Thus our first glance at Fig. 9:3 (p. 260) may not reveal the fact that it is symmetric, but a second inspection, perhaps aided by turning the page partially around so that the line *l* is vertical, shows it to be symmetric. It is furthermore evident that the part of the figure on one side of *l* may be obtained from the part on the other side. We say that this is accomplished by *reflection* in the line *l* or that one half of the figure is the *mirror image* of the other half.

The two figures which are obtained from one another by reflection in a line are sometimes called an *enantiomorphous pair*. The same term may be applied

[1] See topic 3 at the end of Chapter 10.

to physical objects such as a pair of gloves or a pair of shoes. When we speak of reflecting a solid object, we are, of course, using a plane rather than a line to define the image.

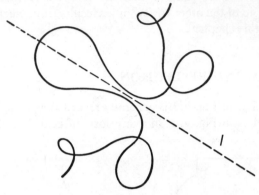

Figure 9:3

It often happens in mathematics that an attempt to understand fully some vaguely comprehended idea results in the discovery of some more fundamental concepts. So it is here. Only certain figures have the property of being symmetric in a line, but *any* plane figure may be reflected in any given line. Symmetry is a descriptive or static property which a figure may or may not have, while reflection in a line is an operation or *transformation* which we may apply to any figure we like.

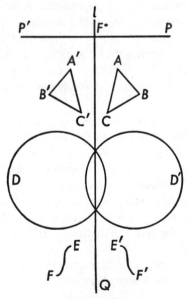

Figure 9:4

It is rather useful to think of a geometric transformation as a rule which enables us to establish corresponding figures in much the same way as the formulas of Chapter 7 enabled us to construct tables of corresponding numbers. Thus the accompanying table describes the relationships shown in Fig. 9:4.

Original figure	point P	triangle ABC	circle D	arc EF	point Q
Figure obtained by reflection in l	point P'	triangle $A'B'C'$	circle D'	arc $E'F'$	point Q

We are now at the point where we can define what we mean by the mirror image of a point in a line.

Definition

The mirror image (or image), P', of a point P in a line l is determined as follows: If P is on l, then it is its own mirror image. If P is not on l, P' is that point with the property that l is the perpendicular bisector of the segment $\overline{PP'}$.

This means that if P is not on l, P' is determined by finding F^*, the foot of the perpendicular from P on l and finding the point P' on this perpendicular such that F is the midpoint of the line segment $\overline{PP'}$.

The definition of mirror image of a point leads immediately to the idea of symmetry:

Definition

A set, S, of points is said to be *symmetric* in a line if the set S is the same as the set of mirror images of the points of S; or, more briefly, if S is its own image.

Since we think of geometric figures as sets of points, the above definition applies to figures as well.

EXERCISES

1. Name some commonly occurring pairs of enantiomorphous objects.
2. When we look in a mirror we see that in the image right and left are interchanged. Why are not up and down interchanged too?
3. In what line or lines are the following figures symmetric?
   a. An isosceles triangle. b. An equilateral triangle.
   c. A circle. d. A rectangle.
   e. A square.
4. Let P' denote the image of P in the line l. Show that $(P')' = P$, that is, the

image of the image of P is P. Let x be a number corresponding to the point X on a line m and denote by x' the coordinate of the image of X in some line perpendicular to m. Suppose $x' = ax + b$; that is, if $x = 1$, then $x' = a \cdot 1 + b = a + b$; if $x = 2$, then $x' = a \cdot 2 + b = 2a + b$; and so on. Find conditions on a and b so that $(x')' = x$.

5. Suppose the line l is the line $y = x$. What will be the images of each of the following points?

a. (1, 1).   **b.** (1, 2).   **c.** (3, −4).   **d.** (r, s).

6. What effects on the x and y coordinates of a point are produced by reflecting the point in the x-axis? In the y-axis?

7. Determine the image in a line l of each of the following:
    a. A straight line parallel to l.
    b. A straight line perpendicular to l.
    c. A straight line neither parallel nor perpendicular to l.
    d. The line l.
    e. A line segment of which l is the perpendicular bisector.
    f. A circle tangent to l.
    g. A circle with a diameter along l.
    h. A triangle with one side along l.

8. State with respect to which lines each of the following is symmetric.
    a. A line.   **b.** A line segment.
    c. A triangle no two sides of which have equal length.
    d. A regular pentagon (see Section 3 of Chapter 6).

9. Would the following definition be equivalent to the one given above? Why? A figure is said to be symmetric in a line l if it can be divided into two parts in such a way that one part is the image of l of the other part.

10. How would you define symmetry with respect to a point?

11. Show that the distance between any two points in the plane is the same as the distance between the images of the points in any given line.

3. SUCCESSIVE REFLECTIONS

Let us see what would happen to a point if we were to reflect it successively in two intersecting lines. In Fig. 9:5 we begin with a point P, reflect it in a line l, obtaining P_1; then reflect this point in line m (why not in l again?) obtaining P_2; then reflect P_2 in line l, obtaining P_3; and so on. Evidently we might also have begun by reflecting P in m, following this by a reflection in l; and so on as illustrated in Fig. 9:6.

While the points constructed in this way from P "jump about" considerably, even a casual inspection convinces us that they do not wander all over the plane. Indeed, if we notice that the intersection, C, of l and m remains in the same position when reflected in either l or m, we can use the result of Exercise 11 above to conclude that all the points P, P_1, P_2, P_3, and so on, are

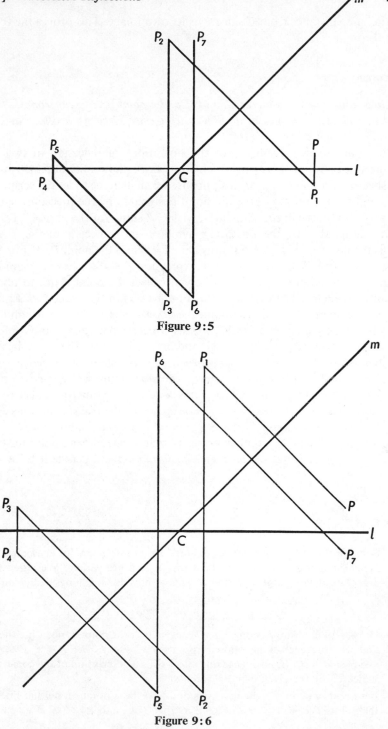

Figure 9:5

Figure 9:6

at the same distance from C since C is its own image. This proves the following theorem.

Theorem

Given a pair of lines intersecting in C *and a point* P *different from* C, *then* P *and its successive images in the lines all lie on a circle with center at* C *and passing through* P.

If we are to deal readily with the operations of reflection in two lines, we need a convenient way in which to describe the points obtained by successive transformations. For this purpose we shall agree to let the script letter $\mathscr{L}$ stand for the phrase "take the point designated by the preceding symbols and reflect it in the line *l*." Similarly, we let $\mathscr{M}$ stand for the phrase "Take the point designated by the preceding symbols and reflect it in the line *m*." Thus the point P_1 of Fig. 9:5 is now called $P\mathscr{L}$ and the point P_1 of Fig. 9:6 is now called $P\mathscr{M}$. When we write the expression $P\mathscr{L}\mathscr{M}\mathscr{L}$ we understand it to stand for the point obtained as follows: Reflect the point P in *l* to obtain a point $P\mathscr{L}$; then reflect the point $P\mathscr{L}$ in *m* to obtain the point $P\mathscr{L}\mathscr{M}$; finally reflect the point $P\mathscr{L}\mathscr{M}$ in *l* to obtain the point $P\mathscr{L}\mathscr{M}\mathscr{L}$. Our shorthand notation is not only more concise, it gives us concrete symbols $\mathscr{L}$ and $\mathscr{M}$ for the *operations* of reflecting in the lines *l* and *m*, respectively. The symbols $\mathscr{L}$ and $\mathscr{M}$ then stand for *geometric transformations*. We shall agree that a symbol like $\mathscr{L}\mathscr{M}$ shall stand for the geometric transformation which carries an arbitrary point P into the point $P\mathscr{L}\mathscr{M}$. Thus $\mathscr{L}\mathscr{M}$ is the geometric transformation which results from the successive performance of the transformations $\mathscr{L}$ and $\mathscr{M}$. We shall agree to call $\mathscr{L}\mathscr{L}$ a *geometric transformation*, but since (by Exercise 4 above) it has the peculiar property that it carries each point into itself, ($P\mathscr{L}\mathscr{L} = P$), we call it the *identity transformation* and denote it by $\mathscr{I}$. Then $\mathscr{L}\mathscr{L} = \mathscr{I}$.

EXERCISES

1. Some of the following expressions stand for points, some for geometric transformations and some are meaningless. Which are which? We agree that P and Q designate points and that $\mathscr{L}$ and $\mathscr{M}$ have the meanings just given them.

    $$P\mathscr{M}\mathscr{M}\mathscr{L}, \quad P\mathscr{M}P\mathscr{L}; \quad \mathscr{L}\mathscr{M}\mathscr{L}\mathscr{L}; \quad \mathscr{L}\mathscr{M}\mathscr{L}P;$$
    $$\mathscr{M}\mathscr{L}\mathscr{M}\mathscr{L}\mathscr{M}\mathscr{L}\mathscr{L}\mathscr{M}; \quad \mathscr{M}\mathscr{L}P\mathscr{M}\mathscr{L}\mathscr{M}Q.$$

2. Draw a pair of intersecting lines *l* and *m*, mark a point P not on either line, and construct each of the following points: $P\mathscr{L}, P\mathscr{M}, P\mathscr{L}\mathscr{M}, P\mathscr{M}\mathscr{L}, P\mathscr{M}\mathscr{L}\mathscr{L}\mathscr{M}$, and $P\mathscr{M}\mathscr{L}\mathscr{M}\mathscr{L}$. In each case put into words a description of the construction indicated by the given expression.

3. Let *l* and *m* be perpendicular lines and let P be a point on neither line. Find the points $P\mathscr{L}\mathscr{M}, P\mathscr{L}\mathscr{M}\mathscr{L}\mathscr{M}, P\mathscr{L}\mathscr{M}\mathscr{L}\mathscr{M}\mathscr{L}\mathscr{M}$, and $P\mathscr{L}\mathscr{M}\mathscr{L}\mathscr{M}\mathscr{L}\mathscr{M}$.

4. Do what is asked in Exercise 3 but interchanging $\mathscr{L}$ and $\mathscr{M}$. How are $P\mathscr{L}\mathscr{M}$ and $P\mathscr{M}\mathscr{L}$ related?

5. Construct as many different points $P\mathscr{L}$, $P\mathscr{L}\mathscr{M}$, $P\mathscr{L}\mathscr{M}\mathscr{L}\mathscr{M}$, and so on, as you can when l and m intersect at an angle of 45°.

6. Reflect the vertices of an equilateral triangle successively in its *three* sides. (Do not attempt to get *all* the points so obtainable.) If you used only two sides and the vertices opposite them, how many points would you obtain and how would they be arranged?

7. Do what is asked in Exercise 6 for a square rather than a triangle.

8. Show that $P\mathscr{L}\mathscr{M}\mathscr{L}\mathscr{L}\mathscr{M}\mathscr{L}$ coincides with P. How can we obtain P from $P\mathscr{M}\mathscr{L}\mathscr{M}$ by successive reflections? What geometric transformation should be inserted in the parentheses in order that $\mathscr{L}\mathscr{M}\mathscr{L}\mathscr{M}$ (   ) shall equal $\mathscr{I}$, the identity transformation?

4. ROTATIONS

If the reader has carried out the preceding constructions with care, he may have been struck by the fact that the points constructed fall into two separate patterns. In Fig. 9:5 we see that the points P, P_2, P_4, and P_6 are all obtained by proceeding counterclockwise around C, the "spacing" between successive points appearing to be about the same at each step. Similarly, we get from P_1 to P_3 by proceeding clockwise around C and to P_5 from P_3 by continuing clockwise an equal amount.

That it is actually true that $\angle PCP_2 \cong \angle P_2CP_4 \cong \angle P_4CP_6 \cong \ldots$ is a consequence of the following theorem, where the symbol $\cong$ means "congruent" or "has the same measure as" or "has the same number of degrees as."

Theorem

If P *is any point other than* C, $P_2 = P\mathscr{L}\mathscr{M}$ *and* θ *is the measure of the angle from*[1] *line* l *to the line* m, *then the measure of the angle* PCP_2 *is equal to* 2θ.

Proof. If α is the measure of the angle from line l to the line through C and P, then $-\alpha$ will be the measure of the angle from line l to the line through C and

---

[1] There are, of course, two positive angles less than 180° *between* lines l and m. We shall agree that the phrase "the angles from line l to line m" is to mean the angle which, when described in the counterclockwise sense, has for its initial side l and its terminal side m. While this description will enable you to "mark in a figure" the correct angle, it is nevertheless true that it would be rather difficult to give a precise definition of "counterclockwise sense" without first giving rather careful proofs of a chain of theorems which you might consider too obvious to require proof! We also agree that the angle PCP_2, for instance, is the angle from PC to PC_2 in the above sense.

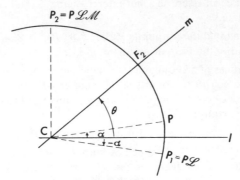

Figure 9:7

$P_1 = P\mathscr{L}$. Then (see Fig. 9:7) $\text{meas}(\angle P_1 CF_2) = \theta + \alpha$, where "meas" denotes the measure of the angle. Hence

$$\text{meas}(\angle(F_2CP_2) = \theta + \alpha,$$

$$\text{meas}(\angle PCP_2) = \theta + \alpha + \theta - \alpha = 2\theta.$$

This completes the proof.

With the aid of this theorem it is very easy to construct as many of the points $P\mathscr{L}\mathscr{M}$, $P\mathscr{L}\mathscr{M}\mathscr{L}\mathscr{M}$, $P\mathscr{L}\mathscr{M}\mathscr{L}\mathscr{M}\mathscr{L}\mathscr{M}$, and so on, as we wish. Thus in Fig. 9:8 we get the points marked by making arcs PP_2, P_2P_4, P_4P_6, ..., all equal to twice the angle from l to m. ($P_2 = P\mathscr{L}\mathscr{M}$, $P_4 = P\mathscr{L}\mathscr{M}\mathscr{L}\mathscr{M}$, ...)

One rather unexpected result follows from this construction. The points P_2, P_4, P_6, ..., would remain in precisely the same position if instead of reflecting P in l and m we were to reflect P successively in two lines l' and m' through C *provided the angle from l' to m' is the same as the angle from l to $m.*

This fact is illustrated in Fig. 9:9.

The geometric transformations $(\mathscr{L}\mathscr{M})$ and $(\mathscr{L}'\mathscr{M}')$ evidently have precisely the same effect on every point in the plane. This effect is briefly described by saying that the entire plane is turned or *rotated* around C through an angle 2θ. We agree to take for granted the additional phrase "in a counterclockwise direction" when θ is positive and the phrase "in a clockwise direction" when θ is negative.

If we write out a description of, say, P_{16} in terms of $\mathscr{L}$ and $\mathscr{M}$, we have $P_{16} = P\mathscr{L}\mathscr{M}\mathscr{L}\mathscr{M}\mathscr{L}\mathscr{M}\mathscr{L}\mathscr{M}\mathscr{L}\mathscr{M}\mathscr{L}\mathscr{M}\mathscr{L}\mathscr{M}$. However, we can let $\mathscr{R}$ stand for the rotation $\mathscr{L}\mathscr{M}$ and write $P_{16} = P\mathscr{R}\mathscr{R}\mathscr{R}\mathscr{R}\mathscr{R}\mathscr{R}\mathscr{R}\mathscr{R}$. Even now we may still damage our eyesight trying to keep track of how many $\mathscr{R}$'s there are in our description of the point. It is natural to count them once and for all and to write $P_{16} = P\mathscr{R}^8$, the "exponent" 8 merely telling us to perform the rotation eight times. A point such as P_7 may now be written

$$P_7 = P\mathscr{L}\mathscr{M}\mathscr{L}\mathscr{M}\mathscr{L}\mathscr{M}\mathscr{L} = P\mathscr{R}^3\mathscr{L}.$$

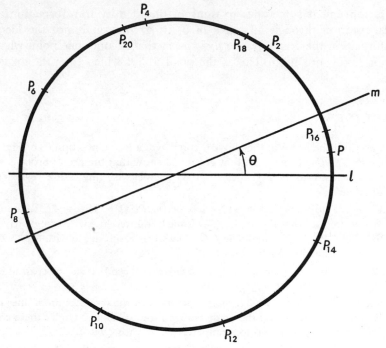

Figure 9:8

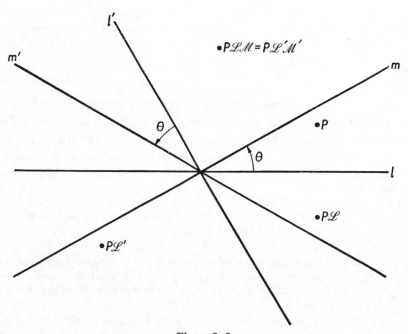

Figure 9:9

It is convenient sometimes to think of the identity transformation as a special rotation through an angle of $0°$. If a rotation is not the identity transformation there is only one *fixed point*, that is, only one point which a rotation takes into itself. This is the point about which the rotation takes place.

EXERCISES

1. Carry out the proof of the above theorem when P is in position P_2 of Fig. 9:7. Compare your proof with that given and show that the above proof is valid for all positions of P (other than C) if clockwise angles are regarded as negative.

2. Using the notation of Fig. 9:7, show that $\angle PCQ = 360° - 2\theta$, where $Q = P\mathscr{M}\mathscr{L}$. (Recall that the angle PCQ is the "counterclockwise" angle with initial side CP and terminal side CQ.) Does this result agree with the theorem proved in Section 4? Explain.

3. Why does the theorem show that the angle between two successive points P_2, P_4, P_6, . . . is the same?

4. Suppose $P\mathscr{L}\mathscr{M} = P\mathscr{M}\mathscr{L}$ for some point P not the intersection of lines l and m. What follows about the angle between the two lines? Under these conditions if Q is another point, compare $Q\mathscr{L}\mathscr{M}$ with $Q\mathscr{M}\mathscr{L}$.

5. Answer the questions of Exercise 4 under the assumption that $P\mathscr{L}\mathscr{M}\mathscr{L} = P\mathscr{M}\mathscr{L}\mathscr{M}$.

6. Show that if $\mathscr{R}^{-1}$ is an abbreviation for $\mathscr{M}\mathscr{L}$, then $\mathscr{R}^{-1}$ is a *clockwise* rotation through an angle 2θ and that $\mathscr{R}^{-1}\mathscr{R} = \mathscr{R}\mathscr{R}^{-1} = \mathscr{I}$, the identity transformation. Letting $\mathscr{R}^{-b}$ stand for the repetition b times of $\mathscr{R}^{-1}$ (b an integer), and letting $\mathscr{R}^0 = \mathscr{I}$, show that
$$\mathscr{R}^a \cdot \mathscr{R}^{-b} = \mathscr{R}^{a-b} = \mathscr{R}^{-b} \cdot \mathscr{R}^a,$$
$$\mathscr{R}^a\mathscr{L} = \mathscr{L}\mathscr{R}^{-a}, \mathscr{M}\mathscr{R}^a = \mathscr{R}^{-a}\mathscr{M}.$$
$$\mathscr{R}^a\mathscr{M} = \mathscr{R}^{a-1}\mathscr{L}, \text{ and } \mathscr{L}\mathscr{R}^a = \mathscr{R}^{-a+1}\mathscr{M}.$$

Are these formulas still true if a is zero or a negative integer?

7. If $\mathscr{R}$ is a rotation, under what conditions will a power of $\mathscr{R}$ be the identity?

8. With the agreements as to notation made in Exercise 6, show that any point obtainable from P by a succession of reflections in l and m must occur in one of the lists
$$P, P\mathscr{R}, P\mathscr{R}^2, P\mathscr{R}^3, \ldots,$$
$$P\mathscr{L}, P\mathscr{R}\mathscr{L}, P\mathscr{R}^2\mathscr{L}, P\mathscr{R}^3\mathscr{L}, \ldots,$$
$$P, P\mathscr{R}^{-1}, P\mathscr{R}^{-2}, P\mathscr{R}^{-3}, \ldots,$$
$$P\mathscr{M}, P\mathscr{R}^{-1}\mathscr{M}, P\mathscr{R}^{-2}\mathscr{M}, P\mathscr{R}^{-3}\mathscr{M}, \ldots.$$

9. Show that if L is a line, and if all points on L are rotated through an angle of 2θ about a fixed point P, not necessarily L, they will then be on a line L' making an angle of 2θ with L. (*Hint:* Use the theorem of Section 3, taking line l parallel to L.)

10. Show that if, in Exercise 9, θ is greater than $0°$ and less than $90°$, then there is only one point Q on line L which is taken by the rotation into another

point on L. Call Q' this latter point and show (1) that the line segment QQ' is bisected by the foot of the perpendicular from P upon L, (2) that angle $QPQ' = 2\theta$, and (3) that Q' is the point of intersection of L and L'.

11. If, in Exercise 9, $\theta = 90°$, can there be a point Q as described in Exercise 10? If so, under what conditions? Whether or not Q exists, how will P be located with reference to lines L and L'?

12. Do all the rotations about a given point form a group? (See Chapter 4, Section 1.) Do the reflections in a given line form a group?

13. If l, m, l', and m' are four lines through a point, show that

$$\mathscr{L}\mathscr{M}\mathscr{L}'\mathscr{M}' = \mathscr{L}'\mathscr{M}'\mathscr{L}\mathscr{M}.$$

5. COMPLETION OR EXHAUSTION?

It is now rather natural to ask ourselves if we can construct the complete set of points obtainable from P by successive reflections in l and m, or if the number of points we can construct is limited only by the point at which we become too exhausted to continue the matter further.

Evidently we can obtain an arbitrarily large number of points P, $P\mathscr{R}$, $P\mathscr{R}^2$, $P\mathscr{R}^3$, ... unless some of these points coincide. For instance, $P\mathscr{R}^2$ might coincide with $P\mathscr{R}^{10}$. Then $P\mathscr{R}^3$ would coincide with $P\mathscr{R}^{11}$, $P\mathscr{R}^4$ with $P\mathscr{R}^{12}$, and so on. But in order for $P\mathscr{R}^2$ to coincide with $P\mathscr{R}^{10}$, it would be necessary that $P\mathscr{R}$ should coincide with $P\mathscr{R}^9$ and that P should coincide with $P\mathscr{R}^8$. Evidently if $P\mathscr{R}^c$ coincides with $P\mathscr{R}^d$, where d is greater than c, P must coincide with $P\mathscr{R}^{d-c}$. Hence, if the number of points obtainable from P by successive rotations is to be finite, we must have $P = P\mathscr{R}^b$ for some integer b. But this is possible only if the angle $b \cdot 2\theta$ through which P has been rotated to obtain $P\mathscr{R}^b$ is an integral multiple of $360°$. That is, $b \cdot 2\theta = a \cdot 360°$ or $\theta = \dfrac{a}{b} \cdot 180°$, where a and b are integers.

Since we may draw the lines l and m so that θ will be any size we please,[1] we may choose to take $\theta = (1/\sqrt{2}) \cdot 180°$. Then the sequence of points P, $P\mathscr{R}$, $P\mathscr{R}^2$, ... will all be different, for otherwise $180°(1/\sqrt{2}) = 180°(a/b)$ or $\sqrt{2} = b/a$, and we saw in Chapter 5 that it is impossible to express $\sqrt{2}$ as the quotient of two integers.

EXERCISES

1. How many rotations through an angle of $30°$ are necessary in order that every point shall return to its initial position? How many if the angle is $72°$? $73°$? $7\frac{1}{2}°$? $37\frac{1}{2}°$?

[1] That this is so depends upon the axiom of completeness of Euclidean geometry. See page 25 of reference **29**.

2. Show that if $\theta = 180°(a/b)$, where a and b are integers, then not more than $2b$ distinct points will be obtained by reflecting P in l and m. Will fewer points be obtained for some positions of P?

3. Show that if P is chosen so that the angle from CP to l equals θ, where 2θ is the angle of rotation of $\mathscr{R}$ and C is the fixed point of the rotation, then $P\mathscr{R}\mathscr{L} = P$. For what positions of P will $P\mathscr{R}^b\mathscr{L} = P$?

4. If exactly $2k$ distinct points can be obtained by reflecting a point P in lines l and m, what can you say about the size of the angle between l and m?

6. PARALLEL MIRRORS

Let us now determine the effect of successive reflections in parallel lines. In Fig. 9:10 we show the effect of reflecting some points in the parallel lines l

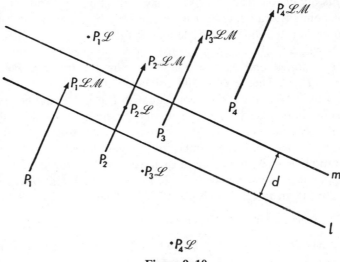

Figure 9:10

and m. The arrows in the figure are drawn to call attention to the fact that the transformation $\mathscr{L}\mathscr{M}$ moves every point a distance $2d$ in a direction perpendicular to l and m. We leave it as an exercise for the reader to prove that this is true for all positions of P.

From the preceding statement it evidently follows that the transformation which carries P into $P\mathscr{L}\mathscr{M}$ may equally well be determined by lines l' and m' parallel to l and m and equally far apart.[1] Thus in Fig. 9:11, $P\mathscr{L}$ and $P\mathscr{L}'$ are different, but $P\mathscr{L}\mathscr{M} = P\mathscr{L}'\mathscr{M}'$.

---

[1] We must also require that m' be on the same side of l' as m is of l. This intuitively evident requirement may be accomplished by demanding that the distance between l and l' shall equal the distance between m and m'.

The transformation $\mathscr{L}\mathscr{M}$ illustrated in Fig. 9:10 is determined by any one of the parallel and equal arrows of that figure. We shall call such a transformation a *translation* and denote it by $\mathscr{T}$. When $d = 0$, the translation is the identity transformation.

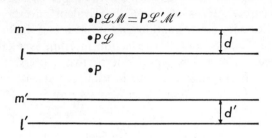

Figure 9:11

EXERCISES

1. Prove that if P is reflected in line l and then in line m, parallel to l, P goes into a point whose distance from P is $2d$, where d is the distance from l to m. What can be said about the direction of the translation?
2. Show that if P and P' are any two points, there is just one translation for which the image of P is P'.
3. Describe what you mean by a "translation."
4. Given the two translations $\mathscr{T}_1$ and $\mathscr{T}_2$, let $\mathscr{T}_1 \cdot \mathscr{T}_2$ (or $\mathscr{T}_1 + \mathscr{T}_2$) denote the result of performing first $\mathscr{T}_1$ and then $\mathscr{T}_2$. Show that $\mathscr{T}_1 \cdot \mathscr{T}_2$ is a translation. Are $\mathscr{T}_1 \cdot \mathscr{T}_2$ and $\mathscr{T}_2 \cdot \mathscr{T}_1$ the same translations?
5. Show that if a translation leaves one point fixed, then it is the identity transformation.
6. How can one obtain from $\mathscr{T}$ the translation $\mathscr{T}^{-1}$ having the property that $\mathscr{T}\mathscr{T}^{-1} = \mathscr{I}$? What is $\mathscr{T}^{-1}$ in terms of $\mathscr{L}$ and $\mathscr{M}$ when $\mathscr{T} = \mathscr{L}\mathscr{M}$ and $\mathscr{L}$ and $\mathscr{M}$ are reflections in parallel lines l and m? Is $\mathscr{T}^{-1}\mathscr{T} = \mathscr{I}$?
7. Show that the set of all translations including the identity transformation, form a group where the operation is $\cdot$ described in the previous exercises. Is it an Abelian group?
8. Is there a translation $\mathscr{T}$ such that $\mathscr{T}^2 = \mathscr{I}$?

7. EUCLIDEAN TRANSFORMATIONS

Suppose we have a transformation of points into points with the property that if P' is the image of P and Q' the image of Q, then the distance PQ is equal to the distance $P'Q'$. If this property holds for all points in the plane, we call the transformation a "rigid motion" or "Euclidean transformation."[1] The latter name is used because Euclidean geometry is a study of properties which remain unchanged by a Euclidean transformation. We call two figures "congruent" if one can be obtained from the other by a Euclidean transformation. We call two transformations "equivalent" if every point has the same image under one transformation that it has under the other.

From Exercise 11 of Section 2 a reflection in a line is a Euclidean transformation. Since a rotation is equivalent to two successive reflections, it follows that a rotation is also a Euclidean transformation.

There is a remarkable theorem which we now prove.

Theorem

Every Euclidean transformation in the plane is equivalent to a succession of three or fewer reflections in lines.

Proof. We divide the proof into several parts.

Part I. First let A, B, C, be three noncollinear points and A', B', C' their respective images and suppose that $AB = A'B'$, $AC = A'C'$, $BC = B'C'$, where AB denotes the distance between A and B, and so on. Then A', B', C' are noncollinear. For suppose they were collinear, then one of the three following distances would be equal to the sum of the other two:

$$A'B', A'C', B'C'.$$

But this would imply that one of the distances AB, AC, BC would be the sum of the other two which denies the noncollinearity of A, B, C. That is, we have shown that if A', B', C' are collinear, then A, B, C are. The contrapositive of this statement is what we set out to prove.

Part II. Suppose A, B, C and A' B' C' are points satisfying the conditions of Part I. Then we prove that there is a sequence of at most three reflections which takes the set of points A, B, C into the set A', B', C'.

First, we want to take A into A'. If $A = A'$ we go on to the next step. If not, we can take A into A' by reflecting the points A, B, C in the perpendicular bisector of $\overline{AA'}$. Call B_1 and C_1 the respective images of B and C under this reflection.

Second, we want to take B_1 into B'. If $B_1 = B'$, we go on to the next step. If not, we can take B_1 into B' by reflecting the points A', B_1, C_1 in the perpendicular bisector of $\overline{B_1B'}$. Let C_2 be the image of C_1 under this reflection. What happens to A'? Now $A'B_1 = AB$ because A' and B_1 are the images of A and B, respectively, under the first reflection and a reflection is a Euclidean

---

[1] Another name for such a transformation is an "isometry."

transformation. Also $A'B' = AB$ by the hypothesis of this part of the proof. Hence

$$A'B_1 = A'B'.$$

This shows that A' is on the perpendicular bisector of $\overline{B'B_1}$ and hence is unchanged by reflection in this line. Hence by our two reflections we have gone from A, B, C into A', B', C_2.

Third, if $C_2 = C'$, we are through this part of the proof. If this is not the case, we can take C_2 into C' by reflecting A', B', C_2 in the perpendicular bisector of $\overline{C_2C'}$. What does this do to A' and B'? As in the previous part, $A'C' = A'C_2$ and $B'C' = B'C_2$, which shows that both A' and B' are on the perpendicular bisector of $\overline{C'C_2}$ and hence are unchanged by the third reflection. This completes the proof of this part.

Since a sequence of reflections is a Euclidean transformation, we have in the process shown that for any two sets of three points satisfying the requirements of Part I, there is a Euclidean transformation taking the first set into the second. We wish to show that *every* Euclidean transformation taking one such set of three points into another is equivalent to a sequence of three or fewer reflections.

Part III. Suppose A, B, C and A', B', C' are two sets of three points satisfying the conditions of Part I of this proof. Let $\mathscr{E}$ be a sequence of three or fewer reflections taking A, B, C into A', B', C', respectively. Suppose $\mathscr{E}'$ is another Euclidean transformation which takes A, B, C into A', B', C'. (It might or might not be a sequence of three or fewer reflections.) We show in this final part of the proof that $\mathscr{E}$ and $\mathscr{E}'$ must be equivalent. Suppose they were not. Then there would be some point R which $\mathscr{E}$ takes into R' and $\mathscr{E}'$ takes into R'', and where R' and R'' are different points. Since both $\mathscr{E}$ and $\mathscr{E}'$ take A, B, C into A', B', C', respectively, the following equations are true:

$$AR = A'R' = A'R'', \qquad BR = B'R' = B'R'', \qquad CR = C'R' = C'R''.$$

This shows that A', B', C' all lie on the perpendicular bisector of the segment $\overline{R'R''}$. This is impossible, since they are not on the same line. This completes the proof of the theorem.

Notice that a given Euclidean transformation is equivalent to many different sequences of reflections. For instance, we might have begun our proof by reflecting in the perpendicular bisector of $\overline{BB'}$ or of any two corresponding points.

EXERCISES

1. Show that if A and B are taken into points A' and B' by a Euclidean transformation, then every point on the line segment $\overline{AB}$ is taken into a point on the line segment $\overline{A'B'}$.

2. Suppose that a Euclidean transformation has two "fixed points," that is, has two points which are their own images. Prove that either this transformation takes every point into itself (that is, is the identity transformation), or it is a reflection in the line through the two fixed points.

3. Suppose that a Euclidean transformation has exactly one fixed point. Prove that the transformation is a rotation through the point.

4. Suppose that a Euclidean transformation has no fixed points. Must it be a translation?

5. Suppose that a Euclidean transformation has exactly one fixed line. What must the transformation be? (By "fixed line" we do not mean that every point on the line is fixed, but that the image of every point on the line is on the line.)

6. Suppose that a Euclidean transformation has no fixed lines. What can be said about the transformation?

7. Show that the Euclidean transformations and the operation · form a group, where, for any two transformations $\mathscr{E}_1$ and $\mathscr{E}_2$, $\mathscr{E}_1 \cdot \mathscr{E}_2$ means "apply $\mathscr{E}_1$ and then $\mathscr{E}_2$."

*8. In Exercise 3 you were asked to show that if a Euclidean transformation has a fixed point and only one, it must be a rotation. Since every rotation except the identity has exactly one fixed point, we say that having a single fixed point "characterizes" a rotation. What fixed points and lines characterize a translation? A reflection?

8. ORIENTATION—AN INTUITIVE APPROACH

In this section we consider intuitively a concept which is hard to pin down rigorously. (It is dealt with more carefully in Section 9.) Suppose A, B, C is a set of three noncollinear points. They will be the vertices of a triangle. Suppose we draw a circle through these three points and move our pencil around the circle from A, to B, and C, and back to A. If the pencil moves in a clockwise direction, we say that the triangle has *clockwise orientation*; if the pencil moves in a counterclockwise direction we say the triangle has *counterclock-*

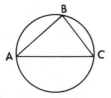

Figure 9:12

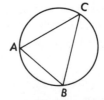

Figure 9:13

wise orientation. In Fig. 9:12 the triangle has clockwise orientation; in Fig. 9:13 it has counterclockwise orientation.

It would be rather hard to prove the following property of orientation and reflection but if you draw a few figures it will at least seem plausible. We shall not attempt to prove it.

Property of reflection and orientation: If one set of three noncollinear points is the mirror image of another set in line *l*, then the orientation of the two sets is different; that is, one set has clockwise orientation and the other set has counterclockwise orientation.

A very important consequence of the property is the following: *An even number of reflections does not change the orientation of a set of three noncollinear points.* This is true because the effect of changing the orientation twice is to leave it unchanged. A consequence of this is that translations and rotations do not change the orientation.

The previous paragraph is connected with another intuitive idea. In testing the congruence of two triangles one often imagines them to be triangular frames which can be moved around. If one can be moved to "coincide" with the other (whatever that means), then the "triangles" are congruent. If the congruent triangles have the same orientation, this may be accomplished by merely sliding one triangle (that is, translating and rotating). If they have opposite orientation, an additional reflection in a line is necessary, that is, one has to move the triangle out of the plane. The reader should be cautioned that in this paragraph we are not talking about triangles in the mathematical sense but the physical models which we sometimes find it convenient to use when we think about them.

Using the ideas of this section and the theorem of Section 7, one may deduce some remarkable properties. Suppose *A*, *B*, *C* are three noncollinear points and we reflect them first in one line, then in another, and so on, until we have reflected them successively in 29 lines. Since each of these reflections is a Euclidean transformation, the sequence is a Euclidean transformation, which we might call $\mathscr{E}_{29}$. By the theorem of the Section 7 we know that $\mathscr{E}_{29}$ is equivalent to some sequence of three or fewer reflections. Since $\mathscr{E}_{29}$ changes the orientation, it must be equivalent to a sequence of one or three reflections.

We defined a transformation to be Euclidean if it leaves distance unaltered. You may have wondered why we did not consider angles also. The point is that if distances are left unchanged so are the measures of angles, because if the corresponding sides of two triangles are congruent, so are the corresponding angles.

EXERCISES

In these exercises, assume the results stated above in this section.
1. Prove that every Euclidean transformation which preserves the orientation of some set of three noncollinear points is either a translation or a rotation and hence preserves the orientation of every set of three noncollinear points. Show how the number of fixed points determines whether such a transformation is a translation or a rotation.

2. Prove that a transformation consisting of first a translation and then a rotation not the identity is equivalent to a single rotation. (*Note:* The fixed point of the second rotation may be different from that of the first.) Can the same conclusion be drawn if the rotation comes first and the translation second?

3. Prove that the set of all Euclidean transformations which do not change orientation forms a group.

4. Does the set of Euclidean transformations which change orientation form a group?

5. Is it possible for a sequence of two rotations to be equivalent to a single translation?

6. Is it possible for a sequence of two translations to be equivalent to a single rotation?

7. Suppose a Euclidean transformation is equivalent to a sequence of three reflections in lines but not to a single reflection. What can be said about orientation and fixed points?

8. Does the set of rotations not necessarily about the same point form a group?

9. Does the set of translations form a group? Does the set of translations and rotations form a group?

10. Let $\mathcal{R}_1$ and $\mathcal{R}_2$ be two rotations, neither the identity. Let C be the fixed point of $\mathcal{R}_1$. Show that if C is also the fixed point of $\mathcal{R}_2$, then $\mathcal{R}_1 \cdot \mathcal{R}_2$ is equivalent to $\mathcal{R}_2 \cdot \mathcal{R}_1$.

11. Let $\mathcal{E}_1$ and $\mathcal{E}_2$ be two Euclidean transformations with the property that $\mathcal{E}_1 \cdot \mathcal{E}_2$ is equivalent to $\mathcal{E}_2 \cdot \mathcal{E}_1$. If P is a fixed point of $\mathcal{E}_1$, show that $P\mathcal{E}_2$ is also.

12. Let $\mathcal{R}_1$ and $\mathcal{R}_2$ be two rotations, neither the identity, with the property that $\mathcal{R}_1 \cdot \mathcal{R}_2$ is equivalent to $\mathcal{R}_2 \cdot \mathcal{R}_1$. Show that $\mathcal{R}_1$ and $\mathcal{R}_2$ have the same fixed point. (*Hint:* The result of Exercise 11 is helpful here.)

*9. ORIENTATION—A LESS INTUITIVE APPROACH

In this section, for those who are interested, we shall indicate how the idea of orientation can be given a less intuitive foundation. We wish to make the following definition.

Definition

Let $\mathcal{E}$ be a Euclidean transformation. If it is equivalent to a succession of one or three reflections in lines, we say it is *orientation-changing*. If it is the identity or equivalent to a succession of two reflections in lines, we say it is *orientation-preserving*.

You may remark that a definition need not be justified. We defined new numbers as we pleased and why do we not have comparable liberty here? Of course we do have such liberty, but we want to have a definition that is sensible. If a given transformation could be both orientation-changing and orientation-preserving, our definition would be both trivial and misleading. Hence we need to prove the following theorem.

Theorem 1

If a Euclidean transformation is equivalent to a succession of one or three reflections in a line, it is neither the identity transformation nor equivalent to a succession of two reflections in lines.

The proof of this theorem will be much easier if we first prove the following auxiliary result, which we call a "lemma."

Lemma

If $\mathcal{K}$, $\mathcal{L}$, $\mathcal{M}$, and $\mathcal{N}$ are four reflections in lines k, l, m, *and* n, *respectively, then $\mathcal{K}\mathcal{L}\mathcal{M}\mathcal{N}$ is equivalent to a succession of two reflections.*

Proof. To prove this lemma we consider three cases:

First, if $m \| n$, then either $k \| l$ or $k \|\!\| l$. In the former case, the given transformation is equivalent to a succession of two translations, $\mathcal{K}\mathcal{L}$ and $\mathcal{M}\mathcal{N}$; by Exercise 3 of Section 6, this is equivalent to a single translation, that is, a succession of two reflections. In the latter case, by Section 4, we can replace $\mathcal{K}\mathcal{L}$ by $\mathcal{K}'\mathcal{L}'$, where l' is any line through the intersection of k and l and k' is another line properly chosen through the intersection of k and l. Then if we choose l' parallel to m and n, we see that $\mathcal{L}'\mathcal{M}\mathcal{N}$ is a succession of reflections in three parallel lines. By Section 6, $\mathcal{M}\mathcal{N} \cong \mathcal{M}'\mathcal{N}'$, where $\mathcal{M}'$ is a reflection in a line m' which may be chosen as we please parallel to m, provided n' is chosen properly. We may choose m' to be the line l'. Then $\mathcal{L}'\mathcal{M}'\mathcal{N}' \cong \mathcal{L}'\mathcal{L}'\mathcal{N}' \cong \mathcal{N}'$ and our given transformation is equivalent to $\mathcal{K}'\mathcal{N}'$.

Second, if $m \|\!\| n$ and $l \|\!\| m$, we can, as in the previous paragraph, replace $\mathcal{L}\mathcal{M}$ by $\mathcal{L}'\mathcal{M}'$, where the line m' is chosen parallel to n. This case is then reduced to the previous one.

Third, if $m \|\!\| n$ and $l \| m$, then either $k \| l$ or $k \|\!\| l$. In the former case, $\mathcal{K}\mathcal{L}\mathcal{M}$ is a succession of three reflections in parallel lines, which as we have shown above, is equivalent to one reflection. In the latter case, we can replace $\mathcal{K}\mathcal{L}$ by $\mathcal{K}'\mathcal{L}'$, where $l' \|\!\| m$; this reduces to the second case.

We have considered all possibilities and proved the lemma. Now, to prove the theorem we have to find some way of distinguishing between the two categories of transformations. We have already seen that this can be done to a certain extent by examining the fixed points, since if two transformations are equivalent, they must have the same fixed points. Let us summarize what we know so far about fixed points:

Transformation	Fixed Points
Identity	All points
Reflection in a line l	Each point of l
Translation not the identity	None
Rotation not the identity	Exactly one

Examination of the table shows that a reflection cannot be equivalent to any one of the other three transformations, since of the list only a reflection has a line of fixed points. To complete the proof of the theorem, we need to show that a succession of three reflections cannot be equivalent to any transformation in the list except perhaps a single reflection. We use an indirect proof. That is, suppose

$$\mathscr{L}\mathscr{M}\mathscr{N} \cong \mathscr{R}\mathscr{S},$$

where each of the letters designates a reflection in a line. The equivalence means that $P\mathscr{L}\mathscr{M}\mathscr{N} = P\mathscr{R}\mathscr{S}$ for every point P. Applying the transformation $\mathscr{N} \cdot \mathscr{M}$ to both sides of this equality and noting that $\mathscr{N}^2$ and $\mathscr{M}^2$ are each equivalent to the identity, we have

$$P\mathscr{L} = P\mathscr{R}\mathscr{S}\mathscr{N}\mathscr{M}.$$

But our lemma shows us that $\mathscr{R}\mathscr{S}\mathscr{N}\mathscr{M}$, being a succession of four reflections, is equivalent to a succession of two reflections, which we might call $\mathscr{U}$ and $\mathscr{V}$. Thus

$$P\mathscr{L} = P\mathscr{U}\mathscr{V},$$

for all points P. But we have already shown that $\mathscr{L}$ cannot be equivalent to $\mathscr{U}\mathscr{V}$, by the table of transformations. This is the contradiction we desired and our theorem is proved.

The theorem affirms that every Euclidean transformation is equivalent to exactly one of the following two types:

(1) The identity, a translation or a rotation (that is, a succession of none or two reflections).
(2) A reflection or a succession of three reflections.

Theorem 1 still leaves something to be desired, because to determine whether a Euclidean transformation is orientation-preserving or orientation-changing, we must first express it as a product of three or fewer reflections. It would be much simpler if we could tell which it is by simply counting the number of reflections. That this may be done is shown by the following theorem, which we shall prove.

Theorem 2

A succession of an odd number of reflections is an orientation-changing transformation and a succession of an even number of reflections is orientation-preserving. (This could be used as an alternative definition of the two terms.)

Proof. We have already shown this result if the number of reflections is three or less. Suppose $\mathscr{E}$ is a succession of n reflections. Then, if $n \geq 4$, we can by the lemma replace the last four reflections by two reflections and see that $\mathscr{E}$ is equivalent to a succession of $n - 2$ reflections. If $n - 2 \geq 4$, we can repeat the process and see that $\mathscr{E}$ is equivalent to a succession of $n - 4$ reflections. So we can continue until, for some positive integer k, $\mathscr{E}$ is equivalent to $n - 2k$ reflections, where $n - 2k$ is less than 4. If $n - 2k$ is 2 or 0, that is, if n is even, $\mathscr{E}$ is equivalent to the identity or a succession of two reflections and hence is orientation-preserving by our definition of these terms. If $n - 2k$ is 3 or 1, that is, if n is odd, $\mathscr{E}$ is equivalent to a reflection or a succession of three reflections and hence, by our definition, is orientation-changing. This proves the theorem.

Perhaps the most important consequence of Thereom 2 is the following corollary, which gives further justification for our terminology.

Corollary

If $\mathscr{E}$ and $\mathscr{E}'$ are Euclidean transformations, then

(1) *$\mathscr{E}\mathscr{E}'$ is orientation-preserving if $\mathscr{E}$ and $\mathscr{E}'$ are both orientation-preserving or both orientation-changing.*

(2) *$\mathscr{E}\mathscr{E}'$ is orientation-changing if one of $\mathscr{E}'$, $\mathscr{E}$ is orientation-changing and the other is orientation-preserving.*

We leave justification of this corollary as an exercise.

In closing, we should remark that perhaps the most important assumption we have made in this chapter has to do with the direction of a rotation. We have assumed that there is such a thing and that we know what it means. It would be difficult to pin down completely the meaning of this concept, but this can also be said about a number of other things in this chapter and throughout the book.

EXERCISES

1. Show why the corollary above follows from the previous theorem.
2. Does the set of rotations form a group?
3. In this section the Euclidean transformations are defined with reference to fixed points. For each of the types of transformations, how many fixed lines are there and how are they related? (See Exercise 5 of Section 7.)
4. Suppose $\mathscr{E}$ is equivalent to a succession of three reflections but is not equivalent to a reflection. Does it have any fixed points?
*5. If $\mathscr{T}$ is a translation and $\mathscr{L}$ a reflection in a line, show that there is a translation $\mathscr{T}'$ and a reflection $\mathscr{L}'$ such that $\mathscr{T}\mathscr{L} \cong \mathscr{L}'\mathscr{T}'$.

10. TOPICS FOR FURTHER STUDY

There are brief treatments of topics in this chapter and allied topics in geometry in references: **3**, **15**, and **48** (Vol. 1, Chap. 4, pp. 671–724).

A longer treatment of geometry including reflection and rotation is given in references **15** and **43**. See also the references in Chapter 10.

10

Lorentz Geometry

"I don't know what you mean by 'glory,'" Alice said.

Humpty Dumpty smiled contemptuously. "Of course you don't—till I tell you. I meant 'there's a nice knockdown argument for you.'"

"But 'glory' doesn't mean 'a nice knockdown argument,'" Alice objected.

"When *I* use a word," Humpty Dumpty said in rather a scornful tone, "it means just what I choose it to mean—neither more nor less."

"The question is," said Alice, "whether you *can* make words mean so many different things."

"The question is," said Humpty Dumpty, "which is to be master—that's all!"

1. INTRODUCTION

The usual approach to Euclidean geometry is to write down at the very beginning certain statements called *axioms*. These axioms are in many ways analogous to the rules of such a game as checkers or chess. Their great age and the general familiarity of mankind with them has won these axioms a veneration which is not their due. Indeed, they are not even so very old—they are younger than the Pyramids—and every teacher of geometry will agree that not all persons are familiar with them.

Even a checker game could hardly proceed smoothly if at every play one of the players insisted on a change in the rules of the game. In just the same way the axioms of Euclidean geometry, *once accepted at the start of the theory*, cannot be contradicted at any later stage in its development. If, however, the players were to grow tired of their game of checkers and begin to play chess, we should not be surprised to see them making moves quite impossible before. In just the same way mathematicians sometimes put away their perpendicular lines, their circles, and their equilateral triangles and play the game of a new and different kind of geometry. They even draw diagrams on a sheet of paper to describe their new game, just as one plays chess on a checkerboard.

The historically important approach to a geometry different from that of Euclid is to modify one or more of the Euclidean axioms and study the consequences of the new rules. This was done in the early nineteenth century by three men, Gauss, Bolyai, and Lobachevsky, working independently. The general recognition that there were other geometric "games" to be played had far-reaching effects in all domains of human thought.

We shall not hope to acquaint the student with this sort of "non-Euclidean" geometry in these pages. (See Section 8.) Instead, we shall attempt to describe very briefly the foundations of a geometry which is different from both Euclid's and that frequently discussed under the title non-Euclidean geometry. In some respects our theory is easier to formulate than either of the historically more important types. It may also lay claim to being of greater interest in contemporary physics. (See Section 6.)

2. LOROTATIONS

In Chapter 9 we considered various transformations: reflections in lines, rotations about points, and, in general, Euclidean transformations. We called two triangles congruent if one could be obtained from the other by a Euclidean transformation. There it would have been fairly difficult to give a description of the transformation by means of equations connecting the coordinates of the original point and its image under the transformation. In the geometry of this chapter, the use of coordinates is much the simplest way of describing the transformations which we shall deal with.

We will play our new game on the same board which we used in Chapter 7. By this we mean that a pair of real numbers (x, y) will determine a point, a straight line will be the set of points determined by solutions of the linear equation $Ax + By + C = 0$ (with not both A and B zero), and a pair of linear equations will determine parallel lines (definition!) if and only if the equations have no common solution.

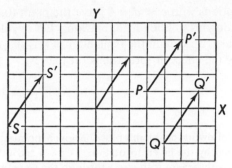

Figure 10:1

But instead of the transformations of Chapter 9 we consider the transformations which take the point (x, y) into the point (x', y') where

$$x' = kx + c,$$
$$y' = y/k + d,$$

with c and d real numbers and k any positive real number.

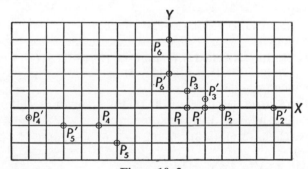

Figure 10:2

For instance, suppose $c = 2$, $d = 3$, and $k = 1$. Then, in Fig. 10:1, P, Q, and S have as their images P', Q', and S', respectively. The arrows suggest that the transformation is our old friend a translation. This is indeed the case whenever $k = 1$. The proof is left as an exercise.

If, on the other hand, we take k different from 1, say $k = 2$, and set $c = d = 0$, we obtain the transformation

$$x' = 2x, \quad y' = \tfrac{1}{2}y.$$

Corresponding points under this transformation are marked in Fig. 10:2, that is, P_1' is the image of P_1, P_2' is the image of P_2, and so on.

P	(1, 0)	(3, 0)	(1, 1)	(−4, −1)	(−3, −2)	(0, 4)
P'	(2, 0)	(6, 0)	(2, $\frac{1}{2}$)	(−8, $-\frac{1}{2}$)	(−6, −1)	(0, 2)

The transformations with $c = d = 0$ leave the point (0, 0) fixed and are in some respects analogous to Euclidean rotations. Let us call them *lorotations*.[1] For simplicity's sake, we shall in the remainder of this chapter take $c = d = 0$ and consider only the effect of lorotations on figures.

EXERCISES

In each of the following problems plot the given points and the points after they are transformed by the given lorotation. Draw a smooth curve through the original points and another smooth curve through the transformed points. Where the curves appear to be straight lines, find their equations.

1. Points: (1, 1), (2, 1), (3, 1), (4, 1), (5, 1), (6, 1).
   Lorotation: $x' = \frac{1}{3}x$, $y' = 3y$.

2. Points: (2, 1), (4, 2), (6, 3), (8, 4), (−4, −2).
   Lorotation: $x' = \frac{1}{2}x$, $y' = 2y$.

3. Points: (2, 0), (4, 1), (6, 2), (8, 3), (−4, −3).
   Lorotation: $x' = \frac{1}{2}x$, $y' = 2y$.

4. Points: (1, 1), (3, −3), (−2, 7), (0, 3).
   Lorotation: $x' = 2x$, $y' = \frac{1}{2}y$.

5. Points: (5, 0), (5, 1), (5, 4), (5, −2).
   Lorotation: $x' = \frac{1}{5}x$, $y' = 5y$.

6. Points: (13, 0), (12, 5), (5, 12), (0, 13), (−5, 12), (−12, 5), (−13, 0), (−12, −5), (−5, −12), (0, −13), (5, −12), (12, −5).
   Lorotation: $x' = \frac{1}{2}x$, $y' = 2y$.

7. Points: (5, 0), (8, 1), (10, 5), (9, 8), (5, 10), (2, 9), (0, 5), (1, 2).
   Lorotation: $x' = \frac{1}{3}x$, $y' = 3y$.

8. Points: ($\frac{5}{2}$, 0), (2, 6), ($\frac{3}{2}$, 8), (0, 10), (−2, 6), (−$\frac{5}{2}$, 0), (−$\frac{3}{2}$, −2), (0, −10), (2, −6).
   Lorotation: $x' = 2x$, $y' = \frac{1}{2}y$.

9. Points: (6, 1), (3, 2), (2, 3), (1, 6).
   Lorotation: $x' = \frac{3}{2}x$, $y' = \frac{2}{3}y$.

10. Points: (12, 1), (6, 2), (4, 3), (3, 4), (2, 6), (1, 12).
    Lorotation: $x' = 2x$, $y' = \frac{1}{2}y$.

11. Points: (14, 7), (8, 8), (6, 9), (5, 10), (4, 12), (3, 8).
    Lorotation: $x' = \frac{1}{2}x$, $y' = 2y$.

12. Points: (1, 13), (2, 8), (3, 7), (4, 7), (6, 8), (12, 13).
    Lorotation: $x' = 2x$, $y' = \frac{1}{2}y$.

[1] Lorotation = lo (an expression of wonderment) + rotation! More seriously, the geometry we are studying is a very much simplified version of the sort of geometry developed by *Lorentz* and now of importance in the theory of relativity.

13. Into what point do the lorotations $x = kx'$, $y = (1/k)y'$ take the point $(2, 3)$ when $k = 2, 3, 4,$ and 5? Draw the graphs of these points.

14. The line connecting $P_1 : (2, 3)$ and $P_2 : (4, 6)$ goes through the origin. Can the same thing be said of the points P_1' and P_2', where P_1' and P_2' are the points obtained from P_1 and P_2 by the lorotation $x = kx'$, $y = (1/k)y'$?

15. Find the inverse of the lorotation of Exercise 14.

16. Show that the lorotations form a group.

17. Prove that any set of points on a line is transformed by a lorotation into a set of points on another (or the same) line.

18. If m and n are two lines and m' and n' their images under the transformation of Exercise 14, show that the image of the intersection of m and n is the intersection of m' and n'.

19. If A and B are two points and their images under a lorotation are A' and B', respectively, show that the line $A'B'$ consists of the images of the points of the line AB.

*20. Prove that under a lorotation, the image of the x-axis is itself, the image of the y-axis is itself, and that no other line is its own image unless $k = 1$ or $k = -1$.

*21. Show that a lorotation is a translation if and only if $k = 1$.

3. MODELS

The fact that Euclidean geometry was invented to describe our physical surroundings makes it very easy to construct models of our figures in this kind of geometry. Thus a long thin rod or a broad "straight" carbon mark on a piece of paper may be used to suggest the abstract concept "straight line." We can suggest the effect of a Euclidean transformation on a figure by drawing the figure on a sheet of paper and then "moving" the sheet around (perhaps turning it over). Of course, if we had happened to draw our figure on a sheet of rubber, we should have had to be very careful not to stretch the rubber in any way, for, if we did, our model of the geometry would not be useful.

It cannot be emphasized too strongly that the use of any model at all in geometry is not only unnecessary, it is positively dangerous *unless we keep clearly in mind at all times that the model is merely a convenient way of describing the abstract system of relationships which constitute the geometry.*

It is very difficult to construct a satisfactory model of our new geometry, chiefly because a lorotation is such a queer "motion." We notice, however, that the lorotation $x' = kx$, $y' = (1/k)y$, multiplies all x coordinates by the same factor. Since $0 \cdot k = 0$, a point on the y-axis is moved into another point on the y-axis. The points on a line $x = 3$ are all moved into points with x coordinate $3k$ and are, therefore, all on the line $x = 3k$. Similarly, the line $x = 7$ goes into the line $x = 7k$ and the line $x = -4$ goes into the line $x =$

$-4k$. Evidently the effect of our lorotation is to "stretch" the plane in the x-direction, keeping the y-axis fixed and moving all lines parallel to it in such a way that their distance from the y-axis is multiplied by k. If k is less than 1, this would be described as a contraction.

From similar considerations we see that the x-axis stays fixed under a lorotation and that lines parallel to it are moved so that their distances from it are multiplied by $1/k$.

A model of our geometry, which we do not seriously propose to use, would be a curtain on a curtain stretcher or a photograph enlarger. Of course, when we stretched the curtain in length we should have to allow it to contract in width—not a recommended procedure in drying curtains.

4. COMMON PROPERTIES

Let us see if we can find any theorems of Euclidean geometry which continue to hold in our new geometry. Looking through the first few pages of Hilbert's *Foundations of Geometry* (reference **29** in the Bibliography), we see the following theorems which could still be proved in our system. (The proofs are not too difficult to be worked out by a student who is well versed in algebra.)

Two distinct points A and B always completely determine a straight line.[1] We, in effect, proved this in Chapter 7 when we showed how to find the equation of a line through two given points.

Any two distinct points of a straight line completely determine that line; that is, if A and B determine a line l and A and C determine the same line l, where B and C are different, then B and C determine the line l.[1]

Two distinct straight lines have either one point or no point in common.[1]

Through a point not on a line l there can be drawn one and only one straight line which does not intersect the line l.[1] (We have agreed to call two such lines *parallel*.)

If two straight lines are parallel to a third, then they are parallel to each other.

5. UNCOMMON PROPERTIES

None of the theorems just stated made any reference to the distance between two points. This is because a lorotation does not preserve what we call in Euclidean geometry the *distance* between two points. For instance, the lorotation $x' = 4x$, $y' = y/4$, carries $(0, 0)$ into $(0, 0)$ and $(3, 4)$ into $(12, 1)$.

---

[1] These statements are taken as axioms in Euclidean geometry. Since we did not take them as basic assumptions, we are obliged to prove them as consequences of our assumptions.

In Fig. 10:3 we see that the *Euclidean distance* from (0, 0) to (3, 4) is 5, while the distance between the transformed points (0, 0) and (12, 1) is $\sqrt{145}$, which is not equal to 5. Notice, however, that the areas of the two triangles are equal. [This latter would not be true if the point (3, 4) were rotated through an angle of 30°, for instance.]

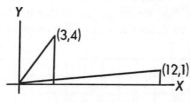

Figure 10:3

Having observed this fact we shall not be surprised if the points on a "circle" about the origin are moved by a lorotation to positions off the circle. Thus the points (0, 5), (3, 4), (4, 3), and (5, 0) are all five Euclidean units of distance from the origin, but the lorotation $x' = 2x$, $y' = \frac{1}{2}y$ moves them to the positions shown in Fig. 10:4. The points on the "circle" C will in fact all go into points on the curve C'.

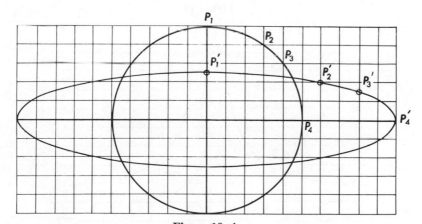

Figure 10:4

The fundamental property of a "circle" in Euclidean geometry is that we can get as many points of it as we please by rotating about the center through different angles any one point on it. Let us see if we can discover a curve having an analogous property under lorotations. We start with the point with coordinates (3, 2) and perform the following lorotations:

$$x' = \tfrac{1}{3}x, \; y' = 3y \quad \text{transforms (3, 2) into (1, 6)}$$
$$x' = \tfrac{1}{2}x, \; y' = 2y \quad \text{transforms (3, 2) into } (\tfrac{3}{2}, 4)$$

$$x' = 2x, \ y' = \tfrac{1}{2}y \quad \text{transforms (3, 2) into (6, 1)}$$
$$x' = 3x, \ y' = \tfrac{1}{3}y \quad \text{transforms (3, 2) into } (9, \tfrac{2}{3})$$

Drawing the graphs of all these points and a smooth curve connecting them, we get Fig. 10:5. The equation $xy = 6$ is evidently satisfied by all the points of the form $(k \cdot 3, 2/k)$, and hence every point obtained from $(3, 2)$ by a lorotation will lie on this curve. However, it is only that part of the curve $xy = 6$ in the first quadrant which we can obtain by "lorotating" $(3, 2)$. (We leave it to the student to draw the rest of the locus of $xy = 6$.)

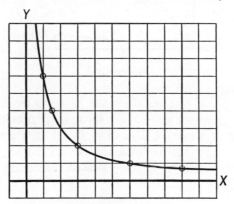

Figure 10:5

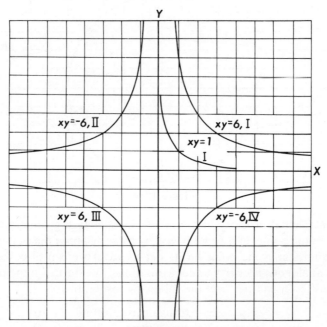

Figure 10:6

Let us call a curve such as that in Fig. 10:5 a *lorcle* (pronounced: "lorkle"). Formally, we define a lorcle to be the locus of all points obtainable from a single point by lorotations about a fixed point called the *center* in the lorcle. In Fig. 10:6 we show some lorcles with center at the origin. Any one of them is completely determined by a single point on it or by the common value of the product of the coordinates of any point on it together with the quadrant in which it lies (I or III and II or IV). A lorcle cannot be drawn in its entirety any more than can a straight line.

In Euclidean geometry all except one of the lines through a point on the circumference of a circle cut it in two points. The one exceptional line touches the circle in only one point and is called the *tangent* to the circle at the point. Is it possible to determine a line tangent to a lorcle in the same way? We shall do this for one numerical example, leaving it to any interested student to carry out the construction for any point on a lorcle with center at the origin.

Let the lorcle be the one shown in Fig. 10:5. It consists of all the points (x, y) which satisfy the equation $xy = 6$ and lie in the first quadrant. To find a tangent line at the point $(3, 2)$, notice that a line $y = ax + b$ will pass through the point $(3, 2)$ if $2 = 3a + b$, that is, if $b = 2 - 3a$. To determine the other point in which the line $y = ax + (2 - 3a)$ cuts the lorcle $xy = 6$, we must solve the equations simultaneously. Substituting in $xy = 6$ we get

$$x[ax + (2 - 3a)] = 6,$$

or

$$ax^2 + (2 - 3a)x - 6 = 0.$$

Factoring, we get

$$(ax + 2)(x - 3) = 0.$$

Hence $x = 3$ or $-2/a$. The line, therefore, intersects the curve $xy = 6$ in the points $(3, 2)$ and $(-2/a, -3a)$. These points are distinct unless $3 = -2/a$, or $a = \frac{2}{3}$. By analogy with Euclidean geometry we call this line (when $a = -\frac{2}{3}$) with equation

$$y = -2x/3 + 4,$$

the *tangent to the lorcle* through the point $(3, 2)$.

We have already seen that the Euclidean distance between two points does not remain the same under lorotations. This might make us suspect that our usual notions of angle need to be abandoned. That this is, indeed, the case we can see as follows.

Let us refer to any figure made up of two half-lines emanating from the same point as an "angle." We must carefully refrain from associating any numerical measure with such a figure for the measurement of an angle in degrees requires a whole sequence of moves in the game of Euclidean geometry, and we are by no means sure that we can make analogous moves in the game we are playing.

A particular angle consists in the "positive" half of the *x*-axis and the half-line from the origin through the point (1, 1). The lorotation $x' = 2x$, $y' = y/2$ or $1/2y$ carries the origin and the positive half of the *x*-axis into themselves, and the line *OB* into the line *OC* through $(2, \frac{1}{2})$ (see Fig. 10:7). Hence the angles *AOB* and *AOC* are congruent!

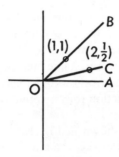

Figure 10:7

This astonishing result should convince the student that the measurement of angles in our new geometry differs radically from the measurement of them in Euclidean geometry. The problem, although not insoluble, is too difficult for us here.

In spite of these difficulties, we can define perpendicular lines in our new system (or should we call them, say, *lopendicular*?). We do it by saying that the tangent to a lorcle is "perpendicular" to the line connecting the center with the point of contact. (What is the analogous theorem in Euclidean geometry?) For instance, by referring to our above discussion, we see that the line $y = -2x/3 + 4$ is "perpendicular" to the line through (0, 0) and (3, 2). (What is the equation of the latter line?)

EXERCISES

1. If P:(4, 5) is carried by a lorotation into a point P', show that the area of the rectangle two of whose sides lie along the *x* and *y* axes and which has P as a vertex is the same as the area of the corresponding rectangle having P' as a vertex. Would a similar statement be true for any other point P?

2. If $P = (2, -3)$ is taken by a lorotation into the point $P' = (3, b)$, what must b be and what is the lorotation?

3. If P is a point not on either axis, Q is the foot of the perpendicular from P upon the *x*-axis, and P' and Q' are obtained from P and Q by a lorotation, show that Q' is the foot of the perpendicular from P' upon the *x*-axis and that the area of the triangle PQO is the same as the area of triangle $P'Q'O'$, O being the origin.

4. Show that the line $y = mx - 2m + 4$ passes through $(2, 4)$ for any value of the constant m. Find, in terms of m, the points in which the line intersects the curve $xy = 8$. For what value of m will these points coincide? Draw the curve $xy = 8$ and the lines obtained by taking $m = -1/2$, $m = -1$, $m = -2$, and $m = -3$.

5. Show that the line $y = mx - am + b$ passes through (a, b) for any value of m and find the equation of the tangent to the curve $xy = ab$ at the point (a, b).

6. Find the perpendicular (in our new sense) at $(5, 3)$ to the line joining $(5, 3)$ to the origin.

7. Find the equation of the perpendicular, in our new sense, to the tangent line in Exercise 5.

*8. Prove that if P_1 and P_2 are two points and O is the origin, then the area of the triangle P_1OP_2 is the same as the area of the triangle $P'_1OP'_2$, where P'_1 and P'_2 are obtained from P_1 and P_2 by a lorotation.

*9. What is the equation of the line obtained by lorotating the line $y = -2x/3 + 4$ by the lorotation $x = kx'$, $y = y'/k$? Is the resulting line tangent to the lorcle $xy = 6$?

10. Show that the lorotations about the origin form a *group* in the sense in which this term was used in Section 3 of Chapter 4. (See reference **53**, Chap. 18.)

*11. Show that the transformations

$$x' = x + c, \qquad y' = y + d,$$

where c and d are arbitrary real numbers, form a group.

*12. Show that the transformations

$$x' = kx + c, \qquad y' = \frac{1}{k}y + d,$$

where c and d are arbitrary real numbers and k is positive, form a group.

*13. In Euclidean geometry we called

$$\sqrt{(x_1 - x_2)^2 + (y_1 - y_2)^2}$$

the distance between two points (x_1, y_1) and (x_2, y_2). Call the "lortance" between these two points

$$(x_1 - x_2)(y_1 - y_2)$$

Show that the lortance between two points is left unchanged by a lorotation. Show that the lortance between two points is left unchanged by any transformation of Exercise 12.

6. ROTATIONS AND LOROTATIONS

In Chapter 9 we considered rotations. We noticed that a rotation is a Euclidean transformation; that is, it leaves distances unchanged. Distance is thus sometimes called an "invariant under a rotation." Algebraically, rotation about the origin of coordinates can be represented by the following transformation:

(1)
$$x' = rx + sy,$$
$$y' = -sx + ry,$$

where $r^2 + s^2 = 1$. Why does this represent a rotation? First notice that the distance from the origin to the point (x, y) is the square root of $x^2 + y^2$, by the Pythagorean theorem, as may be seen from Fig. 10:8. Then

$$
\begin{aligned}
x'^2 + y'^2 &= (rx + sy)^2 + (-sx + ry)^2 \\
&= r^2x^2 + 2rsxy + s^2y^2 + s^2x^2 - 2srxy + r^2y^2 \\
&= (r^2 + s^2)x^2 + (r^2 + s^2)y^2 \\
&= x^2 + y^2,
\end{aligned}
$$

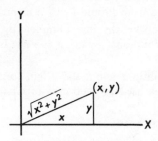

Figure 10:8

since $r^2 + s^2 = 1$. Every circle, $x^2 + y^2 = a^2$, having its center at $(0, 0)$ is left unchanged by the transformation (1); that is, for the transformation (1) the image of each point on the circle is another point on the same circle. We also say that the expression $x^2 + y^2$ is an *algebraic invariant* under transformation (1).

Now consider another transformation. We shall see that this is associated closely with the lorotations which we have been considering.

(2)
$$x' = rx - sy,$$
$$y' = -sx + ry,$$

where now $r^2 - s^2 = 1$. Then

$$
\begin{aligned}
x'^2 - y'^2 &= (rx - sy)^2 - (-sx + ry)^2 \\
&= r^2x^2 - 2rsxy + s^2y^2 - (s^2x^2 - 2rsxy + r^2y^2) \\
&= (r^2 - s^2)x^2 - (r^2 - s^2)y^2 \\
&= x^2 - y^2,
\end{aligned}
$$

since $r^2 - s^2 = 1$. Thus transformation (2) leaves invariant the algebraic expression $x^2 - y^2$. Hence this transformation leaves unchanged each hyperbola (really a lorcle according to the terminology of this chapter) $x^2 - y^2 = a^2$ (see Fig. 7:14); that is, (2) will take each point on such an hyperbola into another point on the same hyperbola. You are asked to show in an exercise below that the hyperbolas (lorcles) $x^2 - y^2 = a$ are the hyperbolas $xy = b$ in another position. Hence in this connection we call $x^2 - y^2$ the lortance between $(0, 0)$ and (x, y) and analogously to Exercise 13 of Section 5, we define the lortance between (x_1, y_1) and (x_2, y_2) to be

$$(x_2 - x_1)^2 - (y_2 - y_1)^2.$$

(See Exercise 3 below.)

The transformation (2) is fundamental in the theory of relativity. Although we do not propose to go very deeply into this in these pages, you can perhaps see a little of what is involved from the following considerations. In Fig. 10:9,

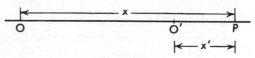

Figure 10:9

let O denote the position of one observer and O' that of another who is moving with a constant velocity relative to O. Let x denote the distance according to O's measuring stick from O to a point P, and x' the distance from O' to P according to the measuring stick of O'. In relativity theory, not only the measuring sticks but the clocks for measuring time are dependent on the motion of O' and the times and distances are related by the equations

(3)
$$x' = rx - sct,$$
$$ct' = -sx + rct,$$

where $r^2 - s^2 = 1$ and c stands for the velocity of light. These are just equations (2) with y replaced by ct. It turns out physically that the ratio s/r is the velocity of O' relative to O divided by the velocity of light.

EXERCISES

1. In transformation (1) let $r = 1/\sqrt{2}$, $s = -1/\sqrt{2}$. Notice that this is a rotation. Show that this rotation takes the hyperbola (lorcle) $x^2 - y^2 = a$ into the hyperbola (lorcle) $x'y' = 2a$. This shows that these two equations represent the same hyperbola in different positions.

2. If instead of the transformation (2) we use $x' = kx$, $y' = y/k$, for k any number not zero, show that for a transformation of this kind the lorcle $xy = r$ is left unchanged. What happens if the transformation is $x' = ky$, $y' = x/k$?

3. What area is left unchanged by transformation (2)?

4. Show that for transformation (2),

$$(x_2' - x_1')^2 - (y_2' - y_1')^2 = (x_2 - x_1)^2 - (y_2 - y_1)^2$$

7. SUMMARY

We could go on almost indefinitely making new definitions and proving new theorems in this geometry. If the student has caught the spirit of what we have already done, any additional proofs would be superfluous. The end we had in view was not so much to develop a body of theorems as to illustrate the possibility of studying mathematical systems distinctly different from Euclidean geometry, yet sufficiently similar to it to be called *geometries.*

In Chapter 11 we return to the familiar surroundings of Euclidean space, but discuss problems which must be attacked by methods not at all like those useful in a first course in Euclidean geometry.

8. TOPICS FOR FURTHER STUDY

1. Geometry of two dimensions: references **1** and **48** (Vol. 4, pp. 2385–2402).
2. Axioms of geometry: reference **29**.
3. Geometry and art: reference **61**; also many numbers of the periodical *Scripta Mathematica.*
4. Four-dimensional geometry: reference **38** (pp. 112–131).
5. Non-Euclidean geometry: references **14**, **38** (Chap. 16), **45**, and **53** (Chap. 17).
6. Finite geometries (geometry with only a finite number of points and lines): reference **34**.

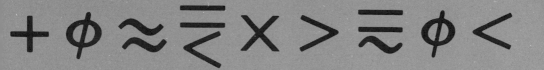

11

Topology

1. THE KÖNIGSBERG BRIDGE PROBLEM

The city of Königsberg is situated on and around an island at the junction of two rivers. At the beginning of the eighteenth century the various parts of the city were connected by seven bridges, as shown in Fig. 11:1. The mathe-

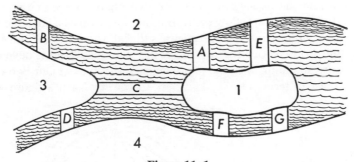

Figure 11:1

maticians of those days were interested in the following problem: Is it possible for a Königsberger to take a walk through the city during which he shall cross each of the seven bridges exactly once? Although a few trials will soon convince you that the answer is no, is it not easy to prove this by trial, for there is a large number of ways of starting the walk. The proof was first given in 1736 by the famous mathematician Euler.

A similar problem that is seen in puzzle books is to determine whether or not one can make a tour of the house in Fig. 11:2, going through each door

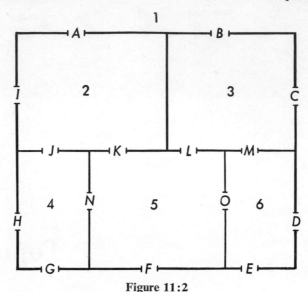

Figure 11:2

once and only once and returning to the same room where the tour began. Here again the answer is no and, owing to the large number of doors, it is very difficult to show this by trial.

The figures for these problems are similar in that they consist of several regions (districts of the city, rooms) connected by passageways (bridges, doors). We can simplify the figures considerably by replacing each region by a single point and the passageway between two regions by a line joining the corresponding points. To follow the transition more easily we label the regions (points) with numbers and the passageways (lines) with letters. If this is done, we have Fig. 11:3 (Notice that in the second example we must have a point labeled 1 corresponding with the region outside the house.) Our two original problems then, in terms of Fig. 11:3, are to traverse such a figure without

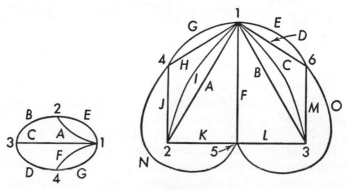

Figure 11:3

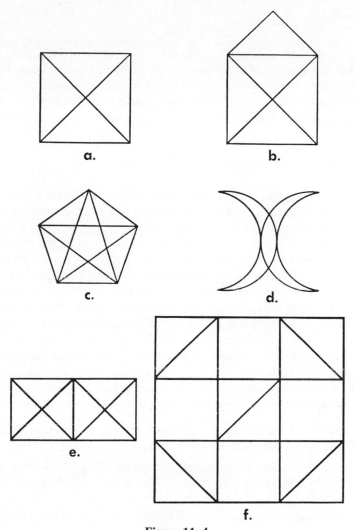

Figure 11:4

covering any line more than once—or to draw the figure without lifting the pencil from the paper and without retracing any line. Other figures are given in Fig. 11:4.

Figures of this sort, consisting of a set of two or more points (finite in number) joined in pairs by lines, no two of which intersect except at a point of the set, we call *networks*. We shall call the points of the set *vertices* and the lines *arcs*. A vertex is said to be *odd* or *even* according as an odd or an even number of arcs touch it. The points which are joined by an arc are called the *end points* of the arc. For example, the Königsberg network has 4 odd and no even vertices; network (f) in Fig. 11:4 has 14 even and 2 odd vertices.

EXERCISES

1. List the numbers of odd and even vertices in each of the networks of Figs. 11:3 and 11:4. Compare the number of odd vertices with your ability to draw the figure in one stroke of the pencil. Can you draw any conclusion from this comparison? If so, test the accuracy of your conclusions with other networks.

2. A wire framework is to be made in the form of the edges of a cube. What is the least number of pieces of wire required? Give reasons for your answer.

2. PATHS IN A NETWORK

By a *path* in a network we shall mean a sequence of distinct arcs of the network with the property that each arc has an end point in common with the next one. No arc can occur twice in the same path. A path is said to be *closed* if it begins and ends at the same vertex. We can think of a path as something one can trace with a pencil without lifting the pencil nor retracting any arc. For some networks it is true that there is a path which includes all the arcs of the network—then we can say that the network is *traversed by a path*.

It is evident that a network cannot be entirely traversed by a single path if it consists of two or more disconnected pieces. A network is said to be *connected* if, for each pair of points *A* and *B* in the network, there is a path in the network joining *A* to *B*. In this section we shall consider only connected networks except in the exercises. We shall see that there are some connected networks which cannot be traversed by a single path. By experimenting with various connected networks we may arrive at the following conclusions:

(1) *The number of odd vertices in any connected network is even.* (Remember that zero is considered to be an even number.)

(2) *If a connected network has no odd vertices, it can be traversed in a single closed path, starting at any vertex of the network and along any arc from that vertex.*

(3) *If a connected network has two odd vertices, it can be traversed in a single path, starting at one of the odd vertices and ending at the other.*

(4) *If a connected network has more than two odd vertices, it cannot be traversed in a single path.*

These are the statements which were proved by Euler. Applying statement (4) to the Königsberg bridge problem, we see that, since there are four odd vertices, the network cannot be traversed in a single path.

The proof of these four statements depends largely on the fact that if *a* is an even number, then *b* − *a* is odd if *b* is odd and even if *b* is even. Here are the proofs.

(1) The number of odd vertices in any connected network is even.

Proof. Let n_1 be the number of vertices of the network which are end points of just one arc, let n_2 be the number of vertices which are end points of just two arcs, and so on. Then the number of odd vertices is

$$N = n_1 + n_3 + n_5 + n_7 + \cdots.$$

Now each arc has two ends, so the number of arc ends is even. The n_1 vertices account for n_1 of these arc ends, the n_2 vertices account for $2n_2$, and so on. Hence

$$S = n_1 + 2n_2 + 3n_3 + 4n_4 + 5n_5 + 6n_6 + 7n_7 + \cdots,$$

the total number of arc ends, is even. Obviously,

$$S - N = 2n_2 + 2n_3 + 4n_4 + 4n_5 + \cdots$$

is an even number (and when *we* say "obviously" we mean "obviously"). Since S is even, N is also even, which is what we desired to prove.

(2) If a connected network has no odd vertices, it can be traversed in a single closed path starting at any vertex of the network and along any arc from that vertex.

Proof. Suppose that our network has no odd vertices. Starting at any vertex, which we may call A, and along any arc from that vertex, we proceed in a path along the arcs. Suppose B is some vertex other than A. Since B is an even vertex, if we enter B by one arc there will be another arc by which we can leave. Moreover, having used up two arcs at B, the number left is still even, and so if later we return to B, the same situation will hold. Hence we can always continue our path until we come back to A. We shall then have a closed path, which, of course, may not traverse the entire network. If it does, we are through; if it does not, we start over again with the unused portion of the network. This portion will still have all even vertices, since our closed path always accounted for the arc ends two at a time. So the process may be continued, and we finally get a number of closed paths which among them use up all the arcs. Our problem now is to connect these closed paths into a single path. This is done as follows. If our first path does not traverse the whole network, there must be at least one of the other paths which has a vertex in common with it since the network itself is connected. Suppose the first path begins and ends at A, the second begins and ends at B, and a common vertex is C. Then we can combine the two paths into one by going from A to C, then around the second path from C to B and back to C, and along the rest of the first path to A. If there are any more paths left, the process can be continued until finally, we have only one path beginning and ending with A and traversing the whole network.

For instance, consider the network of Fig. 11:5. We might have at first two circuits represented by the following sequences of numbers: (12341) and (542615). A common vertex is 2 and we can combine the circuits as follows:

12(61542)341. We could also make use of the common vertex 1 and have 12341(54261).

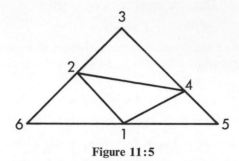

Figure 11:5

(3) If a connected network has two odd vertices it can be traversed in a single path, starting at one of the odd vertices and ending at the other.

Proof. Suppose the network has exactly two odd vertices, A and B. Draw a new arc joining A and B. Then the new network will have no odd vertices, and so it can be traversed by a path starting and ending at B. By statement (2), the path can be chosen so that the first arc traversed is the extra one from B to A. Now remove the extra arc, and we have a path starting at A and ending at B.

(4) If a connected network has more than two odd vertices, it cannot be traversed in a single path.

Proof. If a network can be traversed in a single path, no vertex, except possibly the ends of the path, can be odd. For in going into and out of a vertex the arcs which it touches are used up in pairs, and so we can never use up an odd number. Since a single path has only two ends, such a network can have at most two odd vertices.

A more specific statement of (4) is

(5) *If a connected network has* $2n$ *odd vertices,* n *being a positive integer, it can be traversed in precisely* n *paths and in no less than* n *paths; in other words, the pencil must be lifted* $n - 1$ *times between the beginning of the path and the ending.*

The proof of this is left as an exercise.

Suppose we have a closed path, C, in a plane each of whose vertices is the end point of exactly two arcs. It is intuitively evident (and we shall assume it) that such a path separates the plane into two parts or regions (compare Sections 5 and 6 of Chapter 7). Each region has the property that any two of its points can be joined by an arc or sequence of arcs all of whose points are in the region. One region is finite in extent; we call it the interior of C. The other, infinite in extent, we call the exterior. We call C the boundary of the region.

Similarly, any network separates the plane into regions such that no region contains a point of the network. If the network were thought of as a map, the regions would be the states or countries.

From any network in a plane one may get another, called its *dual*, by replacing each region by a vertex and connecting two vertices of the dual by an arc if the regions of the given network have a common boundary. That is, if regions A and B in a given network have a common boundary c, then in the dual network there will correspond vertices A and B which are end points of an arc corresponding to c. For instance, in Fig. 11:6, the dashed network is

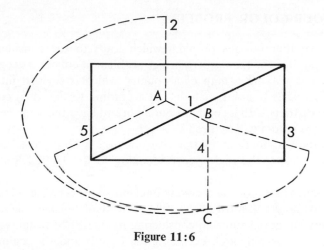

Figure 11:6

the dual of the network of solid lines. The first has three regions A, B, and C and its dual vertices A, B, and C.

EXERCISES

1. Review the exercises of Section 1 in the light of the results of this section.
2. Find the dual networks of parts (a), (b), and (d) of Fig. 11:4. In each case compare the number of regions, vertices, and arcs of a figure with those of its dual. In the dual of a given network to what do the vertices of the given network correspond?
3. Prove statement (5) above.
4. Prove that in a network there is an even number of regions having an odd number of arcs as their boundary, provided that each arc is part of the boundary of exactly two regions. Give an example to show that the qualification is necessary.

5. Which of statements (1), (2), (3), (4), and (5) hold also for all networks which are not connected?

6. Is there a connected network whose dual is not connected?

7. Give an example of a network which is not connected but whose dual is.

8. Consider the figure consisting of a circle with one point on it. Here one can trace the circle starting at any point and returning to the same point. Yet there is just one vertex and it could be interpreted to be odd, because it is the end point of just one arc. Why does this not contradict (1)?

*9. What is the dual of the dual of a network? (A help in finding this is to see what they are for the diagrams considered in Exercise 2 and in Fig. 11:6.) Only an intuitive discussion is required.

3. THE FOUR-COLOR PROBLEM

There is another famous problem which leads to the consideration of networks, and, like the Königsberg bridge problem, it concerns geography. A cartographer is making a map of a country which is divided into several districts. He wishes to color each district according to the usual convention that no two districts with a common border are to have the same color. How many different colors will he need?

An important assumption is that every region is connected. Because of Hawaii and Alaska, the region corresponding to the United States is not connected.

Notice that the map of the the New England states (see Fig. 11:7) is itself a network. Although in most of this section we shall consider the map itself, there are advantages at times in considering its dual. For instance, the dual of the map of New England, excluding the islands and the region outside New England, is shown in Fig. 11:8, where the vertices are labeled with the initials of the states to which they correspond. To color the map of New England would be equivalent to coloring the vertices of the dual map so that no two vertices connected by an arc will have the same color. That is, on the map the condition that no two districts with a common boundary can have the same color corresponds to the requirement on the dual that no two vertices with a common arc can have the same color.

The map of New England can be colored with three colors. (For instance, we could color Maine and Massachusetts red, Vermont and Rhode Island green, and New Hampshire and Connecticut orange.) But what is the least number of colors which will suffice for all maps? The big difference between this question and the ones we asked previously about networks is that this one has never been answered. It has been proved that five colors are enough, and that three are not enough, but whether four will suffice in all cases is still undecided. In any particular network that has ever been investigated four colors have turned out to be sufficient, but this gives us no assurance that

Figure 11:7

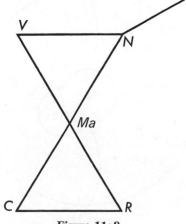

Figure 11:8

there might not be some network, say with several million vertices, which will require five colors. To settle the question one must either give an example of a network which cannot be colored with four colors, or else give a systematic method by means of which any network, no matter how complicated, can be properly colored with four colors. All the mathematicians who have worked at the problem have come to the conclusion that four colors are probably enough in all cases, so that the latter alternative mentioned above is the more promising.

To see that three colors are not always enough, consider the network in Fig. 11:9. Here we have four vertices, and each vertex is joined to each of the

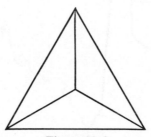

Figure 11:9

other three by an arc. Hence each vertex must be given a different color, for *each pair* of vertices is joined by an arc. This observation naturally causes us to ask, "Why not draw a network with five vertices, each joined to the other four? Would not such a network require five colors to color it properly?" It certainly would, but this network cannot be drawn *on a plane*. A few trials will convince you that this statement is true. Inasmuch as any network obtained from a map can obviously be drawn on a plane, we must rule out

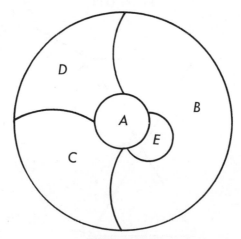

Figure 11:10

this possibility. (Later we shall consider networks on surfaces other than planes.)

Here it is important to notice the restriction mentioned in the first paragraph. Suppose Fig. 11:10 represents an island in the ocean with four counties *A*, *B*, *C*, *D*, and a lake, *E*. If the lake is to be colored blue like the surrounding ocean, five colors are needed, for each of the five regions, considering the ocean and lake as one region in two parts so that the region is not connected, has a common boundary with each of the other four.

In what follows in this section we will consider maps themselves—not their duals.

To indicate how one goes about proving theorems about coloring maps, we shall prove the following theorem.

Theorem
Any map can be properly colored with six colors.

Proof. The proof is quite long, and for convenience we shall break it into several parts.

It will be found useful to consider the entire portion of the plane outside the map as a district of the map. If we can color the map with this added district, we can certainly color it when the district is removed.

A special type of map which we must consider is one in which each vertex is the end point of exactly three arcs. Such a map will be called *regular*. We first prove

(1) *It is sufficient to consider only regular maps.*

Suppose we have proved (as we shall do later) that every regular map can be colored in six colors. If vertex *A* is the end point of only two arcs we can suppress *A* and unite the two arcs into one, without affecting the coloring of the map. Thus we can consider each vertex to be the end point of at least three arcs.

We must now show how to eliminate the vertices having more than three arcs. Suppose, for example, that there is a vertex with five arcs (Fig. 11:11, p. 306). What we do is construct a new map looking just like the old one except that the part in Fig. 11:11 is replaced by Fig. 11:12. We have then added one new district and replaced the irregular vertex by five regular ones. The same thing can be done for each vertex at which four or more arcs come together, and we thus obtain a regular map.

Now we have supposed that this regular map can be colored properly with six colors. In particular, the six districts shown in Fig. 11:12 (p. 306) have colors assigned to them so that no two districts with a common border have the same color. Now if we assign the same colors to the five regions of Fig. 11:11, this still remains the case, for the abolishment of district *f* does not introduce

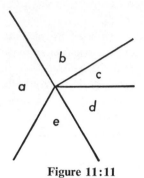

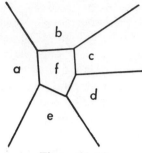

Figure 11:11   **Figure 11:12**

any new borders between districts. (Districts like *a* and *c* in Fig. 11:11 are not regarded as having a common border.) Since these extra districts can be removed simultaneously over the whole map without materially affecting the remaining districts, we shall have a proper coloring of our original map.

(2) *It is sufficient to consider only connected maps.*

Suppose we have a map like Fig. 11:13. Here district *c* is ring-shaped and separates the network into two pieces. However, this is easily avoided, as shown in Fig. 11:14. We have here enlarged district *b* at the expense of *c*, thereby connecting the two parts of the network. Obviously Fig. 11:13 can be colored with six colors if Fig. 11:14 can, and as this process works in any case of this sort, we may suppose that our network is connected.

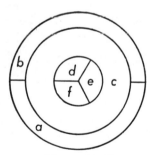

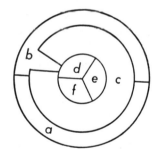

Figure 11:13   **Figure 11:14**

The next thing we prove is a theorem about networks which apparently has nothing to do with maps. However, the connection will be evident later.

(3) *If a connected network on a plane has V vertices and E arcs, and if it separates the plane into F districts, then $F + V - E = 2$. This is called Euler's theorem.*

Any connected network can be built up, or drawn, an arc at a time, each new arc being attached at one or both ends to the part already drawn. Consider the effect on the expression $F + V - E$ of adding such an arc to the network. If the arc is attached at only one end, we do not change the number of districts, but we increase the number of arcs by one, and also the number of vertices by one, for we get a new vertex at the free end of the arc. That is, F is unchanged and V and E are increased by one, so that $F + V - E$ remains the same. If we put a vertex on an arc, we increase V by 1 and E by 1 and do not change $F + V - E$. If both ends of the arc are attached at vertices to the part of the network already drawn, we do not change V, but E is increased by one and so is F, for the addition of such an arc evidently separates one of the previous districts into two. Hence, in this case also, $F + V - E$ is not changed. Now when we start drawing the network we have a single arc, with two vertices, and one district, so that $F + V - E = 1 + 2 - 1 = 2$. Since the value of $F + V - E$ is not changed by adding arcs, it is still equal to 2 for the complete network.

One figure which at first glance appears to violate this result is a circle which has no vertex, one arc, and divides the plane into two regions, but such a figure does not satisfy our definition of a network since it has less than two vertices. If we put two vertices on the circle, Euler's theorem holds. (As it happens, the theorem still holds if just one vertex is put on the circle, although in this case we have no network.)

This property of networks on a plane is very useful in many connections, and we shall have occasion to refer to it later.

(4) *A regular connected map has at least one district which touches five or fewer others.*

Consider the network whose vertices are the points where three of the districts meet and whose arcs are the boundaries between the districts. Using V, E, F as in part (3) of this proof, we have

(A) $$F + V - E = 2,$$

provided this network is connected.

Let n_2, n_3, ... denote the number of districts of the map which have 2, 3, ... sides, that is, which touch 2, 3, ... other districts. A district with k sides has k arcs on its boundary; hence the n_k such districts have kn_k arcs on their boundaries, so the districts have in all $2n_2 + 3n_3 + 4n_4 + \cdots$ arcs on their boundaries. But each arc is on the boundary of two districts, and so

(B) $$2n_2 + 3n_3 + 4n_4 + \cdots = 2E.$$

Now consider the relationship between arcs and vertices. Each of the E arcs has two ends and there are three of these $2E$ ends at each of the V vertices since the map is regular. Hence

(C) $$2E = 3V.$$

Finally,

(D) $$n_2 + n_3 + n_4 + \cdots = F.$$

Multiply equation (A) by 6 to get $6F + 6V - 6E = 12$. Since equation (C) implies $4E = 6V$, we have

$$6F + 6V - 6E = 6F + 4E - 6E = 6F - 2E = 12,$$

and, substituting in this the values of F and $2E$ from equations (D) and (B), we obtain

(E) $$(6 - 2)n_2 + (6 - 3)n_3 + (6 - 4)n_4 + (6 - 5)n_5$$
$$+ (6 - 6)n_6 + (6 - 7)n_7 + \cdots = 12.$$

Now we are trying to prove that there is at least one district with five or fewer sides. If this were false, we should have $n_2 = n_3 = n_4 = n_5 = 0$ and equation (E) would say that

$$(6 - 6)n_6 + (6 - 7)n_7 + \cdots = 12.$$

But this is impossible, for the first term is zero and all the rest are either zero or negative, and so their sum cannot be positive. Hence n_2, n_3, n_4, and n_5 cannot all be zero; this proves statement (4).

Note that up to this point we have made no use of our six colors. Everything we have said so far will apply to the problem of four colors, or five colors.

(5) *Every regular connected map can be properly colored with six colors.*

We have just proved that a regular connected map has a district with five or fewer sides. Suppose there is a five-sided district (Fig. 11:15). Consider the map obtained by removing the boundary between this district and one of the adjacent ones (Fig. 11:16). If this new map can be colored with six colors, so

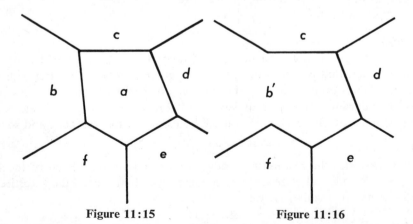

Figure 11:15   Figure 11:16

can the old one, for since the district *a* touches only five others, there is surely
a color available for it when it is put back. The same argument will obviously
work for a district with less than five sides.

So the question now is, "Can the new map, obtained by coalescing districts
a and *b*, be colored?" Well, this new map is still regular (after we eliminate
the two vertices common to just two arcs), and so the same argument can be
applied to it. Continuing in this way we can reduce the number of districts
until we have only six left. We color these all differently, and then work
backward, reintroducing the suppressed districts one at a time, until we have
colored the original map.

There is one possible complication in this process. If a district is coalesced with one which touches it in two places, a ring-shaped district is
created. Before proceeding further we must first eliminate this ring by the process indicated in the proof of statement (2).

Combining statements (5) and (1) we have a proof of the six-color theorem.
Notice that in the first two parts of the proof we replaced the given map by a
more complicated one so that we could apply Euler's theorem. Then part (5)
consists in simplifying the map.

The six colors occur only in proving statement (5), and it is easy to see why
we may need six. As a matter of fact, we shall, in Section 4, by a somewhat
longer argument, prove that even with only five colors the coalescing process
can still be carried out, and thus prove the five-color theorem. But nobody
has yet been able to push the process one step further. The best result that has
been obtained so far is that any map with less than thirty-six districts can be
properly colored with four colors.

For certain special kinds of maps, less than four colors are needed. A
checkerboard is an example of a map requiring only two colors. It can be
shown that two colors are sufficient if the network formed by the boundaries
of the districts has no odd vertices. A little experimenting with such maps will
enable you to see why this is so.

4. THE FIVE-COLOR THEOREM

For those interested we now show the modification of the above argument
necessary to prove that every map may be colored with five colors. Parts (1) to
(4) of the previous argument hold as before and we have given a map containing a five-sided district as in Fig. 11:15 and wish to show: If a map obtained
by eliminating one or more boundaries of the given map can be colored with
five colors, then so can the given map.

In this case we erase the boundary common to districts *a* and *d* as well as
that common to districts *b* and *a* and have Fig. 11:17. Then regions *a*, *b*, and *d*
become one district, *d'*, of the new map. Suppose this map may be colored
with five colors. Then the color of *d'* may be allotted to *b* and *d* of the given

map, leaving a fifth color for district *a*, *unless*, in the given map, *b* and *d* had a boundary in common. If, however, *b* and *d* have a common boundary, the three districts *a*, *b*, and *d* completely surround either district *c* or districts *f* and

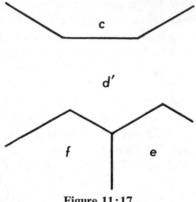

Figure 11:17

e. In both cases it will bc impossible for *c* to have a boundary in common with either *f* or *e*. Then we may go back to Fig. 11:15 and erase the boundaries between *a* and *c* as well as *a* and *f* and carry through our argument as above.

EXERCISES

1. Draw the network which represents the states of the United States in the same way that Fig. 11:7 does the New England states. Use the network to answer the questions:
    a. What is the smallest number of states that must be passed through traveling from Portland, Maine, to Los Angeles, never going outside the country?
    b. How does this compare with the least number of states traversed in going from Miami to Seattle?
    c. Can the map of this country be colored with three colors? If not, where are the exceptional portions?
    d. In how many ways can one remove one state from the country to leave the remaining network disconnected?
    e. What states or state border on the greatest number of other states, where "border on" means have a common boundary which is more than a point?
2. Color with four colors the map in Fig. 11:18.
3. Draw a regular map with ten districts that can be colored with three colors, no two adjacent districts having thc same color.
4. Find all the relationships you can between the *F*, *V*, and *E* for Fig. 11:7 and the *F*, *V*, and *E* for Fig. 11:10.

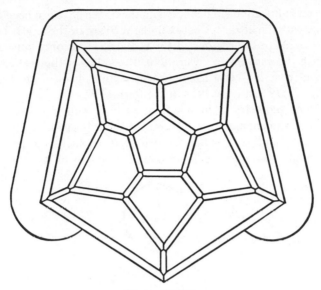

Figure 11:18

5. If all places outside New England are considered part of one vast region (inhabited only by savages), by what should Fig. 11:7 be replaced? Then discuss the questions raised in Exercise 4.

6. Given any connected network A and a network B obtained from A by erasing one of the arcs of A. (It is understood that if the erasure leaves an isolated vertex, the vertex is also erased.) Show that if F, V, and E are the numbers of faces, vertices, and arcs, respectively, of network B, then $F + V - E \doteq 3$ if B is not connected.

7. What is the dual network of the network consisting of two triangles without a common point? What is the dual of the dual?

*8. What are the possible values of $F + V - E$ for a network obtained by erasing two of the arcs of a connected network?

*9. A connected network which contains no closed paths is called a "tree." Thus if A and B are two vertices each of which is the end point of exactly one arc, there is exactly one path along the network from A to B. Prove that a connected network is a tree if and only if the number of vertices is 1 more than the number of arcs. (There are applications of such trees in chemistry and circuit design. See also the references at the end of this chapter.)

*10. What is the form of the dual of a "tree"?

5. TOPOLOGY

Let us stop at this point to consider the kind of problems we have been discussing in connection with networks. Networks are geometrical objects; that is, they are composed of points and lines, and yet we have used none of the usual geometrical methods in dealing with them. We have not been concerned, for instance, with the distances between the vertices, or the angles between the arcs, or the areas of the regions in the networks. Our only concern was the way in which the various arcs were joined together at the vertices. The network arising from the New England states could have been drawn in the form of Fig. 11:19 just as well as in the form of Fig. 11:7.

In our plane geometry in high school we dealt with properties of figures which remained unchanged if we moved them from place to place without changing their shape. A triangle formed of three sticks has the same area whether it is in New York or Timbuktu. The length of its altitudes and its

Figure 11:19

medians are left unaltered by such a journey if the sticks remain *the same size*. Two figures were called *congruent* if by *rigid motion* (a *Euclidean* transformation), that is, moving without distortion, we could make one coincide with the other. As far as our plane geometry was concerned, two congruent figures were identical: They had equal angles, equal sides—everything that we investigated in plane geometry was the same for one as for the other.

In *topology* we allow ourselves more freedom of action. In addition to rigid motions we allow twisting, stretching, and bending but not cutting or welding. We allow ourselves to bend crowbars into U's but not to weld them into hoops. We can make circular hoops into oval hoops or even into the outlines of a square but not into crowbars. Such a transformation we call a *topologic transformation*, as opposed to the transformation of rigid motion which was fundamental in plane geometry. Just as in plane geometry we called two figures *congruent* if we could make them coincide by rigid motion, so in topology we call figures *equivalent* if we can make them coincide by a topologic transformation. Just as in plane geometry we investigated the properties left unchanged by rigid motions, so here in topology we are interested in properties left unchanged by topologic transformations. The traversing of a path and the coloring of a map are topological problems because the shape of the path or the shape of the boundaries of a map make no difference to him who wishes a system for traversing the path or coloring the map. On the other hand, we have now no concern with angles, areas, and such things which can easily be changed by a little twisting.

It must be emphasized that a topologic transformation never brings together two (or more) points that were originally separate, nor does it tear a figure apart so that one point becomes two. Thus the diagrams in Fig. 11:20 are not obtainable from Fig. 11:19 by a topologic transformation. It is evident that these figures do not represent the New England states.

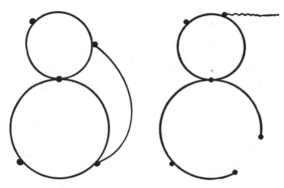

Figure 11:20

While our definition of a topological transformation allows us to round off sharp corners, from some points of view it would not allow us to shake off vertices. However, it is convenient to agree that, before considering equivalence of two figures, we shall *eliminate all vertices which are the ends of exactly two arcs*. For instance, we should say that a line segment with ten points marked off on it would be equivalent to a line segment with no points marked on it; the square with vertices at its corners and the midpoints of its sides would be equivalent to a figure obtained by eliminating these eight vertices, that is, would be equivalent to a circle. That a circle without vertices is not a network is here immaterial, since it is the figure with which we are concerned. In fact, we may apply this definition of topological equivalence to surfaces in three dimensions and say, for instance, that the following surfaces are all equivalent to each other: a sphere, a cube, a pyramid, or a prism. On the other hand, the region bounded by a circle is not equivalent to the ring-shaped region between two concentric circles nor is a sphere equivalent to an inner tube.

EXERCISES

1. Pair off the following figures (see also Fig. 11:21) so that the two figures in each pair are equivalent.

a. The circumference of a circle.
b. A line segment.
c. The surface of a sphere.
d. A hollow sphere.
e. A solid sphere.
f. The surface of a cube.
g. A solid cube.
h. A cube with a hole bored through it.
i. The surface of *h*.
j. The area between two concentric circles.
k. The surface of a doughnut.
l. The edges of a tetrahedron.
m. A piece of gas pipe.
n. A piece of gas pipe with the ends plugged.
o. A swastika.

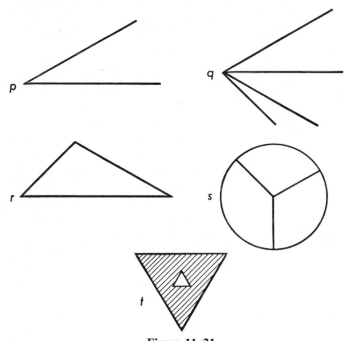

Figure 11:21

2. The following newspaper comment appeared after a popular lecture on topology: "Topology is the branch of mathematics which tells us how to make inner tubes out of billiard balls." What are your comments on the comment?

6. PLANAR NETWORKS

To illustrate some applications of these ideas let us consider in detail a point which arose in connection with the four-color problem. It was stated that the networks arising from maps were such that they could be drawn in a plane. However, the conclusions which one reaches by considering the topological properties of such networks apply to any network which is merely *equivalent* to one in a plane. A network equivalent to a network in a plane, we call a *planar network*. Thus the twelve edges and eight vertices of a cube form a network which does not lie in a plane but which is equivalent to the planar network of Fig. 11:22, where one face of the cube corresponds to the portion

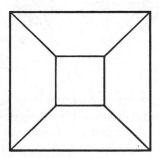

Figure 11:22

of the plane outside the network. Hence the edges and vertices of a cube form a planar network and the conclusions reached in discussing the four-color problem apply to the network on the cube.

The most important of these conclusions is that given in statement (3)— that $F + V - E = 2$. For the cube this gives us $6 + 8 - 12 = 2$.

An important problem in topology is the characterization of planar networks. It was stated that the network with five vertices, each joined to the other four, is not planar. Another network which is not planar arises from the old problem: There are three houses, A, B, C, and three wells, X, Y, Z. The owners of the houses each wish to construct a path to each of the three wells, but being very suspicious of each other they will not permit two paths to cross. (Bridges and tunnels are also barred.) How can the paths be made? Unless you allow one of the paths to go directly over a well, or through one of the houses, the problem has no solution. This is because the network obtained by joining each of three points, A, B, C, to each of three others, X, Y, Z, is not planar.

To determine whether or not a complicated network is planar is not an easy matter. Consider, for example, the network consisting of the edges of a cube plus the lines joining the center to each of the vertices. (Figure 11:23 represents an equivalent network if we assume that the arcs AI, BI, ... hop over EF, FG, ... without intersecting them.) To see that this graph is not planar consider the portion of it drawn in Fig. 11:24. Here we have five vertices joined in all possible ways by arcs, and we have said that such a network is not planar. Hence the complete network cannot be planar either.

This method can be applied to any network. If we can find a portion of the

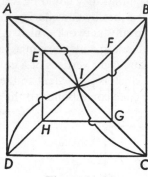

Figure 11:23

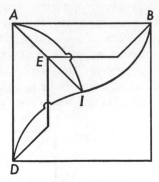

Figure 11:24

network which is equivalent either to five points joined in all possible ways, or
to three points each joined to each of three others, then our network is not
planar. It is interesting that the test also works in the other direction. If no
such portion of the network can be found, then the network is planar. This is
very difficult to prove.[1]

EXERCISES

1. Draw networks in a plane equivalent to:
   a. The edges of a square pyramid.
   b. The edges of a regular octahedron.

2. Show on the basis of statements made above that the network consisting of
   the edges of a cube and one diagonal of the cube is not planar.

3. Which of the following networks are planar? Show that all nonplanar ones
   check with the statements of the paragraph preceding these exercises.
   a. The edge of a tetrahedron and one of its altitudes.
   b. The edges of a box and the diagonals of its faces.
   c. The network whose vertices correspond to the faces of a cube and where
      any edge common to two faces corresponds to a line connecting the
      vertices of the network.
   d. The edges of a tetrahedron, a point within it, and the lines connecting
      the point with the vertices of the tetrahedron.
   e. The network of Fig. 11:9 with the addition of one vertex, E, outside the
      triangle, and an arc connecting E with the vertex inside the triangle.

4. Which of the following networks are planar? Show that all nonplanar ones
   check with the statements of the paragraph preceding these exercises.
   a. The following circles marked on the surface of the earth: the equator,
      the meridian through Greenwich, and the 180° meridian.
   b. The circles in part **a.** plus the axis through the poles.

[1] This is a result of Kuratowski, proved about 1930.

    c.   The circles in part **a.** plus the tropic of cancer.

    d.   The circles in part **a.** plus the tropic of cancer and the tropic of capricorn.

*5.   The simplest finite geometry consists of seven points which we may designate by the seven numerals 1, 2, 3, 4, 5, 6, and 7. There are seven "lines" in this geometry, each consisting of a set of three points as follows: {1, 2, 4}, {2, 3, 5}, {3, 4, 6}, {4, 5, 7}, {5, 6, 1}, {6, 7, 2}, and {7, 1, 3}. (Notice that each pair of lines has exactly one common point and each pair of points is in exactly one line.) Suppose that to depict such a geometry we form a network whose seven vertices are the points of the geometry and which has fourteen arcs formed as follows: The first line determines two arcs with end points two of the three pairs (1, 2), (1, 4), and (2, 4); the second line determines two arcs with end points two of the three pairs (2, 3), (2, 5), and (3, 5); and so on. Is such a network planar? Give reasons for your answer. (See topic 6 in Section 8 of Chapter 10.)

7. POLYHEDRONS

In Section 6 we showed that there is a planar network corresponding to a cube and hence that Euler's theorem gives a relationship among the numbers of its faces, vertices, and edges. Before considering polyhedrons, let us define a polygon in the terminology of this chapter. We could call a *polygon* a connected planar network in which every arc is a line segment and each vertex is an end point of exactly two arcs. Any such network separates the plane into three parts: the inside of the polygon or the region bounded by the polygon, the polygon itself, and the outside of the polygon. We have noticed that a polygon is equivalent (topologically) to a circle. The polygon is the *boundary* of the region which it bounds.

A polyhedron is sometimes thought of as a solid and sometimes as a surface. Here we shall use the term to mean a kind of surface in the same sense that a sphere is not a solid but the set of points in space equidistant from a point called its center. Where in a polygon we have vertices and arcs, in a polyhedron we have vertices, arcs (which we now call edges), and planar regions (which we now call faces). In a polyhedron each face is a planar region bounded by a polygon. Just as in a polygon each vertex is an end point of two arcs, so in a polyhedron each edge is part of the boundary of exactly two faces. Just as a polygon separates the plane into three parts, so a polyhedron separates space into three parts—the inside of the polyhedron, the polyhedron itself, and the outside. Just as a polygon is equivalent to a circle, so a polyhedron is equivalent to a sphere. The latter is intuitively evident if we think of stretching or compressing the polyhedron first so that all the vertices lie on a sphere and then deforming the faces to conform with the sphere.

To show intuitively that any polyhedron may be represented by a planar network, we may think of it as made of some kind of stretchable material, select one of its faces, and cut out this face. Then we could stretch the boundary of this face and so flatten the remainder of the polyhedron that the boundary of the face which we cut out forms the outer boundary of the network consisting of the edges and vertices of the polyhedron. Each face of the polyhedron becomes a region of the network, each edge becomes an arc, and the face which we cut out becomes the region outside the network. The network is connected and hence Euler's theorem holds not only for the network but for the original polyhedron, where now E stands for the number of edges and F for the number of faces. It would also be possible to prove independently Euler's theorem for polyhedrons.

You may recall that a regular polyhedron is defined to be one whose faces are bounded by congruent regular polygons and whose polyhedral angles at the vertices are all congruent. (If you do not remember what a polyhedral angle is, you might look it up. But a knowledge of what it means is not necessary to our discussion here.) There is a remarkable theorem in solid geometry to the effect that there are only five regular polyhedrons with edge, say, 1 inch in length. The proof usually given in solid geometry books depends on the kinds of polyhedral angles. But the restriction to five is really a topological fact and can be proved using Euler's formula with much weaker assumptions.

We can prove the following theorem.

Theorem

If a polyhedron has the following three properties, then there are just five different possible sets of values of F, V, and E, the numbers of faces, vertices, and edges:

(1) *All faces have the same number of edges in their boundaries.*
(2) *All vertices are end points of the same number of edges.*
(3) *Each face has at least three edges in its boundary and each vertex is an end point of at least three edges.*

Note, before starting the proof, that if a polyhedron is regular it certainly has the three properties. But a polyhedron could have these three properties without being regular (for instance, it could be a box not in the form of a cube).

Proof. Assuming property (2), we let r stand for the number of edges having any vertex as a common end point. Then since each edge has two end points, the product rV counts each edge twice. Hence

(A) $$rV = 2E \quad \text{or} \quad V = \frac{2E}{r}.$$

Assuming property (1), we let s stand for the number of edges in the boundary of each face. Then since each edge is on the boundary of two faces, the product sF counts each edge twice. Hence

(B)
$$sF = 2E \qquad \text{or} \qquad F = \frac{2E}{s}.$$

From Euler's formula we know that $V + F - E = 2$. If we replace V and F in this formula by the expressions in (A) and (B), we have

$$\frac{2E}{r} + \frac{2E}{s} - E = 2;$$

that is,

$$E\left(\frac{2}{r} + \frac{2}{s} - 1\right) = 2.$$

Dividing both sides by 2 we get

(C)
$$E\left(\frac{1}{r} + \frac{1}{s} - \frac{1}{2}\right) = 1.$$

Since E is positive, (C) shows

(D)
$$\frac{1}{r} + \frac{1}{s} > \frac{1}{2}.$$

Now if $r \geq 4$ and $s \geq 4$, the left side of (D) would be less than or equal to $\frac{1}{2}$. Hence either $r < 4$, $s < 4$, or both. On the other hand, condition (3) requires that r and s be at least 3. This means that at least one of r and s must be equal to 3.

First suppose $r = 3$. Then (D) becomes

$$\frac{1}{3} + \frac{1}{s} > \frac{1}{2};$$

that is,

$$\frac{1}{s} > \frac{1}{6}$$

or

$$s < 6.$$

Thus if $r = 3$, there are just three possibilities for s: 3, 4, or 5.

Second, if $s = 3$, the same argument would show that there are just three possibilities for r: 3, 4, or 5.

Since we have counted $s = 3 = r$ twice, there are just five different possible pairs of values of r and s. The following table shows that all of these satisfy condition (D).

r	3	3	3	4	5	
s	3	4	5	3	3	
$\dfrac{1}{r}$	$\dfrac{1}{3}$	$\dfrac{1}{3}$	$\dfrac{1}{3}$	$\dfrac{1}{4}$	$\dfrac{1}{5}$	
$\dfrac{1}{s}$	$\dfrac{1}{3}$	$\dfrac{1}{4}$	$\dfrac{1}{5}$	$\dfrac{1}{3}$	$\dfrac{1}{3}$	
$\dfrac{1}{r}+\dfrac{1}{s}$	$\dfrac{2}{3}$	$\dfrac{7}{12}$	$\dfrac{8}{15}$	$\dfrac{7}{12}$	$\dfrac{8}{15}$	

For each pair of values of r and s, we can use (C) to find E, (B) to find F, and (A) to find V. We tabulate the results below, leaving the calculations as exercises.

If $r = 3 = s$, then $E = 6$, $F = 4$, $V = 4$. This polyhedron is called a *tetrahedron*.

If $r = 3$, $s = 4$, then $E = 12$, $F = 6$, $V = 8$. This polyhedron is a cube, sometimes called a *hexahedron*.

If $r = 3$, $s = 5$, then $E = 30$, $F = 12$, $V = 20$. This polyhedron is called a *dodecahedron*.

If $r = 4$, $s = 3$, then $E = 12$, $F = 8$, $V = 6$. This polyhedron is called an *octahedron*.

If $r = 5$, $s = 3$, then $E = 30$, $F = 20$, $V = 12$. This polyhedron is called an *icosahedron*.

We have completed the proof since we have considered all possible cases and found a set of values of E, F, and V for each. In order to show that these polyhedra actually exist we can find a planar network for each. We will here find such a one for the third case and leave the others as exercises. Notice that in each case the name corresponds to the number of faces.

To construct the network for the dodecahedron, notice that $s = 5$ means that each face is a pentagon, that is, has five edges and $r = 3$ means that each vertex is the end point of three edges. To begin the drawing then, we can draw a pentagon (it need not be regular). Since each vertex has three edges, we draw a line segment inward from each of the vertices. Then draw two line segments from the "inside end" of each of these segments. The points of intersection of these in pairs form five pentagons each of which has an edge in common with the pentagon we drew in the first place. Then draw lines inward from each of these intersections and complete the inner pentagon. If you check you will see that the numbers E, F, and V are correct. Actually each time we had no choice in the number of lines to draw and hence the network is unique—that is, any two such networks which we drew would

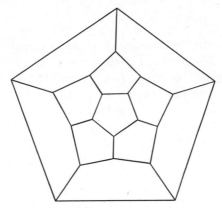

Figure 11:25

have to be topologically equivalent. The network is given in Fig. 11:25 (see also Fig. 11:18).

EXERCISES

1. Calculate the sets of values of E, F, and V for each of the five cases.

2. Draw the networks for the tetrahedron, hexahedron, and octahedron.

3. Compare the hexahedron and the octahedron, noticing that the value of E is the same for both and the values of F and V interchanged. What is the significance of this relationship with regard to the planar networks which represent these polyhedra?

4. Use the results of Exercise 3 to find the network for the icosahedron.

5. Construct a polyhedron as follows. Begin with a cube. Call A a vertex and B, C, and D the midpoints of the edges having A as an end point. If we "cut off" the vertex A by slicing the cube through the plane determined by B, C, and D, we will have a triangular face BCD in the place of the corner of the cube at A. Do this for each of the eight vertices. What kinds of faces do we have in the resulting polyhedron? Does it satisfy the properties of the theorem?

6. Every vertex of a polyhedron is the end point of three edges and each face has three sides. What polyhedron is it?

7. Every vertex of a polyhedron is the end point of r edges and each face has r sides. What possibilities are there for the positive integer r?

*8. The faces of a certain polyhedron are pentagons and hexagons. Each vertex is a vertex of two hexagons and one pentagon. For any given hexagon, three of its sides are in common with other hexagons and the other three with pentagons. Find the number of hexagonal faces, the number of pentagonal ones, the number of vertices, and the number of edges. (If the pentagons and hexagons are regular, this polyhedron is called a semiregular polyhedron. The polyhedron of Exercise 5 is another example of a semiregular polyhedron.)

8. SURFACES

Probably the simplest surface which is encountered in elementary geometry is a sphere. This is defined as the locus of points at a given distance from a fixed point, called the *center*. In solid geometry many things are proved about such surfaces, including properties of tangent planes or great and small circles on the surface. None of these properties, however, is of any interest from the point of view of topology, for if the surface is stretched or deformed in some irregular manner the very notion of tangent plane or great circle may become meaningless. If we think of the sphere as a rubber balloon, the topological properties of the surface are the ones which are not altered when the balloon is stretched, pinched, or deformed in any manner *without tearing*. Since by such a process we can stretch the balloon into the shape of a cube, or of any polyhedron, it is difficult at first to conceive of any properties of the surface which would not be changed by such stretching. However, there are such properties. For instance, the balloon has an inside and an outside, and no amount of stretching or twisting will alter this property. Hence, this is a topological property of a sphere. This property appears so obvious that in elementary solid geometry we just take it for granted without even commenting on it. This is true of a great many topological properties, and it is probably the main reason why topology is such a recent development in mathematics.

Another topological property of a sphere is the fact that any closed curve drawn on the surface divides it into two parts. (By a closed curve we mean any line, curved or partially straight, whose ends are joined but which does not cross itself. Thus Figs. 11:26a and 11:26b are closed curves, but Figs. 11:26c and 11:26d are not. In other words, a closed curve is any figure topologically equivalent to a circle.) This property of a spherical surface is evidently a topological one, and, like the other, seems to be obviously true. However, there are surfaces which do not have this property. The simplest of these is the

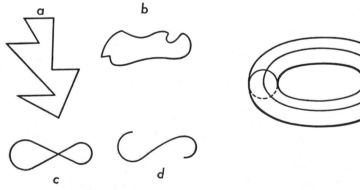

Figure 11:26   Figure 11:27

torus, which is the surface of a doughnut, or an inner tube. In Fig. 11:27 there have been drawn on such a surface two closed curves, neither of which divides the torus into two parts. As a matter of fact, both curves together do not separate the surface, for it is easily seen that we can pass from any point of the surface to any other without having to cross either curve. However, it is fairly evident that no more closed curves can be drawn without spoiling this property.

These observations concerning closed curves lead us to the conclusion that a torus is not equivalent to a sphere. In more familiar terms, a spherical balloon cannot be deformed into the shape of an inner tube. It is not difficult to see how we can get surfaces which are not equivalent to either the sphere or the torus. Two such surfaces are buttons with two and three holes, respectively. Obviously we can construct such examples with as many holes as we wish. It can be proved that on the surface of a button with *n* holes there can be drawn precisely 2*n* closed curves without separating the surface into two or more parts. Hence two buttons with a different number of holes are not topologically equivalent, whereas two with the same number of holes are

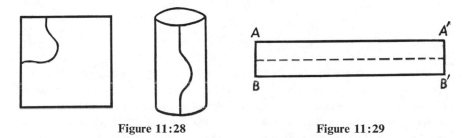

Figure 11:28   Figure 11:29

equivalent. This gives us a complete topological classification of buttons. Of course, the classification extends not only to objects in the shape of a button, but to any object whatever. Thus the surface of one of the chairs in your class-room is equivalent to the surface of a button with holes. (How many?)

So far we have used the word "surface" to refer to the complete surface of some solid object. Let us now use this word for such things as regions bounded by a rectangle or a circle. That is, we consider surfaces with *edges*. In contrast to this usage of the word, we shall refer to the kind of surfaces we considered previously as *closed* surfaces. Surfaces with edges may also be analyzed by considering the curves which can be drawn on them. For example, a rect-angular region is divided into two parts by any curve joining two points of its edge, whereas on a cylindrical surface one such curve may be drawn without separating the surface (Fig. 11:28). Hence a rectangular region and a cylin-drical surface are not equivalent.

A cylindrical surface may be obtained by gluing together the ends of a rectangular strip so that *A* coincides with *A'* and *B* with *B'* in Fig. 11:29. An interesting surface, known as a *Möbius strip*, is obtained by gluing the ends so that *A* coincides with *B'* and *B* with *A'* (Fig. 11:30). This surface has some surprising properties. Like the cylinder we can cut it on a line joining two points of the edge without separating the surface, and yet the resulting edge consists of a single closed curve. Unlike the cylinder, we can cut the surface along a closed curve (the dashed line in the figure) and still have a single

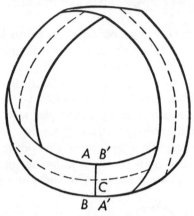

Figure 11:30

piece left. But perhaps the most remarkable property is the following. Imagine an ant crawling along the surface but never crossing the edge. If it starts at point *C* and follows the dashed line it will eventually return to point *C, but on the other side of the surface.* In other words, this surface has the property that you can get from any point of it to any other without crossing over an edge. This is certainly not the case with any closed surface or with the other open surfaces we have considered. Surfaces behaving like the Möbius strip in this respect are said to be "one-sided."

One-sided surfaces have the further interesting property that an orientation cannot be defined on them. In Chapter 9 we defined an orientation in the plane (a two-sided surface) as a distinction between clockwise and counterclockwise rotations, and our point is here that this idea is meaningless on the Möbius strip. To see this draw a directed circle on the Möbius strip and, by a translation, move it along the center line of the surface until it comes back to its original position but on the other "side" of the surface. If you try this, you will see that the arrow on the circle now points in the direction opposite to its original one. Thus we can turn clockwise rotation into counterclockwise rotation by a translation, and hence we cannot distinguish between these two directions of rotation. Surfaces like the Möbius strip for which such a sense of rotation cannot be defined are called "nonorientable."

Returning to Fig. 11:29, let us again glue *A* to *A'* and *B* to *B'*, but giving the line *A'B'* a twist through 360° before gluing. We then get a strip which, although twisted, has most of the properties of the untwisted cylindrical surface, at least as far as the behavior of curves on it is concerned. If we cut this strip along the center line, we obtain two pieces as we should also for the cylinder, but unlike the case of the cylinder the two pieces are linked together so that it is impossible to separate them. Other interesting surfaces can be made by giving the strip several twists before gluing the ends together.

For brevity we will use the term "*n*-strip" to refer to the surface obtained by twisting the strip in Fig. 11:29 through $n \cdot 180°$ before gluing the ends together. Thus a cylinder is a 0-strip, and a Möbius strip is a 1-strip.

EXERCISES

1. Show that an *n*-strip is one-sided if *n* is odd and two-sided if *n* is even.
2. Show that if an *n*-strip is cut down the middle (that is, as in Fig. 11:30), we obtain one piece if *n* is odd and two pieces if *n* is even.
3. If a 1-strip is cut down the middle, we obtain a single piece which is an *n*-strip. Why is this so, and what is the value of *n*, and is the *n*-strip one- or two-sided? If this is cut again down the middle, what happens?
4. If a 2-strip is cut down the middle, each of the two pieces is a two-sided *n*-strip. Why is this so, and what is the value of *n*? If each of these strips is cut again down the middle, what happens?
5. A torus can be made from the strip in Fig. 11:29 by gluing edge *AA'* to edge *BB'* to make a hollow cylinder and then gluing the ends together to form the torus. Show the two curves on the torus of Fig. 11:27 as they would appear on the strip of Fig. 11:29. (Physically this would be hard to manage with paper but not with rubber.)
6. Show how to draw on a button with two holes, four lines which do not separate the surface.

9. MAPS ON A SURFACE

In discussing the four-color problem we considered maps drawn on a plane. It is evident that the same problem can be investigated for maps on a sphere, a torus, or any other surface. As might be expected, maps on different kinds of surfaces require a different number of colors, and the more complicated the surface the more colors required.

It is not hard to see that a map on a sphere will require just as many colors as a map on a plane. For suppose we have a map on a sphere. Thinking of the sphere as a rubber balloon, cut a small hole in one of the districts. By pulling out the boundary of this hole, we can evidently stretch the balloon until it becomes flat. We then have a map on a plane which is equivalent to the original one of the sphere, and hence the same number of colors will be required for each of them.

For the next simplest closed surface, the torus, the situation is quite different. Figure 11:31 is an example of a map consisting of seven districts, each one touching all the remaining six. Such a map will obviously require seven colors. Now the interesting thing about the torus is that we can prove that

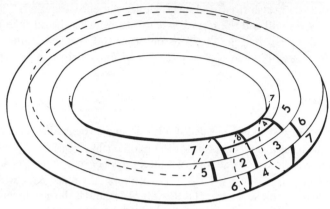

Figure 11:31

seven colors will suffice in all cases. The proof is very much like the one outlined for six colors in the case of the plane. By the same methods we used for maps on a plane we can show that any map on a torus must have at least one district which touches six or fewer others. Since we have seven colors at our disposal, this district can be coalesced with one of the adjacent ones and the process continued until there are only seven districts left. Hence the map-coloring problem for the torus is completely solved: Seven colors is the minimum number.

In the same way it has been proved that for the surface of a button with 2, 3, or 4 holes the number of colors needed is 8, 9, and 10, respectively. It is rather a strange state of affairs that these more complicated surfaces are easy to deal with, while the much simpler plane and sphere cause so many difficulties. (See the last sentence of Section 10.)

The relationship between maps and networks can, of course, be extended to maps on the torus or any other surface. We saw that on a plane—and hence also on a sphere—we could not have a network joining five points in all possible pairs. On a torus, however, we can join as many as seven points in all possible ways. This possibility is assured by the existence of the map in Fig. 11:31. The fact that no map on a torus requires more than seven colors tells us that the network joining eight points in all possible ways cannot be drawn on a torus. Similar statements can be made for the other types of closed surfaces.

A convenient way of dealing with the drawing of a map on the torus is by means of a figure like Fig. 11:32. If we cut a sheet of rubber in the form of the figure and divide it into districts as indicated, then we may (1) roll it into a tube by making the top and bottom edges coincide, and (2) stretch and twist the tube until its ends coincide as indicated in Exercise 5 of Section 8. As a result we will have essentially the torus of Fig. 11:31.

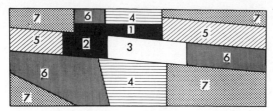

Figure 11:32

EXERCISES

1. Draw a picture of a network on a torus which consists of seven vertices and all the arcs joining them in pairs.
2. Show that the problem of coloring a map on a cylinder is the same as for a sphere or a plane.
3. Construct a map on a Möbius strip which consists of six districts each touching the other five. (*Hint:* Draw the map on the rectangle of Fig. 11:29, remembering how the ends are to be glued together.)
4. Find $F + V - E$ for the network on the surface of the torus in Fig. 11:27; for the curves drawn on the button with two holes in Exercise 6 of Section 8.

10. TOPICS FOR FURTHER STUDY

Most of the topics dealt with in this paper and many allied ones appear in references **6** (Chaps. 8 and 9), **9**, **13** (Chap. 5), **38** (Chap. 8), **52** (pp. 13–17, 73–82, and regular polyhedra in pp. 82–88), and **61**.

For four-color and five-color problems see reference **24**. For "trees" see references **40** (pp. 83–87, 139–144) and **54** (p. 83). For knots see reference **5**.

There is an interesting account of the solution of the map-coloring problem for all surfaces but the plane in a paper by Gerhard Ringel and J. W. T. Youngs, Solution of the Heawood Map-Coloring Problem, *Proceedings of the National Academy*, June 1968.

Bibliography

1. ABBOTT, A., *Flatland, A Romance of Many Dimensions*. Boston: Little, Brown, 1929. (7th ed., New York, Dover, 1952). A popularly written account of life in a world of other than three dimensions. See also reference **48**, Vol. 4, pp. 2385–2402.

2. ALLENDOERFER, C. B., and OAKLEY, C. O., *Fundamentals of College Algebra*. New York: McGraw-Hill, 1967. This is a text for college freshmen.

3. ANDERSON, R. D., Concepts of Informal Geometry, *Studies in Mathematics*, Vol. V. New Haven, Conn.: The School Mathematics Study Group, 1960. A small volume describing for teachers metric and nonmetric geometry.

4. ANDREWS, F. E., Revolving Numbers, *The Atlantic Monthly*, Vol. 155, 1935, pp. 208–211. A very interesting account for nonmathematicians of cyclic or "revolving" numbers. The author's results are stated by means of examples with no attempt at proof.

5. ARTIN, EMIL, The Theory of Braids, *The American Scientist*, Vol. 38, 1950, pp. 112–119. Reprinted in *The Mathematics Teacher*, Vol. 52, 1959, pp. 328–334. Develops some topological ideas related to knots.

6. BALL, W. W. R., *Mathematical Recreations and Essays*, revised by H. S. M. Coxeter. New York: Macmillan, 1956 (Paperback edition, 1962.) Probably the most widely consulted book on mathematical puzzles. In many cases rather complete theories are developed.

7. BELL, E. T., *Men of Mathematics*. New York: Simon and Schuster, 1937. Contains a set of popularly written short biographies of famous mathematicians and describes the ideas they originated.

8. ———, *Numerology*, Baltimore: Williams & Wilkins, 1933. A mathematician's opinion of one of the occult pseudosciences.

9. BING, R. H., Elementary Point Set Topology, *The American Mathematical Monthly*, Vol. 67, 1960 (Slaught Memorial Paper No. 7).

10. BIRKHOFF, G., and MACLANE, S., *A Survey of Modern Algebra*. New York: Macmillan, 1965. A much used textbook in modern algebra which carries much further a number of topics touched on in this book. For a more elementary treatment see reference **39**.

11. BOEHM, G. A. W., *The New World of Math*. New York: Dial, 1959. A "paperback" about mathematics and mathematicians.

12. BREUER, J., *Introduction to the Theory of Sets*, trans. by H. F. Fehr. Englewood Cliffs, N.J.: Prentice-Hall, 1958.

13. COURANT, R., and ROBBINS, H., *What is Mathematics?*, 4th ed. New York: Oxford

University Press, 1947. Assumes a minimum of preparation and presents many unusual and interesting aspects of mathematics. There are many worthwhile exercises and problems.

14. COXETER, H. S. M., and GREITZER, S. L., *Geometry Revisited*. New York: L. W. Singer, 1967 (Volume 19 of the School Mathematics Study Group's New Mathematical Library). Written for high school students and laymen.

15. CURTIS, C. W., DAUS, P. H., and WALKER, R. J., Euclidean Geometry Based on Ruler and Protractor Axioms, *Studies in Mathematics*, Vol. II. New Haven, Conn.: The School Mathematics Study Group, 1959. A brief presentation of Euclidean geometry from a "modern" point of view—intended for teachers.

16. DANTZIG, TOBIAS, *Number, the Language of Science*, 4th ed. New York: Free Press, 1967. A popularly written account of the history of the number concept and number systems, number lore, and the theory of numbers.

17. DROBOT, S., *Real Numbers*. Englewood Cliffs, N.J.: Prentice-Hall, 1964. An elementary description of the system of real numbers as presented to classes of secondary school teachers.

18. DUBISCH, ROY, *Introduction to Abstract Algebra*. New York: Wiley, 1965. Written at the junior-year level of most colleges.

19. DUDENEY, H. E., *Amusements in Mathematics*. London: Thomas Nelson, 1917. This book and the three following are excellent collections of puzzles. Quite often the author raises a question which he does not answer and which can serve as a starting point for investigation by an interested student.

20. ——, *Canterbury Puzzles*. New York: Dutton, 1908.

21. ——, *Modern Puzzles*. London: C. Arthur Pearson, 1926.

22. —— , *Puzzles and Curious Problems*, revision by James Travers. London: Thomas Nelson, 1931.

23. FUJI, J. N., *An Introduction to the Elements of Mathematics*. New York: Wiley, 1961. A text intended for much the same audience as *Elementary Concepts of Mathematics*, with emphasis on sets and logic.

24. FRANKLIN, PHILIP, The Four Color Problem, *Scripta Mathematica*, Vol. 6, 1939, pp. 149–156, 197–210.

25. GARDNER, MARTIN, *Scientific American Books of Mathematical Puzzles and Diversions*. New York: Simon and Schuster, 1959, 1961.

26. GINSBURG, JEKUTHIAL, Gauss's Arithmetization of the 8-Queen's Problem, *Scripta Mathematica*, Vol. 5, 1938, pp. 63–66.

27. GUTTMAN, SOLOMON, Cyclic Numbers, *The American Mathematical Monthly*, Vol. 41, 1934, pp. 159–166. The first part of this article should be understood by one who knows only a little algebra.

28. HARDY, G. H., *A Mathematician's Apology, with foreword by C. P. Snow*. New York: Cambridge University Press, 1967. A fascinating account of a mathematician's outlook on life, written by one of the finest mathematicians of the present century. The foreword is a description of him by a noted close friend.

29. HILBERT, DAVID, *The Foundations of Geometry*, trans. by E. J. Townsend. La Salle, Ill.: Open Court, 1902. A famous work that puts Euclidean geometry on a firm axiomatic foundation.

30. HOGBEN, L. T., *Mathematics for the Million*. London: G. Allen, 1936. (3rd ed., New York: Norton, 1951.) Written by a British economist during a long convalescence—with the object of informing his nonmathematical friends about mathematics. It has been a very widely read book.

31. HOHN, F. E., *Applied Boolean Algebra: An Elementary Introduction*. New York: Macmillan, 1966. Gives some applications of the algebra of sets and logic to switching circuits.

32. JEANS, SIR JAMES, *Mathematics and Music*. New York: Macmillan, 1937. A popularly written account of how mathematics enters into the theory and physics of music.

33. JOHNSON, W. W., Octonary Numeration, *The New York Mathematical Society Bulletin*, Vol. 1, 1891, pp. 1–6.

34. JONES, B. W., Miniature Geometries, *The Mathematics Teacher*, Vol. 52, 1959, pp. 66–72. A brief introduction to finite geometries, that is, geometries with only a finite number of points.

35. ———, Miniature Number Systems, *The Mathematics Teacher*, Vol. 51, 1958, pp. 226–232. Deals with modular systems and some extensions of them.

36. ———, *The Theory of Numbers*. New York: Holt, 1955. A textbook in the theory of numbers which carries further a number of topics in this book.

37. ———, *Introduction to Number Systems*. Stanford, Calif.: The School Mathematics Study Group (Studies, Vol. XIV), 1966. Written for junior high school teachers, the book contains an intuitive development of the real number system at a somewhat lower level than that in reference 17.

38. KASNER, E., and NEWMAN, J., *Mathematics and the Imagination*. New York: Simon and Schuster, 1940. Explores in a fascinating way the beginnings of many of the important mathematical fields of today.

39. KELLEY, J. L., *Introduction to Modern Algebra*. Princeton, N.J.: Van Nostrand, 1960. This was an official textbook for Continental Classroom, a mathematics television program, and is directed toward mathematics teachers. It is accompanied by a student manual by Roy Dubisch.

40. KEMENY, J. G., SNELL, J. L., and THOMPSON, G. L., *Introduction to Finite Mathematics*. Englewood Cliffs, N.J.: Prentice-Hall, 1966. Deals with many novel topics in mathematics for the social scientist.

41. KLINE, MORRIS, *Mathematics—A Cultural Approach*. Reading, Mass.: Addison-Wesley, 1962. A textbook for those interested in mathematics for its cultural values.

42. KRAITCHIK, MAURICE, *Mathematical Recreations*. New York: Norton, 1942. Contains material similar to that in Ball's book.

43. KUTUZOV, B. V., *Geometry*, trans. by L. T. Gordon and E. S. Shater, *Studies in Mathematics*, Vol. IV. New Haven, Conn.: School Mathematics Study Group, 1960. A translation of a Russian text carried out under the auspices of the Survey of Recent Eastern European Mathematical Literature at the University of Chicago.

44. LEVINSON, H. C., *The Science of Chance*. New York: Holt, 1939. A popularly written account of problems in probability and statistics.

45. LIEBER, L. R., and LIEBER, H. G., *Non-Euclidean Geometry*. Lancaster, Pa.: The Science Press, 1931. An amusingly written and illustrated account of a geometry different from that in high school textbooks.

46. McCoy, N. H., *Introduction to Modern Algebra*. Boston: Allyn & Bacon, 1967. A junior-year textbook written at about the same level as reference 18.

47. MOSTELLER, F., ROURKE, R. E. K., and THOMAS, G. B., *Probability and Statistics*. Reading, Mass.: Addison-Wesley, 1960. The second half of the Continental Classroom course (see reference 39).

48. NEWMAN, J. R., *The World of Mathematics*, 4 vols. New York: Simon and Schuster, 1956. An annotated anthology of mathematics.

49. NIVEN, IVAN, *Numbers: Rational and Irrational*. New York: Random House, 1961 (Volume 1 of the School Mathematics Study Group's New Mathematical Library). An account of rational and irrational numbers written for high school students and laymen.

50. POLYA, GEORGE, *Mathematical Discovery*, 2 vols. New York: Wiley, 1961. A book by one of the most interesting, most respected, and most loved of present-day mathematicians, one long concerned with the art and science of problem solving.

51. QUINE, W. V., *Elementary Logic*. Boston: Ginn, 1941. An elementary account of the basis of logic from a modern point of view.

52. RADEMACHER, H., and TOEPLITZ, O., *The Enjoyment of Mathematics*, trans. by H. Zuckerman. Princeton, N.J.: Princeton University Press, 1957. For the nonmathematician who wishes to learn about the types of phenomena considered by mathematicians and their methods of proposing and solving problems.

53. RICHARDSON, MOSES, *Fundamentals of Mathematics*, 3rd ed. New York: Macmillan, 1966. A college freshman textbook which is worth exploring.

54. ROSSKOPF, M. F., and EXNER, R. M., *Logic in Elementary Mathematics*. New York: McGraw-Hill, 1959. An elementary treatment of logic for nonspecialists.

55. SAWYER, W. W., *Mathematicians Delight* and *Prelude to Mathematics*. Baltimore: Penguin, 1954, 1957. "Designed to convince the general reader that mathematics is not a forbidding science but an attractive mental exercise."

56. ———, *A Concrete Approach to Abstract Algebra*, San Francisco: Freeman, 1959. An elementary introduction to abstract algebra.

57. School Mathematics Study Group, *Essays on Number Theory*, Vols. I and II. New Haven, Conn.: Yale University Press, 1960. An elementary account of some of the topics in number theory.

58. ———, *The Golden Measure*. Pasadena, Calif.: A. C. Vroman, 1967 (Number 9 in the Reprint Series). A small pamphlet of short articles on the golden measure with references to articles of a similar nature.

59. ———, *Mathematics and Music*. Pasadena, Calif.: A. C. Vroman, 1967 (Number 8 in the Reprint Series).

60. SMITH, D. E., *The History of Mathematics*, Vol. 2. Boston: Ginn, 1925. A rather complete and authoritative history of mathematics; contains references to still more complete works.

61. STEINHAUS, H., *Mathematical Snapshots*. New York: Stechert-Hafner, 1938. An annotated set of illustrations chiefly concerning topology and polyhedrons.

62. STOLYAR, A. A., *Introduction to Mathematical Logic*. Cambridge, Mass.: M.I.T. Press, 1968. A translation from the Russian of material used in Russia down to the high school level.

63. TERRY, G. S., *Duodecimal Arithmetic*. London: Longmans, 1938. An ardent disciple of F. E. Andrews has prepared tables for the number system to the base twelve.

64. TINGLEY, E. M., Base Eight Arithmetic and Money, *School Science and Mathematics*, Vol. 40, 1940, pp. 503–508.

65. USPENSKY, J. V., and HEASLET, M. A., *Elementary Number Theory*. New York: McGraw-Hill, 1939. A very original and thought-provoking book, although in many places it is too difficult for the college freshman. However, parts of it can be understood without much mathematical experience and background.

66. VAN DER WAERDEN, B., *Science Awakening*, trans. by Arnold Dresden. Groningen, The Netherlands: Noordhoff, 1954. A well-illustrated and well-written history of mathematics through the golden age of Greece.

67. VEBLEN, OSWALD, The Foundations of Geometry, *Monographs on Topics of Modern Mathematics*, by J. W. A. Young. London: Longmans, 1911. Some of the axioms neglected by Euclid are described here.

68. VILENKIN, N. Ya., *Stories about Sets*. New York: Academic Press, 1968. A translation from the Russian which provides a very unusual introduction to the theory of sets by means of fanciful stories. Actually the book treats some rather deep problems.

69. VOROB'EV, N. N., *Fibonacci Numbers*. Waltham, Mass.: Blaisdell, 1961. A paperback book translated from the Russian, it is written for high school students and contains a set of problems which were the themes of several meetings of "the school children's mathematical club" of Leningrad State University in 1949–1950.

70. WEYL, HERMANN, Emmy Noether, *Scripta Mathematica*, Vol. 3, 1935, pp. 201–220. A beautifully written story of the life and works of the most famous woman mathematician of modern times. Anyone who reads it should glean a feeling of what it is that goes to make a great mathematician.

71. WILLERDING, MARGARET, *A Probability Primer*. Boston: Prindle, Weber Schmidt, 1968. Written for high school students and beginning college students.

72. ——, *Mathematical Concepts, a Historical Approach*. Boston: Prindle, Weber & Schmidt, 1967. This paperback book deals with kinds of numerals, number mysticism, figurate numbers, and the calendar.

73. WILLIAMS, J. D., *The Compleat Strategyst* (rev. ed.). New York: McGraw-Hill, 1966. An introduction to the theory of games, a new branch of mathematics. This is very simply written, at least in the beginning, and should give the uninitiated a good idea of what this branch of mathematics is about.

74. WITHERSPOON, J. T., A Numerical Adventure, *Esquire*, December 1935, pp. 83 ff. An account of casting out nines and related phenomena.

Answers to
Odd-Numbered
Exercises

CHAPTER 1

Section 2, p. 4

1. A square, all the lines perpendicular to a given plane, all even integers, or all negative numbers.

3. **a.** All even integers greater than 77 form a set, since we can tell by looking at a number in the decimal notation whether or not it is even; the last digit of every even number is one of 0, 2, 4, 6, 8.

   c. Since there are no days containing 25 hours, sad to say, the set of all such days is the null set.

   e. This does not constitute a set unless one defines accurately what one means by "the most beautiful."

   g. This will be a set even though we do not know who will be shot in the next ten years, since a person is either in this category or not. The set might even be the null set.

5. Each of A and B would have to be the null set.

7. $A \cap B$ will be the same as A only if every element of A is in B; that is, A is a subset of B.

9. Suppose there were two null sets: $\varnothing'$ and $\varnothing$. Then since every set contains the null set $\varnothing$ we would have $\varnothing' \supseteq \varnothing$. Similarly, since every set contains the null set $\varnothing'$ we have $\varnothing \supseteq \varnothing'$. Thus each of the sets $\varnothing$ and $\varnothing'$ is contained in the other and they must be the same.

Section 3, p. 8

1. $(A \cup B) \cup C$ and $A \cup (B \cup C)$ are both the set of elements in one or more of A, B, and C. Also $(A \cap B) = (B \cap A)$ since the set of elements in both A and B is the same as the set of elements in both B and A. The other properties have been shown already. Recall our agreement about the word "or."

3. If A is the set of houses with white blinds, B the set of houses with storm windows, and C the set of two story houses, then $(A \cap B) \cap C$ is the set of two-story houses with white blinds and storm windows. If A is the set of even numbers, B the set of negative numbers, and C the set of numbers different from zero, then $(A \cap B) \cap C$ is the set of all negative even numbers except zero.

5. If the interchange indicated is made in the first table we have

$\cap$	B	$\varnothing$
B	B	$\varnothing$
$\varnothing$	$\varnothing$	$\varnothing$

This is the same as the second table, since except for $B \cap B$, the symbol in the body of the table is $\varnothing$.

7. Let d be an element of $A \cap (B \cup C)$; that is, d is in A and in either B or C. Thus d is either in A and B or in A and C. Thus d is in $(A \cap B) \cup (A \cap C)$. On the other hand, suppose d is in $(A \cap B) \cup (A \cap C)$. If it is in $(A \cap B)$, then it is in $A \cap (B \cup C)$; if it is in $(A \cap C)$, it is in $A \cap (B \cup C)$. Thus we have shown that each element in one of the sets is in the other.

9. First, to show that $A \cup B = B$ implies $A \cap B = A$, note that the first states that A has no elements which are not in B; that is, every element of A is in B, which means that $A \cap B = A$. Second, if $A \cap B = A$, this means that every element of A is in B, which is the same as saying that $A \subseteq B$.

Actually in both these cases we are in a sense not playing fair, for we are really proving $A \subseteq B$ on the way. Another way of proving the first assertion is longer and possibly a little more "honest." Using $A \cup B = B$ we have $A \cap B = A \cap (A \cup B) = (A \cup A) \cap (A \cup B)$ by the distributive property. But $A \cup A = A$, and we have $A \cap B = A \cap (A \cup B) = A$.

11. By property (5), $A \cup B = B$ implies $A \subseteq B$. Interchanging B and A (see Exercise 6) we see that $A \cap B = B$ implies $B \subseteq A$. The two results together imply that $A = B$.

13. By property (5), $B \cup A = B$ implies $A \cap B = A$. Hence, if $A \cap B = \varnothing$, then $A = \varnothing$.

15. If A is the set of rainy days, B the set of Thursdays, and C the set of Saturdays, then $A \cap (B \cup C)$ is the set of rainy days which are either Thursdays or Saturdays. This is the same as the set consisting of rainy Thursdays and rainy Saturdays.

17. Properties (1), (2), (3), and (4b) hold. Property (4a) does not hold. $A \uplus B = B$ implies $A = \varnothing$, and thus (i) and (iii) of property (5) hold. However, $A \subseteq B$ does not imply $A \uplus B = B$. Notice that in this connection the null set is a subset of A, B, and $A \uplus B$. Although it is a common subset, $A \uplus B$ contains no *elements* in common with $A \cap B$.

Section 4, p. 10

1. The set $(A' \cap B')$ will be that region outside both A and B, that is, region 4 of the diagram. The set $(A \cup B)'$ is the set of points outside the union of A and B, that is, region 4, again. The two represent the same region.

3. $(A \cap B)$ is a region 2 and $(A' \cap B')$ is region 4, that is, the region outside both A and B. Thus $(A \cap B) \cup (A' \cap B')$ is the region in both or neither of A and B.

5. Here the symbol $(A - B)$ means the elements in A but not in B and $(B - A)$ those in B but not in A. The union of the two is the set of elements in A or B but not both. This is what is meant by $(A \cup B) - (A \cap B)$. (This is the "exclusive or"; see Exercise 17 of Section 3.)

7. $A - B$ will be the same as A if and only if there are no elements of B which are in A, that is, if the two sets are disjoint.

9. Use the Venn diagram in Fig. 1:3 to see that

$$A \cap (B - C) = (1 \cup 2 \cup 3 \cup 4) \cap (2 \cup 5) = 2,$$
$$(A \cap B) - (A \cap C) = (2 \cup 3) - (4 \cup 3) = 2.$$

Hence the sets are the same.

Section 6, p. 11

1. The relationship "less than or equal to" has the three properties.

3. For property (1) set $B = A$ and see that $A \cup A = A$ and $A = A$.

For property (2) $A \subseteq B$ is equivalent to $B \cup A = B$ and $B \subseteq A$ is equivalent to $A \cup B = A$. Thus $(A \cup B) = (B \cup A)$ implies $A = B$.

For property (3) $A \subseteq B$ is equivalent to $B \cup A = B$ and $B \subseteq C$ is equivalent to $C \cup B = C$. We want to show that these imply $C \cup A = C$. This follows from the sequence

$$C \cup A = (C \cup B) \cup A = C \cup (B \cup A) = C \cup B = C.$$

5. The statements imply that $C \subseteq G \uplus F$, where $\uplus$ denotes the exclusive "or," which excludes elements of both G and F. Thus no element of $G \cap F$ can be in C.

7. **a.** Let the set be the positive integers and zero and denote the relationship by R. Then if a and b are two numbers of the set, we write aRb; that is, a is in the relationship R with b, if $0 \leq b - a \leq 1$. In words, a is in the relationship R to b if the difference $b - a$ is between 0 and 1 inclusive. First, property (1) holds, since aRa means $0 \leq a - a \leq 1$, which is true since $a - a = 0$. Second, property (2) holds since aRb implies $0 \leq b - a$ and bRa implies $0 \leq a - b$. Thus $0 = b - a$ and $b = a$.

Finally, to see that property (3) does not always hold, select $a = 0, b = 1, c = 2$. Then aRb since $0 \leq 1 - 0 \leq 1$; bRc since $0 \leq 2 - 1 \leq 1$. But aRc does not hold since $c - a = 2 > 1$; that is, $c - a$ is not between 0 and 1 inclusive.

CHAPTER 2

Section 3, p. 18

1. **a.** The lake is rough when the wind blows. The wind blowing is sufficient to make the lake rough. If the wind blows, necessarily the lake will be rough.

b. A google is rambunctious when it is flaccid. Being flaccid is sufficient to make a google rambunctious.

c. A necessary condition that a number be a multiple of 6 is that it be a multiple of 3.

d. If you do not call me to tell me you are safe, I shall not sleep. I will sleep only if you call me to tell me you are safe. Your calling me to tell me you are safe is necessary to my sleep.

    e. If he remembers, I shall be surprised. His remembering is sufficient to make me surprised. I shall necessarily be surprised if he remembers.

    f. We smell the stockyards if the wind blows from the east. The east wind is sufficient for the smell of the stockyards where we are. When the east wind blows, necessarily we smell the stockyards.

  3. (Converses for Exercise 2)

    a. If I keep awake at night, it is because of drinking coffee.

    b. Every isosceles triangle is equilateral.

    c. If it is sweet, it is a use of adversity.

    d. If it gathers no stones, it is rolling moss.

    e. If two lines do not meet, they are parallel.

    f. If a number is divisible by 3 and 5, it is divisible by 15.

  5. Statements a and d.

  7. The table would be:

Hypothesis	Conclusion	Implication
True	True	True
True	False	False
False	True	False
False	False	False

Then "p implies q" would be the same as "p and q." Another possible agreement might be that the implication is true when the hypothesis and conclusion are both true or both false but otherwise false.

  9. **a.** By finding a time when the sun is shining through the rain.

    c. By finding a case when A is not the null set and still $A \cup B = \varnothing$. This is not possible.

    e. By finding an example in which the implication is true but the converse is not.

    g. One would try to find a set B containing A and for which $A \cup B = A$ and yet $B \neq A$. This would be impossible.

Section 4, p. 22

  3. Then $(A \cap B)'$ corresponds to $(p \wedge q)'$. We have the following truth table:

p	q	$p' \vee q'$	$p \wedge q$	$(p \wedge q)'$
T	T	F	T	F
T	F	T	F	T
F	T	T	F	T
F	F	T	F	T

The third and last columns are the same.

  5. The symbol $(p' \vee q)'$ means that neither of the following is true: not p, q. This means that both are false. But "not p" is false when p is true. Thus the statement is equivalent to "p and not q." We can also show it by interchanging p and p' in $(p \vee q)' \leftrightarrow (p' \wedge q')$. Finally, by use of the truth table we have

p	q	p'	$(p' \vee q)$	$(p' \vee q)'$	q'	$p \wedge q'$
T	T	F	T	F	F	F
T	F	F	F	T	T	T
F	T	T	T	F	F	F
F	F	T	T	F	T	F

7. Note that the implication $p \rightarrow q$ is true except when p is true and q is false. Thus the negation of the implication is false except when p is true *and* q is false.

    a. Hence the negation of the statement of Exercise 2a in Section 3 is: "I drink coffee in the evening and it does not (always) keep me awake at night."

    b. "The triangle is equilateral and not isosceles." That is, the implication will be denied if we can find an equilateral triangle which is not isosceles.

    c. "There is a use of adversity which is not sweet."

    d. "There is some rolling moss which gathers stones."

    e. "Two lines are parallel and they meet." That is, "there are two parallel lines which meet."

    f. There is a number divisible by 3 and 5 which is not divisible by 15.

9.   **a.** Today is not Wednesday or it is not my birthday.

    b. Today is Wednesday or it is not my birthday.

    c. Today is Wednesday and it is not my birthday.

    d. Today is not Wednesday and it is my birthday.

11.

⊻	T	F
T	F	T
F	T	F

13.

	Contra	Conv	Obv
Contra	i	Obv	Conv
Conv	Obv	i	Contra
Obv	Conv	Contra	i

For instance, the converse of the obverse is: The converse of the contrapositive of the converse. But the (contrapositive of the converse) is, by the previous exercise, the (converse of the contrapositive). Thus the converse of the (contrapositive of the converse) is the converse of the (converse of the contrapositive). But the converse of the converse is the implication (we have an associative property) and thus our answer is: the contrapositive of the implication, that is, the contrapositive.

Section 5, p. 25

1. The diagram would show that the set of Polynesians is contained in the set of brown persons. The following are equivalent to the given one: a, d, and e. The statement would be false if a Polynesian were found who is not brown.

3. The set of Americans is a subset of the set of Canadians. The following are equivalent: b, and d. If we could find an American who is not a Canadian, the given statement would be proved false.

5. The set of days when the sunrise is red is contained in the set of rainy days. The given statement is equivalent to c. The statement would be proved false if we were to find a day with a red sunrise in which it did not rain.

7. Denver is not an element of the set of places which have earthquakes. Statements equivalent are a and e. If there should be an earthquake in Denver, the statement would be proved false.

9. The intersection of the set of soldiers and the set of cruel persons is not the null set. None of the statements is equivalent to the given one. To prove the statement false one would have to show that no soldier is cruel, or, depending on the interpretation of "some," more than one.

11. The intersection of the set of strange persons and those who like mathematics is not the null set. None of the conclusions are warranted. To deny the statement one would have to show that nobody who likes mathematics is strange.

Section 6, p. 29

1. If M stands for the set of men, W the set of white animals, and A the set of Americans, the diagram here would be consistent with the statements given. In that case there need be no men who are Americans. Even if all men were white animals, none need be Americans. It would be possible for all men who are Americans to be of some color other than white.

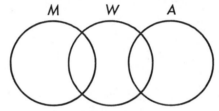

Figure 2:6:1

3. If a hydroplane is called a seagoing vessel, all overlappings would be possible. Thus some of each category would be in each other category. For instance, there could be a hydroplane bomber.

5. If the second statement is put first, we have a syllogism which implies sentence a. Statements b, c, and d do not follow. Statement e is the contrapositive of a and hence does follow. Other statements which would follow are: "If he is not brown, he is not Polynesian," and "If he is not Polynesian, he is not Balinese."

7. Again there is a syllogism if the order of the two given statements is changed. Statement a is implied. Statement c is also implied. One could include, as well, the contrapositive of each of the given statements and of statement a.

9. Only statement b follows.

11. "I do not like cabbages; Brussel sprouts are like cabbage; hence I do not like Brussel sprouts." An unmentioned hypothesis might be that my dislike extends not only to cabbages but to anything like cabbages.

13. The only statement which follows is e.

15. There would be a number of interpretations here but one would be this: If men persecute you and revile you for my sake, your reward in heaven will be great. (Additional evidence for this statement is that the prophets were similarly persecuted and their reward in heaven is great.) The conclusion is not valid because of the omission of the phrase "for my sake."

17. Statement a follows but not statement b.

19. The two assumptions are: "Men never tell the truth to a girl" and "He has not told me he loved me." These do not form a syllogism.

21. The two statements would be: "If a man is in love, he does not tell the truth," and "He told me he loved me." The conclusion given is warranted. For, if he were in love the first statement would say that he did not tell me the truth, which contradicts the statement that he told me he loved me. Hence the only conclusion is that he does not love me.

23. To abbreviate the hypotheses, we have: a. The government must not diminish opportunities for work of its citizens. b. We must not be needlessly dependent on other countries. Then it follows that to exchange one hour of American labor for two or three of foreign labor acts against both of the hypotheses (except for the condition "needlessly" in part b). In addition, the American workman is deprived of purchasing power. Given the validity of the two hypotheses, this is a logical argument, except for the doubt left by the word "needlessly."

25. The two hypotheses are: If a car shows evidence of being in a wreck it is a doubtful purchase. A welded or straightened frame is evidence that it has been in a wreck. Hence a car with such a frame is a doubtful purchase. This is logical.

27. The first three lines have a certain logic. One assumes that losing money and/or a friend is bad. The key word is "oft." If it happens more often that a loan makes a friend, the conclusion would be different. The last three lines are merely a statement which many of us doubt.

29. The principal conclusion would be that each statement implies the other, that is, that all three are equivalent. For an example of this, see property (5) of Section 3 of Chapter 1.

31. If T stands for "It is Tuesday" and S for "Susie stops in for tea," the first statement may be expressed $T \rightarrow S$. If C stands for "I have cinnamon rolls" and if B stands for "I go to the little bakery around the corner," the second statement may be expressed $C \rightarrow B$. The third statement can then be written (not C) $\rightarrow$ (not S). The contrapositive of the last is $S \rightarrow C$. Thus we have $T \rightarrow S \rightarrow C \rightarrow B$; this tells us that on Tuesdays I always go to the little bakery around the corner.

Section 8, p. 37

1. The mere fact that a depression occurred during Hoover's administration does not indicate that it was his fault. One would need more information. Of course statement b is even less tenable than a.

3. Those who go to graduate school may choose Reed College in larger numbers than those who do not go on. The facts mentioned constitute evidence but more is needed.

5. Merely the incidence is not enough, although again it is evidence. In this and many other cases, one needs a controlled experiment. If two individuals are alike except that one smokes and the other not, then if the former developed cancer and the latter did not, one might assume the conclusion stated, especially if the experiment were repeated with the same results.

7. Again the conclusion is not justified. Feeblemindedness might lead to excessive drinking.

9. There is some evidence for the conclusion. But longer compulsory education might change statistics without changing the number of criminals.

11. Conclusion a would seem to be justified if in all other respects John's diet and mode of living remained the same. Further evidence would be furnished if he stopped eating yeast and began to lose weight. Of course, what makes *John* gain might not work for other persons.

Section 9, p. 40

1. Even if everyone born in March had bad feet, it would of course not mean that such trouble could not be eliminated.

3. "One cannot control supply by controlling production. It is unwise to reduce production toward the danger point." These are apparently the two hypotheses, but one can certainly form no syllogism from them. The last sentence is certainly not a conclusion from the two previous ones.

5. "The object of reading a book is to learn from the author. It is easier to learn from the author if you put yourself in a receptive mood. Hence put yourself in a receptive mood." The other statements elaborate on this theme. The argument seems a logical one. Although Ruskin and Emerson seem to take opposite points of view, self-reliance is not incompatible with learning from someone else.

7. One does not need to belabor the fallacy here.

9. There is really no argument here. The paragraph consists of examples supporting the first sentence.

11. "Low temperatures may cause condensation on the walls. Condensation contains acid. If the chimney is not tight this causes corrosion. Therefore, to avoid corrosion use a tile lining." More accurately, "to avoid corrosion from this source a tile lining can be helpful." This seems a logical argument.

13. "To have an understanding of the war Americans must see how the fighting goes. To see why we fight, one must see the manifestations of democracy. *Life* shows these things." These are the three hypotheses. The tacit assumption is that Americans want to see and be shown and also that they agree with the hypotheses. On these assumptions, how can *Life* fail to be read!

15. The parent could say: "You will not give back my child." Then, if the kidnapper keeps the child, the parent would be truly saying what would happen and the kidnapper would be breaking his word. Hence the kidnapper must return the child. Notice that the kidnapper did not say that he would keep the child if the parent did not tell him truly what would happen.

Section 10, p. 45

1. The answers for this question are the diagrams for Exercise 3.

3. The answers for this question are the expressions for Exercise 1.

5. There is a slight simplification for part a in that the parentheses may be omitted.

Part c is equivalent to $s + v$ because, from the diagram, the circuit is closed if S or V is closed, while if neither S nor V is closed there is no closed circuit through T. Also 4e is a simplification of 3e.

7. Let T stand for the switch at the top of the stairs and B that at the bottom and let both be two-way switches, as in Example 5. Denote the two positions by 1 and 0. Suppose that when both switches are in position 1, the light is on. Then it is to be turned off from either T or B by flipping the switch. This will be so if 1, 0 or 0, 1 corresponds to the light being off. Then, if the light is off, it must be possible to turn it on by flipping the switch at the top of the stairs. Suppose 1 is the position for T and 0 for B. If the top switch is flipped, it comes into the 0 position, and hence the light is to be on when both switches are in the zero position. If, on the other hand, the position for T is 0 and that for B is 1, flipping the switch for T will result in positions 1 and 1, when the light is on. One can deal similarly with the switch at the bottom of the stairs. Hence we wish the following:

Light on: 1,1 or 0,0;   light off: 1,0 or 0,1

The diagram shown here will serve our purpose. There is another answer to this exercise which is just as good.

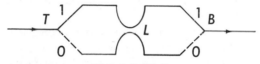

Figure 2:10:7

Section 11, p. 49

1. The expressions are Exercises 4e and 3e of Section 10 and diagrams for Exercises 2e and 1e, respectively. If we make the replacements given we would have

$$s + (t \cdot v) \stackrel{?}{=} (s + t) \cdot (s + v).$$

This is not true in general numbers. For example,

$$3 + (4 \cdot 5) \neq (3 + 4) \cdot (3 + 5),$$

since the former is equal to 23 and the latter to 56.

2. No. Yes.

3. Here we would use two double switches and obtain the result illustrated. Notice that this is essentially the same diagram as for the answer of Exercise 7 of Section 10.

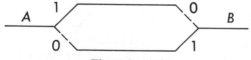

Figure 2:11:3

5. If we replace $\varnothing$ by 1 and B by 0 we have the table for a series connection. If we replace $\varnothing$ by 0 and B by 1 we have the table for a parallel connection.

7.

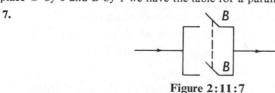

Figure 2:11:7

9. This would correspond to a switch permanently open.

11. The table will be

	1	0
1	0	1
0	1	0

This would correspond to the exclusive "or"—see the answer to Exercise 3.

CHAPTER 3

Section 1, p. 54

1. Order is important in all except perhaps the shopping list, writing to one's sweethearts, and putting gasoline and water into a car.

3. Triplets, quintette, couple, brace, triad.

5. If set S of people has fewer people than set T, then after one sets up a one-to-one correspondence between every person in S with a person in T, there are some in T which are left out of the pairing.

7. "Twice as many" is associated with a two-to-one correspondence, for example, the eyes and the nose, the two sides of a coin and the coin.

9. Certainly the first and third properties hold. The second holds if it is understood that the equality sign is also replaced by "equivalent." In place of this property is the stronger one: If A is equivalent to B, then B is equivalent to A.

11. The union of two finite sets is a finite set, since if $n(S)$ and $n(T)$ are numbers, then the number of elements in $S \cup T$ is a number not more than $n(S) + n(T)$. (See Exercise 8 in Section 2.)

13. The union of a finite set and an infinite one is always an infinite one, since the union of two sets has at least as many elements as either.

15. We let n correspond to $2n$, for $n = 0, 1, 2, 3, 4, \ldots$.

Section 2, p. 57

1. The table need contain only the sums $a + b$, where a is less than or equal to b. The number of entries is $10 + 9 + 8 + \cdots + 2 + 1 = 55$ instead of 100.

3. If s is an integer, there is a set S such that $n(S) = s$ and, for t, a set T such that $n(T) = t$ and $T \cap S = \varnothing$. If $s = 0$, then $S = \varnothing$. Now $s + t = n(T \cup S)$, and since $T \cup S$ is a finite set, it has a number associated with it, that is $s + t$.

5. Let s, t, and v be three nonnegative integers and S, T, and V the sets whose numbers are s, t, and v, respectively, such that no two sets have any element in common. Then

$$(s + t) + v = n((S \cup T) \cup V) = n(S \cup (T \cup V)) = s + (t + v),$$

since the union of three sets is associative.

7. We use the associative and commutative properties of addition.

9. Let B be the set of blond persons and G be the set of girls. Then

$$n(B \cup G) = n(B) + n(G) - n(B \cap G) = 25 + 16 - 10 = 31.$$

Hence there are 31 persons in the room.

Section 3, p. 61

1. 60.

3. **a.** $(5 + 1) \cdot 0 + 17 = 17.$ **b.** $8 + 12 + 55 = 75.$ **c.** $0 \cdot (7 \cdot 6) = 0.$

5. **a.** "*e* added to the sum of *c*, and *a* plus *b*."

b. "*d* multiplied by the product of *ab* and *c*."

c. "*t* multiplied by the sum of *z* and the product of *x* and *y*."

7. One may omit the second pair of brackets and the second pair of parentheses in Exercise 2b and all the brackets and parentheses in Exercises 5a and 5b. Of course, more may be omitted if one uses the convention that, for instance, $6 \cdot 3 + 5$ means $18 + 5$; but this is not by the associative property.

9. $ab = b + b + \cdots + b$, where there are *a* *b*'s in the sum. Since each *b* is a positive integer, the sum will be. In fact, the set of positive integers is closed under addition.

11. From Exercise 9, $(a + b)c = 0$ implies that $c = 0$ or $a + b = 0$. In the latter case, $a + b = 0$ implies that $a = b = 0$.

Section 4, p. 63

1. $(3 + 4) + 5 = 3 + (4 + 5)$; that is, $7 + 5 = 3 + 9 = 12$.

3. If $a = 0$, then $0 \cdot c = 0(bc) = 0$; if $b = 0$, then $0 \cdot c = a \cdot 0 = 0$; if $c = 0$, then $(ab) \cdot 0 = a \cdot 0 = 0$.

5. **a.** $(3 + 5)(4 \cdot 5)[3 + 4 \cdot 5] \cdot [4 \cdot 5 + 3 \cdot 4] \cdot [3 + 5 + 6 \cdot 2] =$
$$8 \cdot 20 \cdot 23 \cdot 32 \cdot 20 = 2,355, 200.$$

b. $(2 + 3)[2 + 1 \cdot 7] \cdot 3 \cdot (7 + 6) \cdot [4 + 3 \cdot 2] \cdot [7 \cdot 5 + 6 \cdot 3] =$
$$5 \cdot 9 \cdot 3 \cdot 13 \cdot 10 \cdot 53 = 930,150.$$

7. There are fifteen possibilities for placing one pair of parentheses on $1 + 2 \cdot 3 + 4 + 5$. All are equal to 16 except the following:

$$(1 + 2) \cdot 3 + 4 + 5 = 18, \qquad 1 + 2(3 + 4) + 5 = 20, \qquad 1 + 2(3 + 4 + 5) = 25.$$

9. The ( ) in $a(b + c) = ab + ac$ cannot be omitted because, without them, the statement would mean $ab + c$, which is equal to $ab + ac$ only when $c = ac$; that is, $a = 1$ or $c = 0$.

11. $(2 + 4)(3 + 5) = 6 \cdot 8 = 48 = 6 + 12 + 10 + 20$. Also

$$(a + b)(x + y) = (a + b)x + (a + b)y = ax + bx + ay + by.$$

13. The correct right-hand sides are as follows (reasons omitted): the first two parts of (1), the last two of (2), the first two of (3), the first part of (4), and the first part of (5).

15. $1 \cdot 5 + 3 = 5 + 3 = 1(5 + 3)$. But $4 \cdot 5 + 3 = 23 \neq 4(5 + 3) = 32$. See the answer to Exercise 9.

17. The student described would be using a kind of distributive property for multiplication alone. We know from the associative property for multiplication that $3 \cdot (4 \cdot 5) = (3 \cdot 4) \cdot 5 = 12 \cdot 5$, which is not equal to $12(3 \cdot 5)$.

19. (ii): $A \cup B = B$ corresponds to $a + b = b$, which holds only if $a = 0$.

(iii): $A \cap B = B$ corresponds to $a \cdot b = a$, which holds only if $a = 0$ or $b = 1$.

Section 7, p. 70

1. $(234)(56) = (200 + 30 + 4)(50 + 6) = (200 + 30 + 4)50 + (200 + 30 + 4)6$
$$= 200 \cdot 50 + 30 \cdot 50 + 4 \cdot 50 + 200 \cdot 6 + 30 \cdot 6 + 4 \cdot 6.$$

3. a. eee in the dozal system is equivalent to $11 \cdot 144 + 11 \cdot 12 + 11 = 1727$ in the decimal system. Also one can note that eee $= 1000 - 1$ in the dozal system, which is equal to $12^3 - 1 = 1728 - 1$ in the decimal system.

b. t5e is, in the decimal system, equal to $10(144) + 5 \cdot 12 + 11 = 1511$.

c. tete $=$ te(100) $+$ te $=$ te(101) in the dozal system. In the decimal system this is equal to the product $131 \cdot 145 = 18{,}995$.

5.

Decimal system	327	4576	17,560
Base twelve	233	2794	tle4

7. a.

Base twelve	Decimal system	**b.**	7te
t6e	1523		80
420	600		t19
6e1	997		————
————	————		1688
1980	3120		

c. 240e. The given sum is equal to $95e + 670 + 1000$ in the system to the base twelve.

9. a.

Base twelve	Decimal system	**b.**	(eee)(te) $=$
3t4	556		$(1000 - 1)$te $=$
23	27		te000 $-$ te $=$ tte11.
———	———		
e70	3892		
788	1112		
———	———		
8830	15012		

c. 2eee84 in base twelve $= 746{,}452$ in the decimal system.

11. $12^3 - 1 = 1727$.

Section 8, p. 72

1. 101011001; 2340; 180.

3. The first card will have all the odd numbers less than 32, that is, all whose last digit is 1 when written in the binary system. The second will contain those numbers whose next to the last digit is 1 in the binary system. The third contains numbers whose second from the last digit is 1 in the binary system, and so on for five cards. For instance, 22 is 10110 in the binary system and appears on cards, 10000, 100, and 10, that is, 16, 4, and 2 in the decimal system.

5. $51 = 1 \cdot 49 + 0 \cdot 7 + 2$, that is, 102 in the system to the base seven. $51 = 1 \cdot 32 + 1 \cdot 16 + 0 \cdot 8 + 0 \cdot 4 + 1 \cdot 2 + 1$, that is, 110011 in the binary system.

7. A number written in the binary system is divisible by 8 if and only if its last three digits are zero, just as for the decimal system a number is divisible by 10^3 if its last three digits are zero.

9. $(55)^2 = 4444$ and $(26)^2 = 1111$ in the system to the base seven.

11. 4210 in the system to the base five is equal to $4 \cdot 125 + 2 \cdot 25 + 1 \cdot 5 + 0$, which is 555 in the decimal system.

13. The symbols 3, 4, and 5 represent the same numbers in the decimal and dozal systems and hence the equation holds in both. But 12 and 13 represent different numbers in the two systems. These in the system to the base twelve are the same numbers as 14 and 15 in the decimal system and

$$5^2 + 14^2 \neq 15^2.$$

Hence the given equation does not hold if the numbers are written in the system to the base twelve.

15. $101 = 1 \cdot 64 + 1 \cdot 32 + 0 \cdot 16 + 0 \cdot 8 + 1 \cdot 4 + 0 \cdot 2 + 1$
$$= 2(1 \cdot 32 + 1 \cdot 16 + 0 \cdot 8 + 0 \cdot 4 + 1 \cdot 2 + 0) + 1.$$

Hence when we divide 101 by 2 we have the remainder 1, which is its last digit written in the binary system, and the quotient

$$1 \cdot 32 + 1 \cdot 16 + 0 \cdot 8 + 0 \cdot 4 + 1 \cdot 2 + 0 = 50.$$

When 50 is divided by 2, the remainder 0 is the next-to-last digit in the binary representation of 101 and the quotient is

$$1 \cdot 16 + 1 \cdot 8 + 0 \cdot 4 + 0 \cdot 2 + 1 = 25.$$

And so one continues to find the digits in the binary representation one by one.

17.   The first card is headed by 1 on one side and carries the numbers 1, 4, 7, 10, 13, 16, 19, 22, 25, whose remainders are 1 when divided by 3, that is, whose last digit in the number system to the base three is 1. The opposite side of this card is headed by 2 and contains the numbers whose last digit in the system to the base three is 2, that is 2, 5, 8, 11, 14, 17, 20, 23, 26. The second card is headed by a 3 on one side and contains all those numbers whose next-to-last digit in base 3 is 1, that is, 3, 4, 5, 12, 13, 14, 21, 22, 23. The opposite side is headed by a 6 and contains 6, 7, 8, 15, 16, 17, 24, 25, 26. The third card is headed by a 9 on one side and has the numbers 9, 10, 11, 12, 13, 14, 15, 16, 17, and on the other side is an 18 with the numbers from 18 to 26 inclusive.

19.   The weights are 1, 2, 4, 8, 16, 32. To find the weights to be used for any amount not greater than 63, write the number in the binary system and choose the weights corresponding to the digits which are 1s. For instance 11 is 1011 in the binary system and we use weights 1, 2, and 8, corresponding to the numbers 1, 10, and 1000 in the binary system.

Section 9, p. 75

1.   Suppose it is A's turn to play and he is faced with 1 2 3. If he takes the 1, his opponent can take 1 from the 3 pile and leave him with a 0 2 2, which is losing. In fact, whatever A does, B can leave him with two piles having the same number of counters, which is losing. Hence 1 2 3 is a losing combination. The story is about the same for 1 4 5 and 2 1 1. In fact, the only winning combination in the list is 3 4 5, since, faced with this, A can take two from the 3 pile and leave his opponent with 1 4 5, which is losing.

Section 10, p. 77

1. **a.**   Remove 5 from the pile containing 9 counters.
     b.   Remove 14 from the pile containing 29.
     c.   Remove 2 from the pile containing 8.
     d.   This is a losing combination.

3.   The first column, counting from the left, whose sum is odd must have either three or one 1s. In the former case one may choose any pile from the three and there will be one and only one way to remove counters from this pile to make all column sums even. In the latter case there is no alternative but to remove counters from the pile whose binary representation has a 1 in this column. So there can never be just two choices.

Section 12, p. 80

1. Calling one's child, burning of paper, and sailing a boat do not have inverses.

3. $3/(4 + 5) \neq 3/4 + 3/5$ and, in general, $a/(b + c) \neq a/b + a/c$. But in the other order $(4 + 5)/3 = 4/3 + 5/3$ and, in general, $(b + c)/a = b/a + c/a$. Hence in the second order the distributive property does hold.

5. $5 + (3 - 2) = 5 + 1 = 6$ and $(5 + 3) - 2 = 8 - 2 = 6$. The equality holds in the following: $a + (b - c) = (a + b) - c$ because $a + (b - c) + c = a + b$ and $(a + b) - c + c = a + b$, using the associative property.

7. $7 - (3 - 2) = 7 - 1 = 6$ but $(7 - 3) - 2 = 4 - 2 = 2 \neq 6$. Hence the two are not equal.

9. $723 - 345 = 700 + 20 + 3 - (300 + 40 + 5)$. 723 is also equal to the sum $600 + 120 + 3 = 600 + 110 + 13$ and hence

$$723 - 345 = (600 + 110 + 13) - (300 + 40 + 5)$$
$$= (600 - 300) + (110 - 40) + (13 - 5)$$

from the commutative and associative properties and the definition of $a - b$. The last is equal to $300 + 70 + 8 = 378$.

11. Since a succession of any three numbers with solidi between is ambiguous, one must have two parentheses or brackets and parentheses. The possibilities are

$$256/[(16/8)/2] = 256/1 = 256,$$
$$256/[16/(8/2)] = 256/4 = 64,$$
$$(256/16)/(8/2) = 16/4 = 4,$$
$$[256/(16/8)]/2 = 64,$$
$$[(256/16)/8]/2 = [16/8]/2 = 1.$$

13. If a is any positive integer $a \cdot 0 = 0$ and, since 0 is an integer, 0 is thus a multiple of a. Also $0/b = 0$, since $0 = 0 \cdot b$.

17. The numbers 3240 and 24 in the system to the base twelve are 5520 and 28, respectively, in the decimal system; the former is not divisible by the latter.

Section 14, p. 83

1. The seventh prime number is 17, which is less than B for $n = 3$ since, for $n = 3$, $B = 2 \cdot 3 \cdot 5 + 1 = 31$. For all greater n, B is larger. Hence for no $n \geq 3$ is B the $(n + 1)$st prime.

3. Suppose k is composite and $k = rs$, where neither r nor s is 1. If $r^2 > k$ and $s^2 > k$, then $r^2 s^2 > k^2 = r^2 s^2$, which is a contradiction. Hence $r^2 \leq k$ or $s^2 \leq k$.

5. $2 \cdot 3 \cdot 5 \cdot 7 + n$ is divisible by 2 if n is, divisible by 3 if n is, divisible by 5 if n is, or divisible by 7 if n is. But every number of the list 2, 3, 4, 5, 6, 7, 8, 9, and 10 is divisible by at least one of 2, 3, 5, 7. Hence $2 \cdot 3 \cdot 5 \cdot 7 + n$ is divisible by one of these for n between 2 and 10 inclusive. This shows that no such number is a prime number.

7. The set of prime numbers and that of positive integers are both infinite sets, just as the positive integers and the positive even integers are. In neither case do the two sets coincide.

9. From 4 to 998 inclusive there are $1 + \frac{1}{2}(998 - 4) = 498$ even numbers. None of these are prime numbers. The numbers 9, 15, 21, 27, and 33 are five more which are not prime. Hence there are at least $498 + 5 = 503$ composite numbers less than 1000. The same kind of argument could be used for 1,000,000 in place of 1000.

11. Vinogradof's theorem is implied by the theorem quoted. For suppose every even number greater than M is the sum of two prime numbers. If D is an odd number $>M + 3$, then $D - 3$ is an even number greater than M and would be the sum of two primes by the theorem quoted. Then $D = 3 + a + b$ for two prime numbers b and a.

CHAPTER 4

Section 3, p. 92

1. Addition is commutative since the table is symmetric about the diagonal from the upper left corner to the lower right, the closure property holds since only the numbers 0, 1, 2, 3, 4, 5, 6 appear in the body of the table, and there is an inverse always since each row contains a zero.

3. The answer depends partially on what one means by "round". Certainly the voyage did not rule out the possibility of an elliptical cross section or a cylindrical shape. It probably did show that the cross section is not a square.

5. (Partial answer) Circle with 6 divisions:

x	0	1	2	3	4	5
0	0	1	2	3	4	5
1	1	2	3	4	5	0
2	2	3	4	5	0	1
3	3	4	5	0	1	2
4	4	5	0	1	2	3
5	5	0	1	2	3	4

The group is Abelian, since it is symmetric about the diagonal. For 8 divisions the results are analogous.

7. (Partial answer) The product of two numbers can be zero without either being zero; for example, $3 \cdot 4 = 0$. There is no 1 in row 3 and hence no multiplicative inverse for 3. The set does not form a group under multiplication.

9. a. $x = 7$.   **b.** $x = 11$.   **c.** $2x = 5$ has no solution since the only values of $2x$ are 0, 2, 4, 6, 8, and 10.   **d.** $10x = 2$ has two solutions: $x = 5$ and $x = 11$.

11. Yes, for the multiplication table is

x	1	5	7	11
1	1	5	7	11
5	5	1	3	7
7	7	3	1	5
11	11	7	5	1

13. To solve $y \,\#\, a = b$, let a' be the inverse of a and add a' to both sides on the right: $b \,\#\, a' = (y \,\#\, a) \,\#\, a' = y \,\#\, (a \,\#\, a')$ by the associative property. But $a \,\#\, a' = e$ and

$y \mathbin{\#} e = y$. Hence if there is a solution, $y = b \mathbin{\#} a'$. Then, by substitution, one can show that this is a solution.

Section 4, p. 95

1. On the circle with 12 divisions there is no multiplicative inverse for the number 3, for instance.

3. Let $a = 3$, $b = 5$. Then $(-a) = 4$, $(-b) = 2$, and $ab = 3 \cdot 5 = 1$, $(-a)(-b) = 4 \cdot 2 = 1$.

5. By property (7), $0 \cdot x = 0$ no matter what x is.

7. All the properties of a field hold except for the existence of a multiplicative inverse of every number not zero.

9. There is an infinite number of integers. If two integers are not equal, one must be less than the other, whereas on a circle we have no inequalities.

Section 5, p. 98

1. $a \equiv b(\bmod m)$ means $a - b = mx$ and $b \equiv c(\bmod m)$ means $b - c = my$, where x and y are integers. Hence $a - c = (a - b) + (b - c) = mx + my = m(x + y)$. This implies that $a - c$ is divisible by m; that is, $a \equiv c(\bmod m)$.

3. If an element A were in two different classes L and M, then we could have B in class L and C in class M, with B and C not in the same class. This would deny property (3c). Also we could not speak properly of *the* class of A if there were two.

5. By property (5), $a \equiv b(\bmod m)$ implies $ac \equiv bc(\bmod m)$, and $c \equiv d(\bmod m)$ implies $bc \equiv bd(\bmod m)$. Then by property (3), $ac \equiv bd(\bmod m)$. One can also prove this result using the methods of Exercise 1.

7. Using the results of Exercise 5, we see that the two given congruences imply $a \cdot a^r = b \cdot b^r(\bmod m)$; that is, $a^{r+1} \equiv b^{r+1}(\bmod m)$. This result shows that, for $r = 1$, $a \equiv b(\bmod m)$ implies $a^2 \equiv b^2(\bmod m)$. Then, taking $r = 2$, we have that the last congruence implies $a^3 \equiv b^3(\bmod m)$. So we can continue up to any positive integer r.

9. $a^5 \equiv b^5(\bmod m)$ by Exercise 7 with $r = 5$; $7a^2 \equiv 7b^2(\bmod m)$ by Exercise 7 and property (5). The rest follows from property (4).

11. Here we can use the argument for result 2 in Section 2, as follows: Write the ath row of the multiplication table $(\bmod m)$:

(i) $a, 2a, 3a, \ldots, (m - 1)a$.

If two are congruent, call them $ra \equiv sa(\bmod m)$, which implies that m is a factor of $ra - sa = (r - s)a$. But m and a have no common factor greater than 1, and hence m is a factor of $r - s$. This is impossible unless $r = s$ since r and s are chosen from the set $1, 2, 3, \ldots, m - 1$. Thus the numbers of (i) are congruent in some order to the numbers $1, 2, 3, \ldots, m - 1(\bmod m)$. One of these must be b, and for this one we have a solution of $ax \equiv b(\bmod m)$. Notice that there is only one solution.

13. The system will work for any odd number of teams, since if m is the number of teams, the congruence $2x \equiv c(\bmod m)$ is solvable for any c if m is odd, using the result of Exercise 11. Thus, on day c, the opponent of a is determined by $a + x \equiv c(\bmod m)$ and the only team which plays with itself on that day is team x, where $2x \equiv c(\bmod m)$.

15. We know that for any integers a, b, and c we have $a(b + c) = ab + ac$. This implies $a(b + c) \equiv ab + ac(\bmod m)$. Thus the remainders when $a(b + c)$ and $ab + ac$

are divided by m are the same. Hence the numbers on the circle corresponding to the two sides of the congruence are the same.

Section 7, p. 102

1. $9^{30} \equiv (-1)^{30} \equiv 1 \pmod{10}$ and hence the last digit of 9^{30} is 1. Also $9^{30} \equiv 1^{30} \equiv 1 \pmod 4$.

3. We can compute the following table:

x	0	1	2	3	4	5	6
$x^3 \pmod 7$	0	1	1	-1	1	-1	-1

5. Notice that the "first barrel" corresponds to the sequence of numbers 1, 9, 17, 25, ..., that is, the numbers congruent to $1 \pmod 8$. Now $497 \equiv 1 \pmod 8$. Then counting 3 barrels to the right we see that the "500th" barrel is number 4.

7. If a number is greater than 9, the sum of its digits will be less than the number. So the process described decreases the number at each step until we have a number less than ten. If it is 9, we divide by 9 to get a remainder 0; if it is less than 9, what we have is the remainder. Hence in most cases we do not even have to divide by 9.

9. If we write the multiples of 9 from 9 to 90 every possible last digit occurs. On the other hand, the numbers from 10 to 17 inclusive and 28 and 29 have all possible last digits and none are divisible by 9. The same results holds for divisibility by 7.

11. Recall that for the decimal system $10 \equiv 1 \pmod 3$ and $\pmod 9$; thus for divisibility by 3 or 9 one may add the digits. In the number system to the base twelve, one seeks an x such that $12 \equiv 1 \pmod x$. Thus x must be a factor of $12 - 1 = 11$ and hence $x = 11$. Thus in this system we can check for divisibility by 11 by adding the digits.

13. Here $13 \equiv 1 \pmod x$ is true if x is a factor of $13 - 1 = 12$. Thus x can be any one of 2, 3, 4, 6, 12.

15. Now $1000 \equiv -1 \pmod 7$, since 7 is a factor of 1001. Hence $b + 1000a \equiv b - a \pmod 7$. Thus $324{,}457 = 457 + (1000)(324) \equiv 457 - 324 \pmod 7$.

17. Since $3 \equiv 1 \pmod 2$, if the sum of the digits of the number written in the system to the base 3 is divisible by 2, the number is divisible by 2. For instance, the number 1210 in this number system is divisible by 2, as well as 2101.

19. If a set of numbers is correctly added, it will check by casting out the nines since the remainders are involved. But it can check by this means without being correct. For instance, $12 + 14 = 8$ is not correct, but it checks.

21. If we had merely divided by 9, we would have seen the result only for the particular numbers involved.

Section 8, p. 109

1. From the last paragraph in the text, we need check only for closure and the existence of the identity in the subset. First, the set $\{a\}$ forms a subgroup, by itself, since a is the identity. If a subset contains b, it contains $b^2 = c$ and hence contains $\{a, b, c\}$, which is a subgroup. Since, from the table, d, e, and f are the only elements whose squares are the identity, we have just three groups with two elements: $\{a, d\}$, $\{a, e\}$, and $\{a, f\}$. If a subset contains d and e, it contains $de = c$; hence $c^2 = b$ and $db = f$ and it is the whole group. Similarly, if it contains d and b, it must be the whole group. Hence the only subgroups are $\{a\}$, $\{a, d\}$, $\{a, e\}$, $\{a, f\}$, $\{a, b, c\}$, and the whole group G(3).

3. Both groups have six elements and are Abelian. Call L the group of integers mod 6 with addition. 1 of H must correspond to 0 of L. Let x of H correspond to 1 of L and form the following table:

L	0	1	2	3	4	5
H	1	x	x^2	x^3	x^4	x^5

since $x \cdot x$ must correspond to $1 + 1$, $x \cdot x \cdot x$ to $1 + 1 + 1$, and so on. Notice that since $L = 0$ and $L = 6$ are the same (mod 6), we must have in H, $x^6 \equiv 1 \pmod 7$. Now, to have a one-to-one correspondence between the elements of H and those of L, the entries for H must be distinct. This is not the case for $x = 2$ since $x^3 \equiv 1 \pmod 7$. But the powers of 3 are distinct (mod 7), as may be seen from the following table:

n	1	2	3	4	5	6	
3^n	3	2	6	4	5	1	(mod 7)

Thus using $x = 3$ we have a one-to-one correspondence between the elements of H and L with the property that if t_1 and t_2 of L correspond to h_1 and h_2, respectively, of H, then $t_1 + t_2$ corresponds to $h_1 h_2$. Thus we have an isomorphism.

5. For (1) note that in b, 1 and 2 as well as 3 and 4 are interchanged. Hence if we do this twice we come back to the identity. We have the table

	a	b	c	d
a	a	b	c	d
b	b	a	d	c
c	c	d	a	b
d	d	c	b	a

For (2) note that $b^2 = c$, $bc = b^3 = d$, and $bd = b \cdot b^3 = b^4 = a$. So the table is

	a	b	c	d
a	a	b	c	d
b	b	c	d	a
c	c	d	a	b
d	d	a	b	c

Both are Abelian since the tables are symmetric about the diagonal, each row contains the identity element a, and closure holds. We assume associativity.

7. Let g represent the rotation of the square through 90° and see that the group consists of g, g^2, g^3, $g^4 = e$. It will be isomorphic with group (2) of Exercise 5 under the correspondence $b \leftrightarrow g$.

9. Here set up as in Exercise 3:

H	0	1	2	3	4	5	
K	1	x	x^2	x^3	x^4	x^5	(mod 9)

Here the value $x = 2$ works. We leave the details to the student. K is not isomorphic to $G(3)$ since the latter is not Abelian.

11. The table here is

	i	c	v	o
i	i	c	v	o
c	c	i	o	v
v	v	o	i	c
o	o	v	c	i

The isomorphism is with group (1) of Exercise 5, since the square of each element is the identity. Here we set up the correspondence:

a	b	c	d
i	c	v	o

13. Since $4^2 \equiv 16 \pmod{17}$ and $4^3 \equiv 13 \pmod{17}$, the isomorphism is with group (2) in Exercise 5 with the correspondence $b \leftrightarrow 4$.

CHAPTER 5

Section 3, p. 114

1. **a.** -5.   **b.** 15.   **c.** 4.   **d.** 3.

3. $-3°$.

5. $a + b + (-a) + (-b) = a + (-a) + b + (-b) = 0 + 0 = 0$, using the commutative and associative properties of addition and the definition of $(-a)$ and $(-b)$. Thus the additive inverse of $a + b$ is $(-a) + (-b)$, which is $-(a + b)$. If in (4) we replace a by $(-a)$ we get

$$(-a) + (-b) = -(b - (-a)) = -(b + a).$$

Section 4, p. 115

1. If $a = 5$ and $b = -7$, then $(-a)b = (-5)(-7) = 35$, $b(-a) = (-7)(-5) = 35$, and $-(ab) = -(-35) = 35$.

(2) If $a = (-3)$, $b = (-2)$, then $(-a)b = 3(-2) = -6$, $b(-a) = (-2)3 = -6$, and $-(ab) = -((-3)(-2)) = -6$.

3. $-3(2 - 7) = (-3)(-5) = 15$ and $(-3)2 + (-3)(-7) = (-3)(2 + (-7)) = (-3)(-5) = 15$.

5. **a.** The first is not equal to $r(S - 1)$ unless $r = S$. $Sr - S = S(r - 1)$ by the distributive property. $r(S - 1) + r - S = rS - r + r - S = rS - S = Sr - S$.

b. $(a + b)(x - y) = (a + b)x - (a + b)y = ax + bx - ay - by$, which is equal to the given expression. The other two are not.

c. The first and last are equal to the given expression and the other two are not.

7. Case I. a is negative and b and c are not. Then $(-ab)c = ((-a)b)c = (-a)(bc) = -(a(bc))$. Case II. b is negative and a and c are not. Then $(-ab)c = (a(-b))c = a((-b)c) = a(-bc) = -(a(bc))$. Case III. c is negative and a and b are not. This is shown similarly.

9. $(-a)(b - c) = (-a)(b + (-c)) = (-a)b + (-a)(-c) = -ab + ac = ac - bc$.

Section 5, p. 118

1. a. $1 > -5$.   **b.** $-4 > -7$.

c. If $a < b$, then $a - 5 < b - 5$ by property (1i), and by two parts of property (2i) $7a < 7b$ and $-8a > -8b$.

3. If $a > b$, then $a + c > b + c$. Also $ac > bc$ if c is positive, $ac = bc$ if $c = 0$, and $ac < bc$ if c is negative.

5. If $a \geq b$, then $a + c \geq b + c$. Also $ac \geq bc$ if c is positive, $ac = bc$ if $c = 0$, and $ac \leq bc$ if c is negative.

7. If $|a - b| + |b - c| = |a - c|$, one can see intuitively that, since under these conditions the distance between a and c is the sum of the distances between a and b and between b and c, then b must be between a and c. To prove this result we can show that if b is not between a and c, then the equality does not hold. If b is not between a and c, then a and c are on the same side of b. Suppose b is less than both a and c. Then $|a - b| = a - b$ and $|b - c| = c - b$ and their sum is $a + c - 2b$. This cannot be equal to $|a - c|$, since if $a + c - 2b = a - c$ we would have $2c = 2b$ and $c = b$, while if $a + c - 2b = c - a$ we would have $a = b$. Either of these contradicts our assumption that b is less than both a and c. The case when b is greater than both a and c can be dealt with similarly.

9. It is not necessary to consider subtraction, since if we add $(-c)$ to both sides of the inequality $a + c < b + c$ we get $a < b$.

11. $|a| < |b|$ does not always imply $|ac| < |bc|$, for if $c = 0$, we have $|ac| = |bc| = 0$.

13. Suppose $A \leq B$; then by property (3i) $B < A$ is impossible and thus $B \leq A$ implies $B = A$.

15. Property (3i) does not hold for inclusion of sets. For instance, let S represent the set of all odd integers and T the set of all even integers. Then neither set is contained in the other nor are they equal.

17. First, if $a \geq 0$, $b \geq 0$ we have $|ab| = ab$, $|a| = a$, and $|b| = b$. Second, if $a \geq 0, b < 0$, we have $|ab| = -ab, |a| = a, |b| = -b,$ and $-ab = a(-b)$. The other two cases are dealt with similarly: $a < 0, b \geq 0$ and $a < 0, b < 0$.

Section 8, p. 125

1. $29/42$, $-13/77$.

3. None of the equalities hold. For instance,

$$\frac{a + 3b}{3d} = \frac{(a/3) + b}{d}.$$

5. a. $\dfrac{ad + bc}{bd}$.   **b.** $\dfrac{adf + cbf + ebd}{bdf}$.

c. $\dfrac{a}{b} - \dfrac{c}{d} \dfrac{eh - fg}{fh} = \dfrac{a}{b} - \dfrac{ceh - cfg}{dfh} = \dfrac{adfh - bceh + bcfg}{bdfh}$.

7. If $a \neq 0$, $ab = 0$ implies $[(1/a)]$ $ab = (1/a)0$, $[(1/a)a]b = 0$, and hence $b = 0$. Similar reasoning shows that $b \neq 0$ implies $a = 0$.

9. $(r/s)(b/b) = rb/sb$. But $rb/sb = r/s$, since $rbs = sbr$.

11. (See Chapter 4, Section 5) Reflexive property: $a/b = a/b$, since $ab = ba$. Symmetric property: $a/b = c/d$ if and only if $ad = bc$, $c/d = a/b$ if and only if $cb = ad$, and

$ad = bc$ if and only if $cb = ad$. Transitive property: $a/b = c/d$ implies $ad = bc$; $c/d = e/f$ implies $cf = de$. We want to show that $a/b = e/f$, that is, $af = be$. Now $cb = ad$ implies $fcb = fad$. Replacing fc by de in the last equation gives us $deb = fad$. Since $d \neq 0$, we can multiply both sides by $1/d$ (or divide by d) and get $eb = fa$, which is what we wanted.

Section 10, p. 128

1. a. $1/2$. **b.** $-35/12$. **c.** $-34/15$. **d.** $(xad - xbd - yc)/x$.

3. $7/(-3) = -7/3 < (-3)/7 < 8/13 < 5/8$.

5. Since b and $b + 1$ are positive, $a/b > (a + 1)/(b + 1)$ is equivalent to $a(b + 1) > b(a + 1)$, that is, $ab + a > ba + b$, which is equivalent to $a > b$.

7. If a and b are rational numbers, they can be written in the forms r/s and t/u, respectively, where r, s, t, and u are integers and s and u are not zero. Then $ab = (rs/su)$, where rt and su are integers.

9. Using the notation in the answer to Exercise 7, where now neither r nor s is zero, we have $b/a = (t/u)(s/r) = ts/ur$.

11. We know that if b and d are positive, then $a/b < c/d$ if and only if $ad < cb$: this serves as a basis of comparison of the denominators in the given fractions. Hence if b and d are both positive, the larger of the fractions ac/bc, ca/da, is that with the smaller denominator.

13. If $b > c > 0$, the inequality $b > c$ is equivalent to $b/bc > c/bc$, that is, $1/c > 1/b$, which is the same as $1/b < 1/c$.

Section 11, p. 132

1. Five gross inches is 12 feet; four doza inches is 4 feet. Hence 542 inches in the dozal system is 16 feet and 2 inches.

3. A pint is half a quart and hence .1 quart in the binary system. A gill (one-eighth quart) is .001 quart in the binary system.

5. $7(.428571) = 2.999997 = 3 - 3/1,000,000$. Hence $.428571 = 3/7 - 3/7,000,000$.

7. $\frac{1}{11} - .090909 < .0000001$.

9. Since 3 is the first remainder when 1 is divided by 7, if we began the division at the second step, we would have $\frac{3}{7} = .428571 \ldots$. Similarly, 2 is the second remainder and the repeating part begins at the second place of the original decimal. Third, 6 is the fourth remainder. The answer to the last question is "not quite."

11. $\frac{1}{37} = .027027 \ldots$.

13. $\frac{1}{2}(\frac{1}{6} + \frac{5}{31}) = \frac{61}{372}$ is halfway between the given numbers.

15. $\frac{4}{11} < \frac{3}{8}$, since $4 \cdot 8 < 11 \cdot 3$; $\frac{23}{11} > \frac{27}{13}$, since $23 \cdot 13 > 11 \cdot 27$.

17. $\frac{1}{2} < \frac{2}{3}$, $\frac{2}{3} > \frac{3}{5}$, $\frac{3}{5} < \frac{5}{8}$, $\frac{5}{8} > \frac{8}{13}$. Thus it appears that successive differences are alternately positive and negative. To see that this is true in general, notice that three successive terms are

$$(c - b)/b, b/c, c/(b + c).$$

Then

$$b/c - (c/b)/b = (b^2 + bc - c^2)/cb,$$
$$c/(b + c) - b/c = (c^2 - bc - b^2)/c(b + c).$$

Both denominators on the right sides are positive and the numerator of the first is the negative of that of the second. Hence one difference must be positive and the other negative.

Section 12, p. 134

1. If p/q is a fraction in simplest form and $(p/q)^2 = 3$, then $p^2 = 3q^2$, which implies that 3 is a factor of p. Write $p = 3r$ and have $9r^2 = 3q^2$, which implies that $3r^2 = q^2$. This implies that 3 is a factor of q and the fraction p/q would not be in simplest form.

3. If p and q have no factor in common greater than 1, then this is also true of p^2 and q^2. Then, from the statement in the exercise $p^2 = 2q^2$ implies that p^2 is a factor of 2. The equation also shows that 2 is a factor of p^2. Since they are both positive, it follows that $2 = p^2$. This is impossible, since 2 is not the square of an integer. Hence our original supposition that $p^2 = 2q^2$ is false.

5. Suppose r is a prime number. Then $(p/q)^2 = r$ implies $p^2 = rq^2$ and we can use the same argument as in Exercise 3.

Section 13, p. 136

1. Here we use the Pythagorean theorem, which affirms that $a^2 + b^2 = c^2$, where c is the length of the hypothenuse of a right triangle whose other sides have lengths a and b, respectively. Then $\sqrt{2}$ inches will be the length of the diagonal of a square of side 1 inch. And $\sqrt{3}$ will be the length of the hypothenuse of a right triangle whose other sides are of lengths 1 and $\sqrt{2}$, respectively.

3. $(1.5)^3 = 3.375$ and $(1.4)^3 = 2.744$. This shows that $\sqrt[3]{3}$ lies between 1.4 and 1.5, and each of these decimals is closer than .1 to $\sqrt[3]{3}$.

5. Yes. For example, $\sqrt{3} + (2 - \sqrt{3}) = 2$.

Section 15, p. 139

1. $(1 + i)(2 + i)(3 + i) = (1 + 3i)(3 + i) = 10i$.

3. $\dfrac{3 + i}{2 - i} = \dfrac{3 + i}{2 - i}\dfrac{2 + i}{2 + i} = \dfrac{5 + 5i}{5} = 1 + i$.

5. $|3 + i| = \sqrt{10}$, $|2 - i| = \sqrt{5}$, $|1 + i| = \sqrt{2}$ and $\sqrt{10}/\sqrt{5} = \sqrt{2}$.

7. The vector corresponding to the sum of the numbers is the vector sum of the vectors corresponding to the given numbers.

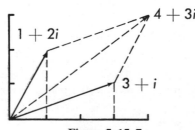

Figure 5:15:7

9. $(2 + 3i)^2 - 4(2 + 3i) + 13 = 4 + 12i - 9 - 8 - 12i + 13 = 0$.

11. The number 5, for instance, is a complex number which is rational. No imaginary number can be rational—all rational numbers are equal.

13. The number $\frac{3}{2}$ is rational, real, and complex only; 8(mod 9) is none of these since the number systems on a circle are completely different; $3i$ is imaginary and complex only; $\sqrt{2}$ is irrational, real, and complex; $\sqrt{-3}$ is imaginary and complex.

15. This is demonstrated by expanding each of

$$(a + bi)[(c + di) + (e + fi)] \quad \text{and} \quad (a + bi)(c + di) + (a + bi)(e + fi).$$

Use of the distributive property for real numbers and the definition of the product of two complex numbers shows the two expressions equal.

17. The imaginary numbers are not closed under multiplication, since, for instance $i^2 = -1$, which is not imaginary. There are other reasons also.

19. $a + bi = c + di$ implies $a - c = (d - b)i$. The left side is real and hence the right must be. This is impossible unless $d - b = 0$. This implies that $a - c = 0$ and hence $d = b$ and $a = c$.

CHAPTER 6

Section 2, p. 146

1. All positive numbers a.

3. For the geometric proof use Fig. 6:2 with 1 replaced by b. Algebraically,

$$(a + b)^2 = (a + b)(a + b) = a^2 + ba + ab + b^2 = a^2 + b(2a + b).$$

Hence $2ab + b^2$ must be added to a^2 to get $(a + b)^2$.

Section 3, p. 147

1. No answer needed.

3. **a.** $l = 85$, the sum $S = 946$.   **b.** 366.
     c. $-17 = 7 + (-3)(n - 1)$ implies $n = 9$, $S = -45$.
     d. $l = \frac{1}{4} + 23(-\frac{3}{8})$ implies $l = -67/8$, $S = \frac{195}{2}$.

5. The eleventh triangular number is 66 and the twelfth is 78.

7. $\dfrac{(a - 1)a}{2} + \dfrac{a(a + 1)}{2} = \dfrac{a}{2}(a - 1 + a + 1) = \dfrac{a}{2}(2a) = a^2.$

The geometric proof may be seen from figures such as Fig. 6:4.

9. The first may be seen from Section 2. That the answer to the second equation is no can be seen from Exercise 7 of Section 2.

11. $S = \dfrac{a + l}{2} n$, $l = a + (n - 1)d$ implies that

$$S = \dfrac{2a + (n - 1)d}{2}n = \dfrac{(2a - d)n + dn^2}{2} = \dfrac{dn^2}{2} + \dfrac{2a - d}{2}n.$$

13. The two differences are 3 and 13. Hence $13 - 3 = 10$. Thus the third and fourth differences are $13 + 10 = 23$ and 33 and the fourth and fifth terms of the sequence are $21 + 23 = 44$ and $44 + 33 = 77$. Using the formula of Exercise 11, with $d = 10$ the common difference for the arithmetic progression of the differences, we have $d/2 = 5$, $a = 3$, and, since $(6 - 10)/2 = -2$, we have

$$S = 5n^2 - 2n$$

as the formula for the sum of the arithmetic progression. Since each term after the first of the sequence is obtained by adding a sum of the arithmetic progression to the first term, we have the formula

$$T = 5 + 5n^2 - 2n$$

as the formula for the nth term, counting 5 as the zeroth term. This may be checked.

15. His thirty-sixth check is $300 + 10 \cdot 35 = \$650$. Thus his total salary over the 36 months is $\frac{1}{2}(300 + 650)36$, that is, 36 times the average of his smallest and largest check. The amount is \$17,100.

To give the answers to Exercises 17 and 19 would spoil the problems.

Section 4, p. 153

1. $0 < a < b$ means that $0 < a^n < b^n$ is true for $n = 1$. Then we assume the inequalities for n and use this to prove them for $n + 1$. That is, we assume that $0 < a^n < b^n$ and seek to prove $0 < a^{n+1} < b^{n+1}$. Here, recalling Section 5 of Chapter 5, we see that a positive and $a^n < b^n$ implies that $a^{n+1} = a \cdot a^n < a \cdot b^n$, and b^n positive implies that $a \cdot b^n < b \cdot b^n = b^{n+1}$. These two inequalities and the transitive property complete the proof.

3. Since $x^1 - y^1$ is divisible by $x - y$ (in fact, equal to it), the result holds for $n = 1$. Then we assume that $x - y$ is a factor of $x^n - y^n$ and try to prove that it is a factor of $x^{n+1} - y^{n+1}$. Now

$$x^{n+1} - y^{n+1} = x(x^n - y^n) + xy^n - y^{n+1}$$
$$= x(x^n - y^n) + y^n(x - y).$$

This has $x - y$ as a factor since it is a factor of $x^n - y^n$.

5. If k is odd, $k + 1$ is even and hence $k(k + 1)$ is even. If k is even, the product $k(k + 1)$ is of course also even.

7. To show the result by induction notice that it holds for $n = 1$ since $1^3 - 1 = 0$, which is divisible by 6. We assume that $n^3 - n$ is divisible by 6 and want to prove that $(n + 1)^3 - (n + 1)$ is also. But

$$(n + 1)^3 - (n + 1) = n^3 + 3n^2 + 3n + 1 - n - 1$$
$$= (n^3 - n) + 3n(n + 1).$$

Now $3n(n + 1)$ is divisible by 6, since, by Exercise 5, $n(n + 1)$ is even. It is easier to prove this without mathematical induction.

9. If $n = 1$, $(n - 1)(2n - 1)n = 0$, which is divisible by 6. Assume that $(n - 1)(2n - 1)n$ is divisible by 6. Then $n(2n + 2 - 1)(n + 1)$ will be divisible by 6 if 6 divides

$$n(2n + 1)(n + 1) - (n - 1)(2n - 1)n = n[2n^2 + 3n + 1 - (2n^2 - 3n + 1)] = n(6n),$$

which is clearly divisible by 6. This again can be proved a little more simply without mathematical induction.

11. The formula holds for $n = 1$. Call S_n the sum of the first n terms and assume $S_n = \frac{1}{6}n(2n + 1)(n + 1)$. We want to show that $S_{n+1} = \frac{1}{6}(n + 1)(2n + 3)(n + 2)$. Now

$$S_{n+1} = (1^2 + 2^2 + 3^2 + \cdots + n^2) + (n + 1)^2$$
$$= \frac{1}{6}n(2n + 1)(n + 1) + (n + 1)^2$$
$$= \frac{1}{6}(n + 1)(2n^2 + n + 6n + 6)$$
$$= \frac{1}{6}(n + 1)(2n^2 + 7n + 6).$$

We see that this is what we want by multiplying $(2n + 3)$ and $(n + 2)$ to get $2n^2 + 7n + 6$.

13. The first four sums are $\frac{1}{2}$, $\frac{5}{4}$, $\frac{17}{8}$, and $\frac{49}{16}$. We can write these $\frac{1}{2}$, $(1 \cdot 4 + 1)/4$, $(2 \cdot 8 + 1)/8$, and $(3 \cdot 16 + 1)/16$ and from these guess the formula to be $[(n - 1)2^n + 1]/2^n$. This holds for $n = 1, 2, 3, 4$. Assuming it for n and adding the $(n + 1)$st term we have

$$\frac{(n - 1)2^n + 1}{2^n} + \frac{2^{n+1} - 1}{2^{n+1}} = \frac{(n - 1)2^{n+1} + 2 + 2^{n+1} - 1}{2^{n+1}}$$

$$= \frac{n \cdot 2^{n+1} + 1}{2^{n+1}}.$$

Section 5, p. 157

1. $121.90.

3. $82.03.

5.

No. of times during year	2	3	4	6	12
Amounts	112.36	112.49	112.55	112.62	112.68

7. The percentages of increase in successive years are 6%, 5.7%, 5.4%, 5.1%, 4.8%.

9. No, except for the first year. For Exercise 7 the percentages decrease since the numerical increase is the same while the population grows. For Exercise 8 it is just the opposite.

11. Let P stand for the present population. Then $\frac{1}{2}P$ will represent the number of persons under 16 years of age. At the end of 16 years these and their children will amount to a population of $\frac{1}{2}P + 2(\frac{1}{2}P) = 3P/2$. At the end of 16 years the population of the other half will become $(6/5)(P/2)$. The sum of these two amounts is $21P/10$. This means that the population will more than double in 16 years.

Section 6, p. 163

1. a. 726. **b.** 2046. **c.** 3069. **d.** $\frac{341}{512}$.

3. At the end of a minute the number of cells would be 2^{60}. Since 2^{10} is about 1000, 2^{60} is about $(1000)^6 = 10^{18}$, that is, 1 with eighteen zeros after it. In 59 seconds the thimble would be exactly half full.

5. a. The common difference is $\frac{1}{6}$. Hence the sum is $\frac{65}{6}$.
  b. The common ratio is $\frac{1}{2}$ and the sum is $\frac{341}{512}$.
  c. It is neither an arithmetic nor a geometric progression.
  d. Here the common ratio is $-\frac{2}{3}$ and the sum is

$$\frac{1}{10}\left(3 - \frac{1024}{3^9}\right).$$

  e. The common difference is $-\frac{1}{3}$ and the sum is -11.

7. If a, b, and c are in an arithmetic progression $b = a + d$ and $c = a + 2d$. If it is also a geometric progression, we can equate the ratios:

$$\frac{a + d}{a} = \frac{a + 2d}{a + d}.$$

The first fraction is equal to $1 + d/a$ and the second $1 + d/(a + d)$. The two can be equal only if $d = 0$ and in this case the progression is

$$a, a, a, \ldots$$

with difference 0 and ratio 1.

9. Under plan A of Example 5 of this section the quarterly interest payments are $6.00, $4.50, $3.00, and $1.50, a total of $15. This is much less than the $24 required by the bank. For a total of $24 under plan A, the interest rate would be $(24/15)6 = 9.6\%$.

11. $56.33, $54.16.

13. $182.87.

15. $1295.05.

17. $9,471.30.

19. $500 + $833.

21. The five payments are $260, $248, $236, $224, and $212, and the sum is $1180.

23. If we write the geometric progression as $a, ar, ar^2, ar^3, \ldots$, the differences will be $ar - a = a(r - 1)$, $ar^2 - ar = ar(r - 1)$, $ar^3 - ar^2 = ar^2(r - 1)$, and the ratio is again r. For this to be the same progression we must have $ar - a = a$, that is, $ar = 2a$, which implies $r = 2$.

25. If a, b, and c form an arithmetic progression we must have $b - a = c - b$ and $2b = a + c$, or $b = \frac{1}{2}(a + c)$, the average or arithmetic mean.

27. To show $\frac{1}{2}(a + b) \geq \sqrt{ab}$ is equivalent to showing $\frac{1}{4}(a + b)^2 \geq ab$, since all letters stand for positive numbers. The second inequality is equivalent to $a^2 + 2ab + b^2 \geq 4ab$, which is equivalent to $a^2 - 2ab + b^2 \geq 0$, or

$$(a - b)^2 \geq 0.$$

This is true for all real numbers a and b and the equality holds only when $a = b$.

Section 7, p. 169

1. $122.01.

3. $106.17.

5. Suppose the result is true down to the nth row and let a and b be two successive numbers in this row. Then b will be the sum of the numbers above a in the diagonal of a, indicated by D in the figure. Then for the $(n + 1)$st row of the Pascal triangle, between a and b we will have $(a + b)$, which is in the same diagonal as b. Then $a + b$ will be a plus the sum of the numbers above a in diagonal D, that is, the sum of the numbers above the $(n + 1)$st row in this diagonal.

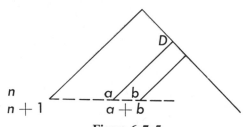

Figure 6:7:5

7. Notice that the first pyramidal number is 1, the first triangular number. The second pyramidal number is the sum of the first two triangular numbers; the third pyramidal number the sum of the first three triangular numbers; and so forth. Since the numbers on the third diagonal of the Pascal triangle are triangular numbers, we see by the answer to Exercise 5 that each number in the fourth diagonal is the sum of those above it in the third diagonal: the sum of the triangular numbers above it. For instance, 20 is the fourth pyramidal number and is the fourth coefficient in $(x + y)^6$. It is the sum of $10 + 6 + 3 + 1$ above it on the third diagonal.

9. For odd exponents one can see the result by inspection, but for an even exponent it is better to notice that the sum specified is the sum of the coefficients in $(1 - 1)^n$, which is, of course, zero. This also holds for n odd, of course.

Section 8, p. 171

1. If S_n denotes the sum on the left side of the equality,

$$S_n = \frac{(\frac{1}{3})^{n+1} - 1}{\frac{1}{3} - 1} = (\frac{3}{2})[1 - (\frac{1}{3})^{n+1}] = \frac{3}{2} - \frac{1}{2}(\frac{1}{3})^n.$$

Hence $\frac{3}{2} - S_n = \frac{1}{2}(\frac{1}{3})^n$. This will be less than $1/1,000,000$ if $(\frac{1}{3})^n < 1/500,000$, or $3^n > 500,000$. Now $3^6 = 729$ and hence $3^{12} = 729^2 = 531,441 > 500,000$. Thus for all $n \geq 12, \frac{3}{2} - S_n \leq 1/1,000,000$. In fact, from $\frac{3}{2} - S_n = \frac{1}{2}(\frac{1}{3})^n$ we can see that the right side can be made as small as we please by making n large enough. That is, by choosing n large enough we can make S_n as close as we please to $\frac{3}{2}$. This means that the limit of S_n as n becomes infinite is $\frac{3}{2}$.

3. Here $S_n = \frac{(\frac{1}{10})^{n+1} - 1}{.1 - 1} = \frac{10}{9}\{1 - (.1)^{n+1}\}$ and the limit as n becomes infinite is $\frac{10}{9}$, whose decimal value is $1.111111 \ldots$ This $\frac{1}{9} = .111111 \ldots$

5. If we let $x = .123123123 \ldots$, $1000x - x = 123$ and $x = \frac{123}{999} = \frac{41}{333}$.

7. The sum within the brackets is $S_n = \frac{1}{2}\frac{(-\frac{1}{2})^n - 1}{-\frac{1}{2} - 1}$. As n becomes infinite, that is, arbitrarily large, $(-\frac{1}{2})^n$ approaches 0. Hence

$$\lim_{n \to \infty} S_n = \frac{1}{2}\frac{0 - 1}{-\frac{1}{2} - 1} = \frac{1}{2}\frac{1}{\frac{3}{2}} = \frac{1}{3}.$$

9. The sum becomes large without limit.

11. $\lim_{n \to \infty} (2n + 1)/n = \lim_{n \to \infty} (2 + 1/n) = 2.$

13. The distance covered is the sum of the following infinite series:

$$12 + 12(\tfrac{3}{4}) + 12(\tfrac{3}{4})^2 + 12(\tfrac{3}{4})^3 + \cdots.$$

The limiting value of this sum is $12\frac{0 - 1}{\frac{3}{4} - 1} = 48$ feet. However, it is a property of a pendulum that each swing takes just as long as the previous one. Hence the time required is infinite under the assumptions of the problem.

15. Since $n > B, S_n > S_B$ and $-S_n < -S_B$. Thus $A - S_n < A - S_B < c$.

Section 9, p. 177

1. $x = \frac{9}{5}$.

3. $x = 84$.

5. $x = \frac{13}{5}$.

7. All x.

9. No x.

11. $P = 1000/2.357$, which is approximately 424.3.

13. $x = 0$.

15. If $f^2 = g^2$, then $0 = f^2 - g^2 = (f - g)(f + g)$ and thus $f - g = 0$ or $f + g = 0$. Thus every solution of the given equation is either a solution of $f = g$ or $f = -g$.

Section 10, p. 178

1. **a.** $x < -1$.   **b.** $x > -1$.   **c.** $x \geq 3$.   **d.** $-\frac{1}{2} \geq x$.

   e. If $x > 2$, the inequality is equivalent to $1 \leq 3(x - 2)$ and thus $x \geq \frac{7}{3}$, which implies $x > 2$. If $x < 2$, the inequality is equivalent to $1 \geq 3(x - 2)$, that is, $x \leq \frac{7}{3}$, which is implied by $x < 2$. Hence $x \geq \frac{7}{3}$ or $x < 2$.

3. If $ab > 0$, $a < b$ is equivalent to $a/ab < b/ab$, that is, $1/b < 1/a$. If $ab < 0$, $a < b$ is equivalent to $1/b > 1/a$.

5. Notice that $-3 < 2$ and yet $(-3)^2 > 2^2$. The answer to all three questions is no.

Section 11, p. 180

1. $x + .44(15 - x) = .51(15)$ implies $x = 1.88$.

3. The formula is $x = V\dfrac{S - s}{1 - s} = \dfrac{VS - Vs}{1 - s}$. Using the table, $x = \dfrac{7.7 - 6.6}{1 - .44} = 1.9$.

5. If b stands for the bonus for those in class I, $2b$ and $3b$ will represent, respectively, the bonuses for the other two classes. Then

$$20b + 50(2b) + 40(3b) = 2000 \qquad \text{implies} \qquad b = \$8.33.$$

7. If t is the amount of the tax, we have $t = .04(I - t)$, or $t = I/26$.

Section 12, p. 182

1. $4x + 6 = 28$ implies $x = \frac{11}{2}$ and thus there is no integer solution.

3. In dividing by x we assumed $x \neq 0$ and hence lost this solution.

5. Solved.

7. Let d be the length of the hill in miles. Then $d/1.5$ represents the time to go up the hill and $d/4.5$ that to go down. Then $d/1.5 + d/4.5 = 6$ implies $d = 27/4$ miles.

9. Let s be the length of the side of the square field and w the width of the rectangular field. Then $2s - w$ is the length of the rectangular field and the difference of the areas is $s^2 - w(2s - w) = s^2 - 2sw + w^2 = (s - w)^2 = 10,000$. Hence $s - w = 100$ feet.

11. Let p be the speed of the propeller plane. Then the time for the jet plane is $1200/2p$ and that for the propeller plane is $1200/p$. Thus $1200/2p + 2 = 1200/p$ or $2 = 1200/2p$, which implies $p = 300$ miles per hour, and the speed of the jet plane is 600 miles per hour.

13. If d represents the distance between towns, we have that $2d$ is the distance traveled and $d/10 + d/15$ the time required. The average rate is the quotient $2d/(d/10 + d/15)$. This is 12 miles per hour.

15. 42 pence. This can be solved working from the end backward.

17. Worked out.

19. 28 years.

21. Louisa's age is 10, Christopher's 8, and Phyllis's 4.

23. Christopher is 25 and David 2 years old.

Section 13, p. 190

1. 12 cocks, 4 hens, 84 chickens.

3. Chickens: 2 shillings; ducks: 4 shillings; geese: 5 shillings.

5. 28 jelly beans, 3 licorice strips, 9 caramels.

7. $x = 9, y = 3$.

9. $x = 2, y = 3$.

11. An equivalent pair of equations is $6x + 9y - 21z = 15$, $6x + 4y + 10z = 36$. Subtracting the second from the first we have $5y - 31z = -21$, which has a solution $z = 1, y = 2$. Then the second of the given equations becomes $3x + 4 + 5 = 18$, or $x = 3$. For $z = 2$ and 3, the number y in $5y - 31z = -21$ is not an integer, and for $z > 3$, the second given equation shows that x and y cannot both be positive. Hence the only solution in positive integers is $x = 3, y = 2, z = 1$.

13. Aside from those of form aa/aa the only possibilities are $\frac{16}{64}$, $\frac{26}{65}$, $\frac{19}{95}$, and $\frac{49}{98}$. For the solution see reference 36, pp. 73–75.

CHAPTER 7

Section 1, p. 196

1. Objects can be ordered according to all the properties mentioned, perhaps the least obvious being color and place of manufacture. For the former the ordering could be on a rainbow or, more technically, by the spectrum or wavelength. The place of manufacture could be ordered alphabetically.

3. Weight, scales; loudness, at least partially by energy output; color, spectrophotometer.

Section 2, p. 200

1. For parts **a.**, **b.**, and **c.** the independent variable is: for part **a.**, rate of interest; for part **b.**, height of the triangle; and for part **c.** the variable b. In part **d.** there is no apparent relationship which would lead us to select one as the independent variable.

3. The points corresponding to the tables of values are indicated on the figures (pp. 362–363). The connecting curve is drawn to indicate what other pairs of values might give.

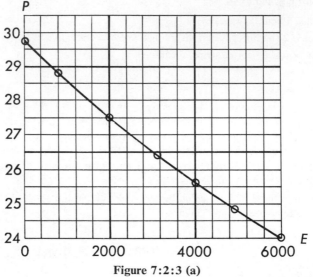

Figure 7:2:3 (a)

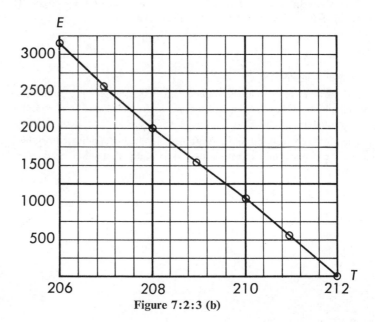

Figure 7:2:3 (b)

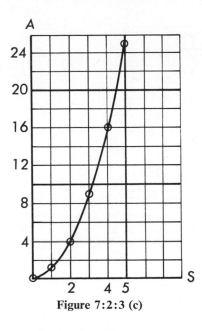

Figure 7:2:3 (c)

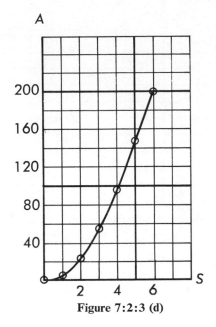

Figure 7:2:3 (d)

Section 3, p. 203

1.

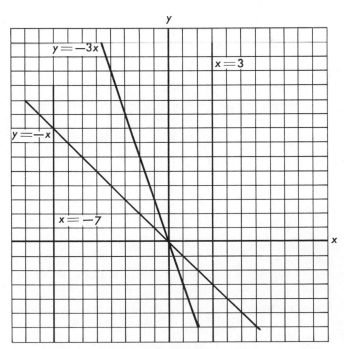

Figure 7:3:1

3.

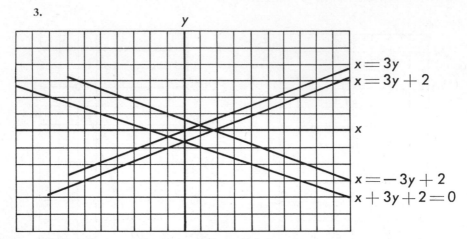

Figure 7:3:3

5.

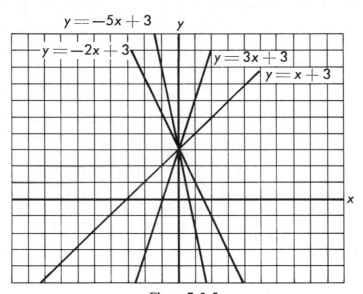

Figure 7:3:5

Section 5, p. 208

1. Here we omit the graphs but describe the results:
   a. Below the line $3x + 2y = 6$.
   b. Below $3x + 2y = 6$ and above $3x + 3y = 6$ (that is, above $x + y = 2$).
   c. Below $-4x + 7y = 8$ and above $x + y = 6$.

3. The new requirement is $m \geq 2d$. Then line L is $m = 2d$, $N = (\frac{14}{3}, \frac{7}{3})$, $L = (6, 3)$. The graph of her requirements is the portion inside the quadrilateral $ALNR$, including all its boundary except NR.

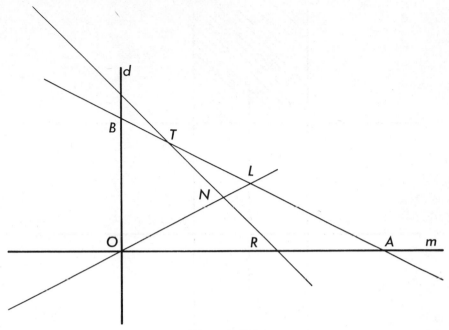

Figure 7:5:3

5. The points between $x = 2$ and $x = -2$ and either above $y = 4$ or below $y = -4$. None of the boundaries are included.

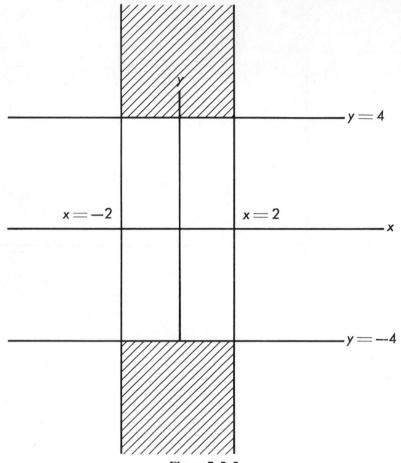

Figure 7:5:5

Section 6, p. 211

1. **a, b.** See the figure.
   c. A table of values is

x	-1	0	1	2	3	4
y	6	2	0	0	2	6

   d. Notice that there is no value of y for $x = 1$. For other values of x we have the table

x	-2	-1	0	2	3
y	$\frac{2}{3}$	$\frac{1}{2}$	0	2	$\frac{3}{2}$

e. A table of values is

$$\begin{array}{c|ccc} x & 1 & \frac{5}{4} & \frac{7}{4} \\ \hline y & 0 & \frac{3}{4} & \sqrt{\frac{33}{4}} \end{array}$$

and also those pairs obtained by changing the sign of one or both of x and y.

f. This is the equation of a circle with center at the origin and radius 5.

h. A table of values is

$$\begin{array}{c|cccccc} r & -2 & -1 & 0 & \frac{1}{2} & 2 & 3 \\ \hline S & -5 & 0 & 1 & \frac{15}{8} & 15 & 40 \end{array}$$

Notice that though the expression as given has no value for $r = 1$, yet for $r = 1$ the sum of the progression is $1 + 1 + 1 + 1 = 4$.

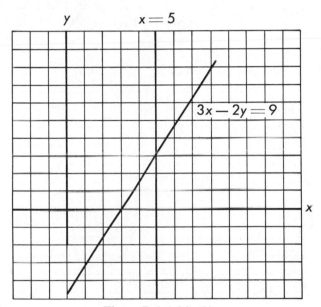

Figure 7:6:1 (a), (b)

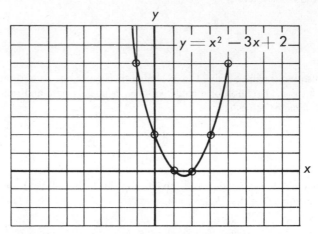

Figure 7:6:1 (c)

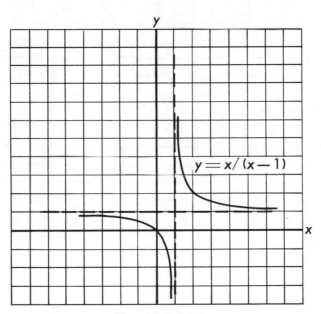

Figure 7:6:1 (d)

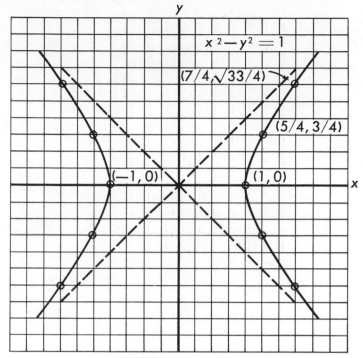

Figure 7:6:1 (e)

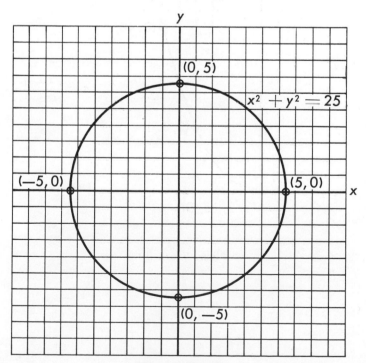

Figure 7:6:1 (f)

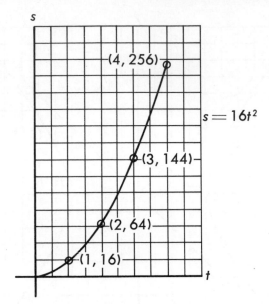

Figure 7:6:1 (g)

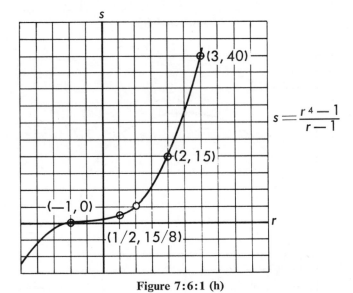

Figure 7:6:1 (h)

3. Let g denote the girth and l the length. Then maximum allowable $l = 100 - g$.

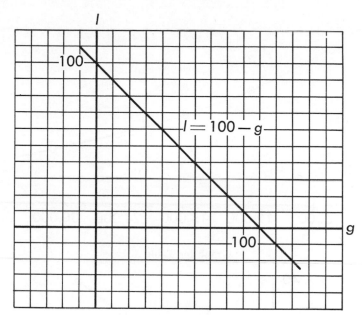

Figure 7:6:3

5. To find algebraically the point of intersection multiply the first equation by 3 to get $3x + 6y = 9$ and subtract the second equation from it. This gives $8y = 8$ or $y = 1$. Then $x + 2y = 3$ implies $x = 1$. Thus $(1, 1)$ is the only pair of coordinates which satisfies both equations, that is, the only point which lies on both lines. Two lines have a point in common if they are in the same plane unless they are parallel. Notice that two distinct lines are parallel if and only if their slopes are equal.

7. In Fig. 7:6 give C the coordinates (x_2, y_2) and A, (x_1, y_1). Then $CB = y_2 - y_1$, $AB = x_2 - x_1$, and the slope is thus $(y_2 - y_1)/(x_2 - x_1)$.

9. **a.** $x > 2$ or $x < 1$.

   b. The points inside the circle but not on it.

   c. The points on the right of the right branch or on the left of the left branch.

   d. The points inside the circle and either to the right of the line $x = 2$ or to the left of the line $x = 1$.

   e. x greater than zero *and* less than 1; that is, $0 < x < 1$.

Section 7, p. 213

1. The graph for Table 7 will not be a smooth curve because property values do not change continuously; Table 9 will not be smooth because of the calibration of the camera. Parts **g.** and **h.** are not smooth because they make sense only for integer values of n and part **i.** changes only at intervals of one-half ounce. The others have smooth graphs: See the figures except for Exercise **1j.**, whose graph may be found in the answer to Exercise **1g.** in Section 6.

Graph for Table 4

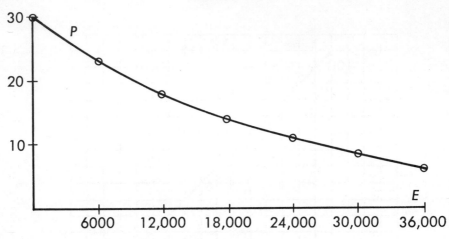

Graph for Table 5

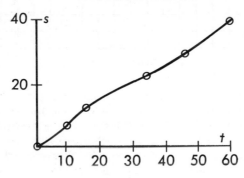

Graph for Table 6

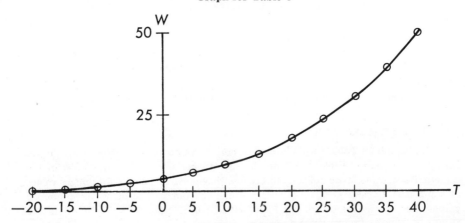

Graph for Table 8

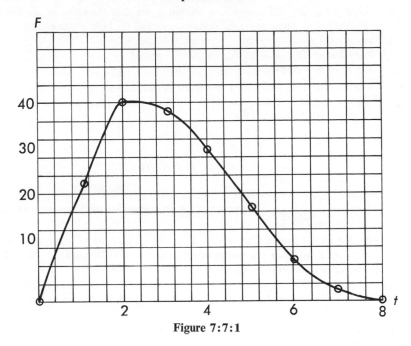

Figure 7:7:1

3. 3200 pamphlets can be published for $180. Although here, strictly speaking, the graph is not a smooth curve, it may be considered so for approximation purposes.

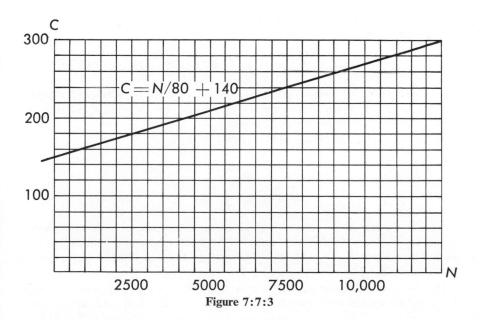

Figure 7:7:3

Section 8, p. 217

1. **a.** Using the result of Exercise 7 of Section 6, see that the slope of the line is $[1 - (-2)]/(5 - 3) = \frac{3}{2}$. Also, using the points (x, y) and $(5, 1)$, the slope is $(y - 1)/(x - 5)$. Then, from $(y - 1)/(x - 5) = \frac{3}{2}$ we get $2y = 3x - 13$.

    c. The equation is $x = -2$, since the line is vertical.

    d. The slope is 8 and the equation is $y = 8x$.

3. Here notice that the values of T are given at intervals of $20°$. Our theory was developed for intervals of 1. Thus we set $T/20 = T'$ and see that the values of T' will be 0, 1, 2, 3, 4 and we will have the table

T'	0	1	2	3	4
W	54	64	74	84	94

Since the values of W are in an arithmetic progression the formula will be $W = aT' + b$ with $b = 54$ and a, the common difference of the W's equal to 10. Thus $W = 10T' + 54 = T/2 + 54$. When $T = 23$, $W = 65.5$.

5. To adjust h to the progression 0, 1, 2, 3, ... we subtract 61 to make the first h' zero and divide by 2 to make the common difference 1; that is, let $h' = \frac{1}{2}(h - 61)$. Notice that the differences for W are all 8 and thus $W = 8h' + 118 = 4(h - 61) + 118 = 4h - 126$. The differences for M are three 8s and two 9s. Choosing the difference 8, we have $M = 8h' + 124 = 4h - 120$, which gives values for M close to that of the table.

7. Here we notice that the differences for S are 2, 3, 4, and 5, which are in an arithmetic progression. In this case it is simpler to compute what S should be for $n = 0$ by computing the D previous to 2 and from that the S previous to 1 as indicated in the table, where the new values are in parentheses:

n	(0)	1	2	3	4	5
S	(0)	1	3	6	10	15
D		(1)	2	3	4	5

Thus referring to the discussion in the section for the quadratic formula we have $c = 0$, $a + b = 1$, $2a = 1$, which implies that $a = \frac{1}{2} = b$ and $S = \frac{1}{2}n^2 + \frac{1}{2}n = (n/2)(n + 1)$, which is just the formula for triangular numbers.

9. Here, to adjust the values, let $f' = (\frac{1}{3})(f - 2)$ and have

f'	0	1	2	3	4
s	0	15	48	99	169
D		15	33	51	70

Except for the last D, the values are in an arithmetic progression with common difference 18; and the last difference is close—19. So we take the common difference to be 18 and have $c = 0$, $a + b = 15$, $2a = 18$, which gives $a = 9$ and $b = 6$. Thus $D = 9f'^2 + 6f' = (f - 2)^2 + 2(f - 2) = f^2 - 2f$.

11. Here the differences are 2.2, 10.2, 18.2, and 26, which are approximately in an arithmetic progression with common difference 8. So $c = 10$, $a + b = 2.2$, $2a = 8$. Thus

$a = 4$, $b = -1.8$, and, if I stands for the investment and T the time in months, we have

$$I = 4T^2 - 1.8T + 10.$$

When $T = 10$, $I = 392$ thousands of dollars and the contractor's loss would be $142,000 if he finished the job.

13. There will be just two second differences and two numbers are always in a unique arithmetic progression.

15. Here we work backward as in Exercise 7 and have

n	(0)	1	2	3	4	5
S	(0)	1	4	10	20	35
D_1		(1)	3	6	10	15
D_2			(2)	3	4	5

From this we have $d = 0$, $a + b + c = 1$, $6a + 2b = 2$, $6a = 1$, and hence $a = \frac{1}{6}$, $b = \frac{1}{2}$, $c = \frac{1}{3}$, and

$$S = (\tfrac{1}{6})n^3 + (\tfrac{1}{2})n^2 + (\tfrac{1}{3})n = (n/6)(n^2 + 3n + 2)$$
$$= \frac{n(n + 1)(n + 2)}{6}.$$

Notice that the values of D_1 are those of S in Table 13 except for the first S. The S's are the pyramidal numbers (see Section 7 of Chapter 6).

17. The differences of the S's are 0, 1, 1, 2, 3, ..., which, except for the initial 0, are the same as the S's themselves. Hence no matter how many times we compute the set of differences of differences, and so on, we always come out with the same except for the first few numbers. Thus no matter how many times we take the differences we cannot come out with an arithmetic progression. This means that no formula like the following can fit the table:

$$y = ax^n + bx^{n-1} + cx^{n-2} + \cdots + rx + s,$$

where n is a positive integer.

Section 9, p. 222

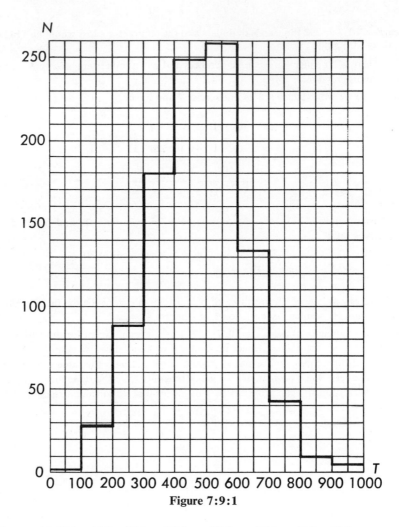

Figure 7:9:1

1.

A	12.5	17.5	22.5	27.5	32.5	37.5	42.5	47.5	
D	3715	3648	3605	3591	3619	3716	3933	4369	

3.

A	52.5	57.5	62.5	67.5	72.5	77.5	82.5	87.5	92.5
D	5241	6646	8576	10,772	12,332	11,763	8989	4638	844

5. A histogram would be appropriate or merely a line graph, as shown.

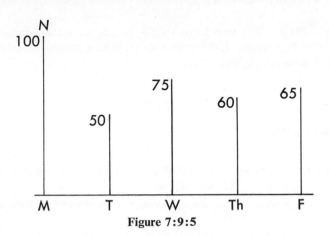

Figure 7:9:5

7. Histograms would be appropriate in all except part b, which requires a smooth curve. One could also use line graphs like those in Exercise 5 for *a, c, e, f,* and *g.*

Section 10, p. 228

1.

	Table 21	Table 23	Table 25
Median	450	5	7
Mode	550	5	8

3. The average for the class is $(12 \cdot 72 + 15 \cdot 70)/27 = 70.9$, approximately.

5. If nine students have a grade of 10 and the tenth student has a grade of 100, the arithmetic mean of the grades is $(90 + 100)/10 = 19$, which is almost twice as much as the grade of the nine students.

7. The batting average is the "average" number of hits per time at bat. It is an arithmetic mean.

9. One must know the median. If it is below the arithmetic mean, then at least half the class have grades less than the arithmetic mean.

11. The person drives $30 \cdot 2 + 70 = 130$ miles in 3 hours. Thus his average speed is $\frac{130}{3} = 43\frac{1}{3}$ miles per hour. This is the arithmetic mean.

13. Let $b = \sqrt{ac}$. Then $\frac{b}{a} = \frac{\sqrt{ac}}{a} = \frac{c}{a}$ and $\frac{c}{a} = \frac{c}{b} = \frac{c}{\sqrt{ac}} = \sqrt{\frac{c}{a}}$. Thus the two ratios are equal and *a, b,* and *c* form a geometric progression.

15. $\sqrt[4]{2 \cdot 6 \cdot 9 \cdot 12} = \sqrt[4]{2^4 \cdot 3^4} = 6$. The arithmetic mean is $7\frac{1}{4}$. The harmonic mean is $4/(\frac{1}{2} + \frac{1}{6} + \frac{1}{9} + \frac{1}{12}) = 4.6$, approximately.

17. Let *d* denote the distance through the towns and 9*d* the distance outside. If Mr. Thims drives at the legal limit all the way, his time would be $d/15 + 9d/50$. This is thus the minimum time in which he could legally make the trip. At 40 miles per hour average, his time would be $10d/40 = d/4$. So the question is: Is the following inequality true?

$$d/15 + 9d/50 \leq d/4.$$

Notice that this is independent of d. Calculation shows that $\frac{1}{15} + \frac{9}{50} = \frac{37}{150} = .247$ approximately, which is barely less than $\frac{1}{4}$. Thus he could barely average 40 miles per hour legally.

19. If $b = 2/(1/a + 1/c)$, then $1/b = \frac{1}{2}(1/a + 1/c)$. So $1/b$ is the arithmetic mean of $1/a$ and $1/c$ and, by Exercise 12, the reciprocals are in an arithmetic progression.

21. This follows directly from Exercise 19.

CHAPTER 8

Section 1, p. 234

1. $3 \cdot 4 = 12$.

3. For the first place there are 6 possibilities, for the second 5, and so on. Thus in all there are $6 \cdot 5 \cdot 4 \cdot 3 \cdot 2 \cdot 1 = 720$ possibilities.

5. If no digit is used twice, there will be $7 \cdot 6 \cdot 5 \cdot 4 = 840$ four-digit numbers. If repetitions are allowed it will be $7^4 = 2401$.

7. There are n possibilities for the first place, $n - 1$ for the second, and so on. Hence the total number is

$$n(n - 1)(n - 2)\ldots3 \cdot 2 \cdot 1 = n!$$

11. We know from Chapter 4, Section 6, that a number is divisible by 3 or 9 if and only if the sum of its digits is divisible by 3 or 9, respectively. Thus this divisibility is independent of the order of the digits. Since the sum of the digits is 21, they are all divisible by 3 but none by 9. To be divisible by 6 they must be divisible by 3 and 2. Thus we count only half of the $6! = 720$, and our answer is 360.

13. The president can be any one of 15, the vice president then anyone of 14, and so on. Thus the number of sets of officers is $15 \cdot 14 \cdot 13 \cdot 12 = 32,760$.

15. To alternate sexes in this case, men must sit at the ends. Then any one of 4 men can sit at the left end, any one of 3 women at his right, any one of 3 men at her right, and so on. So we have $4 \cdot 3 \cdot 3 \cdot 2 \cdot 2 \cdot 1 \cdot 1 = 144$ possible seating arrangements.

Section 2, p. 238

1. (Partial answer) There are 15 subsets of four elements and an equal number of two elements.

3. $2,598,960$.

5. The number of bridge hands is $_{52}C_{13}$, which is approximately $635,000,000,000$.

7. The values are the same.

9. If $r = 0$ or $r = n$, formula (1) would contain the symbol $0!$, which had not yet been defined.

11. This is best seen by drawing or using a checkerboard. Notice that, under the restrictions given, unless the piece is on the top row or one of the edges, there are always two possible places the piece can move. For the first question there must be three moves to the left (and up) and three to the right (and up). We may pick the left moves in any one of $_6C_3 = 20$ ways and then the right ones are determined. For the second question there would be four to the left and three to the right. Thus the number here is $_7C_3 = 35$. For the last the answer is 34. Why?

Section 3, p. 241

1. The coefficient of $x^8 y^5$ will be the number of sets of 5 y's we can choose from 13 parentheses $(x + y)$. This is $_{13}C_5$.

3. Consider the 13 symbols 1, 2, 3, . . . , 12, 13. The number of subsets of 7 symbols is $_{13}C_7$. The number including the number 1 is $_{12}C_6$, the number of subsets not containing 1 is $_{12}C_6$, and the sum of these is $_{13}C_7$.

5. Consider here the numbers 1, $\overset{\cdot}{2}$, 3, . . . , $n - 1$, n and the number of subsets of r of them. This number is $_nC_r$. We may also count them in two parts by counting first those which contain the number 1: There are $_{n-1}C_{r-1}$ of these, and second, those which do not contain the number 1: There are $_{n-1}C_r$ of these. The sum of these two will be $_nC_r$.

7.
$$_{n-1}C_{r-1} + {}_{n-1}C_r = \frac{(n-1)!}{(r-1)!(n-r)!} + \frac{(n-1)!}{(r!)(n-1-r)!}$$
$$= \frac{(n-1)!}{(r-1)!(n-r-1)!}\left(\frac{1}{n-r} + \frac{1}{r}\right)$$
$$= \frac{(n-1)!(r+n-r)}{(r-1)!r(n-r-1)!(n-r)} = \frac{n!}{r!(n-r)!} = {}_nC_r.$$

9. From (5), if we replace x and y by 1, we see that the sum given in the exercise is equal to $(1 + 1)^n = 2^n$. This means that the total number of subsets of a set of n elements including the null set, is 2^n. This is sometimes called "the power of the set."

Section 4, p. 248

1. The respective probabilities are $\frac{1}{36}$, $\frac{2}{36}$, $\frac{3}{36}$, $\frac{4}{36}$, $\frac{5}{36}$, $\frac{4}{36}$, $\frac{3}{36}$, and $\frac{2}{36}$.

3. The number of sets of 4 heads from 8 is $_8C_4$ and the number of different tosses is 2^8.

5. The probability is $_nC_m/2^n$.

7. It can have no heads in just one way and 1 head in any one of four ways. Hence it can roll at most 1 head in five ways and the probability is $\frac{5}{16}$.

9. The numerator of the probability is $_6C_3 + {}_6C_4 + {}_6C_5 + {}_6C_6 = 20 + 15 + 6 + 1 = 42$. The probability is $42/2^6 = 21/32$.

11. If none shows a 6, there are five possibilities for each die, hence 5^4 for the four. If exactly one shows a 6, there are four possibilities and for each of these 5^3 for the other three. Hence the probability is $(5^4 + 4 \cdot 5^3)/6^4$. Referring to Example 2, this is equal to $1 - \frac{171}{1296}$.

13. On the assumption stated at the beginning of the exercises, the answer is $\frac{1}{2}$.

15. The number of possibilities for exactly 3, 4, 5, and 6 fives are, respectively, $_6C_3 \cdot 5^3$, $_6C_4 \cdot 5^2$, $_6C_5 \cdot 5$, and 1. The sum is 2906 and the probability 2906/6^6.

17. $(1 \cdot 5)/(2 \cdot 36) = \frac{5}{72}$.

19. $_{10}C_4/2^{10} = \frac{210}{1024} = \frac{105}{512}$.

21. **a.** Since there are four suits and only one hand for each suit, the probability is $4/_{52}C_{13}$, which is about 1 of 160 billion.

b. If 4 cards are the aces, the remaining 9 cards are chosen from 48, and hence the number is $_{48}C_9$ and the probability $_{48}C_9/_{52}C_{13} = \frac{11}{4165}$.

c. There are four possibilities for the suit of the ace, king, queen, jack, ten, and $_{47}C_8$ for the remaining 8 cards. Hence the probability is $4 \cdot {}_{47}C_8/_{52}C_{13}$.

d. $4 \cdot {}_{47}C_8/{}_{48}C_9 = \frac{3}{4}$. Hence "honors" in a suit is $\frac{3}{4}$ as likely as honors in no-trump.

23.   The denominations of the two pairs can be chosen in any one of ${}_{13}C_2$ ways, that is, 78. For a given denomination there are ${}_4C_2 = 6$ possibilities. Thus for the two pairs the number of possibilities is $78 \cdot 6 \cdot 6$. The fifth card can be any one of 48. Hence the probability is $78 \cdot 6 \cdot 6 \cdot 48/{}_{52}C_5 = \frac{216}{4165}$.

25.   If three are guessed correctly, the fourth must be, since he knows what the 4 cards are. Hence, if 2 are guessed correctly, there is only one way in which the other 2 can be wrong. The ones which he guesses correctly could be any 2 from the 4, for which there are 6 possibilities. So the probability is 6/4!, since there are 4! ways in which designations could be made. Thus the probability is $\frac{1}{4}$.

27.   Think of the men as fixed and the women permuted. There is only one permutation in which they will match, but $6! = 720$ permutations for the women. Hence the chance of each man drawing his wife is $\frac{1}{720}$. As for being at the same table, the number of permutations of the wives would be 2^3, since there are two places each wife can be at a given table. So in this case the probability is $\frac{8}{720} = \frac{1}{90}$.

29.   The probability of getting a full house is the same as before, $\frac{4}{47}$. Also, for a flush there are, as before, 45 different possiblities. For a straight he could draw an ace and a king, a king and a nine, or a nine and an eight. Thus there are $16 \cdot 3 = 48$ possibilities. The number of possibilities for a straight or a flush or both is $45 + 48 - 3 = 90$, and the probability is $90/(47 \cdot 23)$. This is still less than $\frac{4}{47}$, but just barely.

Section 5, p. 251

1.   a.   38,569/100,000.   **b.**   26,237/49,341.   **c.**   14,474/78,106.

   d.   In the first case it merely means that out of 100,000 alive at age 10, 38,569 are alive at age 70.

3.   a. $120/2^{10}$.   **b.** $252/2^{10}$.

   c. $(45 + 10 + 1)/2^{10} = 56/2^{10} = 7/128$.   **d.** $165/2^{10}$.

5.   a. $\frac{247}{995}$.   **b.** $\frac{29}{995}$.   **c.** $\frac{58}{995}$.   **d.** $\frac{16}{995}$.

7.   The advertising is very misleading. Not all children are equally accident-prone. Even if one child were as likely to have an accident as any other, the chance of one or more of a given three having an accident would be $1 - 2^3/3^3 = \frac{19}{27}$, which is about $\frac{2}{3}$.

Section 6, p. 256

1.   a.   If he draws once, his chance of drawing a red ball is $\frac{2}{10} = \frac{1}{5}$, and hence he should pay 20 cents to play.

   b.   There are $2 \cdot 2 \cdot 8 = 32$ different possibilities for drawing one red and one white and thus the probability for this is $32/{}_{10}C_2 = \frac{32}{45}$. For this he should pay $\frac{32}{45}$ dollar. There is only one possibility for two reds and for this he would pay $\frac{2}{45}$ dollar. Hence his payment to play should be $\frac{34}{45}$ dollar or about 75 cents.

   c.   If the ball is replaced, we can think of him as playing the game of part **a.** twice and he should pay 40 cents.

3.   If he is to break even, his number must appear exactly once. This can be in any one of six places and there are five ways in which his number would not occur. Thus the probability of breaking even is $6 \cdot 5^5/6^6 = (\frac{5}{6})^5$, approximately .4. To lose he would have to get a 6 none of the times and the probability is $(\frac{5}{6})^6$, which is approximately .34. The probability of winning is $1 - (\frac{5}{6})^5 - (\frac{5}{6})^6$, which is about .26.

5. From the mortality table, out of 69,804 alive at age 50, 64,543 are alive at age 55; hence 5261 die in the five-year period. Hence the premium should be (5261/69,804) (1000) or about \$75.37.

7. The premium should be (1/5000)(1000) = 20 cents.

9. The probability of his number appearing on exactly one of the three is $3 \cdot 5^2/6^3$ and the expectation $(7.50)/6^3$. The probability that it will appear on exactly two is $3 \cdot 5/6^3$ and the expectation $(3.00)/6^3$. The probability that it will appear on all three is $\frac{1}{6}^3$ and the expectation $.30/6^3$. Hence what he should pay is the sum of these three, that is, 5 cents.

11. Out of 300 cars, 2 will burn in section A and 1 in section B. So the rate for section A will be $\frac{2}{300}$ times the average claim and in B, $\frac{1}{300}$ times the average claim.

13. Here the answer depends on what you expect of the score. If one counts 1 for each right answer and 0 for wrong answers, it is clear that if one answers by chance, say 12 questions, 3 should be right and the score of 3 is for a random guess. If one wishes to adjust the score so that a random guess will result in a zero score, then one scores 1 point for each right answer and $-\frac{1}{3}$ for each wrong one, or at least in this proportion.

CHAPTER 9

Section 2, p. 261

1. One's eyes; sides of a street. (Approximately.)

3. a. The perpendicular bisector of the base.
   b. The perpendicular bisectors of each side.
   c. Every line through the center.
   d. The perpendicular bisectors of the sides.
   e. Part d and the lines through opposite vertices.

5. a. (1, 1) is its own image since it is on the line $x = y$.
   b. $(1, 2)' = (2, 1)$. **c.** $(3, -4)' = (-4, 3)$. **d.** $(r, s)' = (s, r)$.

7. a. The line parallel to l such that l is halfway between the given line and its image.
   c. The line l is the bisector of the angle between the given line and its image.
   e. The line segment is its own image.
   g. The circle is its own image.

9. The definitions are equivalent if we allow the two "parts" to overlap and assume the property mentioned in Section 5 of Chapter 7. Then we can choose as one part the portion of the figure on l together with all points of the figure on one side of l, and as the other part the same portion of the figure on l together with all points on the other side of l.

11. Let m be the line of the reflection and A' and B' the images of A and B, respectively. Let C be any point on m different from A and B. Then $CA = CA'$, since m is the perpendicular bisector of the segment AA' and, similarly, $CB = CB'$. Also angles ACB and $A'CB'$ are congruent. This shows $AB = A'B'$, since triangles ACB and $A'CB'$ are congruent.

Section 3, p. 264

1. The first symbol stands for a point, the third and fifth for transformations, and the others are meaningless.

3. If P is the point (x, y), l the x-axis and m the y-axis, then $P\mathcal{L}\mathcal{M} = (-x, -y)$, $P\mathcal{L}\mathcal{M}\mathcal{L}\mathcal{M} = P$, $P\mathcal{L}\mathcal{M}\mathcal{L}\mathcal{M}\mathcal{L}\mathcal{M} = P\mathcal{L}\mathcal{M}$, and $P\mathcal{L}\mathcal{M}\mathcal{L}\mathcal{M}\mathcal{L}\mathcal{M}\mathcal{L}\mathcal{M} = P$.

5. If one chooses l to be the x-axis and m the line $x = y$ and $P = (x, -y)$, then $P\mathcal{L} = (x, y)$, $P\mathcal{L}\mathcal{M} = (y, x)$, $P\mathcal{L}\mathcal{M}\mathcal{L}\mathcal{M} = (-x, y)$, $P\mathcal{L}\mathcal{M}\mathcal{L}\mathcal{M}\mathcal{L}\mathcal{M} = (-y, -x)$, and $P\mathcal{L}\mathcal{M}\mathcal{L}\mathcal{M}\mathcal{L}\mathcal{M}\mathcal{L}\mathcal{M} = P$.

7. Here the images form a kind of spiral, as can be seen most easily by drawing a figure. Analytically, we can choose two sides to lie along the x and y axes and P the point $(1, 1)$. Then, reflecting in the four sides in clockwise order beginning with the x-axis, we have $P_1 = (1, -1)$, $P_2 = (-1, -1)$, $P_3 = (-1, 3)$, $P_4 = (3, 3) \ldots$, $P_{4r} = (1 + 2r, 1 + 2r)$, $P_{4r+1} = (1 + 2r, -1 - 2r)$, $P_{4r+2} = (-1 - 2r, -1 - 2r)$, and $P_{4r+3} = (-1 - 2r, 1 + 2(r + 1))$.

Section 4, p. 268

1.

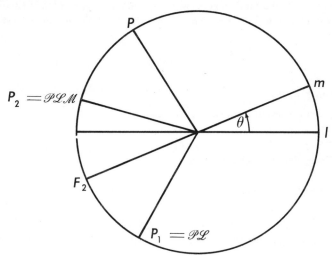

Figure 9:4:1

3. The theorem informs us that the measure of the angle from any point P to its image is 2θ and P_2 is the image of P, P_4 the image of $P_2, \ldots$.

5. If $P\mathcal{L}\mathcal{M}\mathcal{L} = P\mathcal{M}\mathcal{L}\mathcal{M}$, then $P\mathcal{L}\mathcal{M}\mathcal{L}\mathcal{M}\mathcal{L}\mathcal{M} = P\mathcal{M}\mathcal{L}\mathcal{M}\mathcal{M}\mathcal{L}\mathcal{M} = P$. Thus if 2θ is the angle of the rotation $\mathcal{L}\mathcal{M}$, we have $6\theta = 360°$ and $\theta = 60°$ is the angle from l to m. Thus $Q(\mathcal{L}\mathcal{M})^3 = Q$ for all points Q.

7. If $\mathcal{R}$ is a rotation through 2θ, then $(2\theta)x = 360y$ for some integers x and y. Thus $\theta = (y/x)180°$, that is, some rational multiple of $360°$.

9. Let R and S be two points, different from P, on L and R' and S' their images on L'. Then (see Exercise 11 of Section 2) $RS = R'S'$, $PR = PR'$, $PS = PS'$, and triangles PRS and $PR'S'$ are congruent. (We leave to the reader the case when P, R, S are collinear.) Choose l parallel to L and the line m so that the angle from l to m is θ. Let N be the foot of the perpendicular from R (or S) to l, T the intersection of $R'S'$ with l, and $\alpha = \sphericalangle NPR$. Then $\sphericalangle PRS = 180° - \alpha = \sphericalangle PR'S'$. Hence $\sphericalangle TR'P = \alpha$ and in triangle $TR'P$, $\sphericalangle PTR' + \alpha = \alpha + 2\theta$, the exterior angle at P. Hence $\sphericalangle PTR' = 2\theta$. (We also leave to the reader the case when line $R'S'$ is parallel to l.)

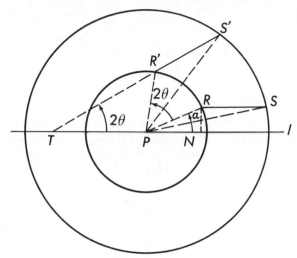

Figure 9:4:9

11. If $\theta = 90°$, the rotation angle is $180°$ and the image of any point $Q = (x, y)$ on L is $(-x, -y)$, which is on the line PQ. Thus the image of a point Q on line L is on L if and only if $PQ = L = L'$.

13. Let P denote the point of intersection of the four lines, θ the angle from l to m, and θ' the angle from l' to m'. Then $\mathcal{LML'M'}$ denotes a rotation about P first through an angle of 2θ and then through $2\theta'$, that is, $2\theta + 2\theta'$. Then $\mathcal{L'M'LM}$ will be the same rotation.

Section 5, p. 269

1. The rotations which take a point into itself, except for the point which is fixed, are integral multiples of $360°$. Hence for any rotation θ the number of repetitions of this rotation necessary to bring the point back to its original position is the least positive integer s such that $s\theta = n360°$, where n is also a positive integer. For three of the values given we have the following table:

θ	30°	72°	73°
n	1	1	73
x	12	5	73

3. If the angle of rotation is 2θ, it may be seen from the figure that the rotation takes P into P' and the reflection in l takes P' into P. Similarly, if the rotation is $2\theta b$, then the rotation and then the reflection will return P to the starting point if the angle from CP to l is θb.

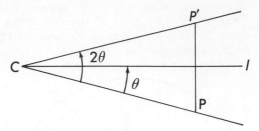

Figure 9:5:3

Section 6, p. 271

1. Let P' be the image of P in l and P'' the image of P' in m. Then S, the midpoint of the segment PP' is on l and R, the midpoint of the segment $P'P''$ is on m. Now, if P' is between m and l, $SP' + P'R = d$ and $PP' + P'P'' = 2(SP' + P'R) = 2d$. We leave to the reader the cases when P' is not between l and m.

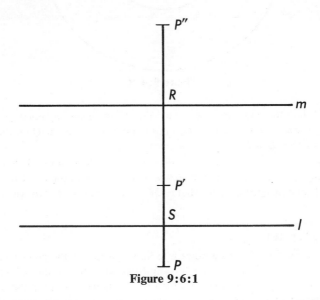

Figure 9:6:1

3. A translation is a transformation with the property that if P' and Q' are the images of any two points P and Q, then the segments PP' and QQ' are parallel and have the same length.

5. Using the definition in the answer to Exercise 3, we see that if P is left unchanged by the translation, the length of the segment PP' is 0. Then this must also be the length of the segment QQ' and $Q' = Q$.

7. Let $\mathcal{T}_1$ and $\mathcal{T}_2$ denote two translations and $P\mathcal{T}_1 = P'$ and $P'\mathcal{T}_2 = P''$, while $Q\mathcal{T}_1 = Q'$ and $Q'\mathcal{T}_2 = Q''$. Then if P, P' and P'' are not collinear, the triangles $PP'P''$ and $QQ'Q''$ are congruent (why?) and have parallel sides. Thus PP'' and QQ'' are parallel segments of the same length and $\mathcal{T}_1\mathcal{T}_2$ is a translation. The group is Abelian.

Section 7, p. 273

1. Notice that the points of a segment $\overline{AB}$ are those points P such that $AP + PB = AB$. If we let P' be the image of P, since the transformation is Euclidean, $AP = A'P'$, $PB = P'B'$, and $AB = A'B'$, which implies that $A'P' + P'B' = A'B'$.

3. Let C be the fixed point of the Euclidean transformation $\mathscr{E}$. Since a reflection has infinitely many fixed points and a translation none, $\mathscr{E}$ can be neither of these. Hence, except for a rotation, the only possibility is that $\mathscr{E}$ is equivalent to a succession of reflections in three lines l, m, and n. If all three are parallel we may, by Exercise 1 of Section 6, replace m and n by l and n', respectively, and see that $\mathscr{E}$ would be a reflection in n', which is impossible. Otherwise we may assume that l and m intersect in a point T. Call $\mathscr{R}$ the rotation $\mathscr{L}\mathscr{M}$ and $\mathscr{N}$ the reflection in n. Since $C\mathscr{R}\mathscr{N} = C$, we have $C\mathscr{R} = C\mathscr{N} = C'$, which implies $TC = TC'$; thus T is on n, which shows that the three lines are concurrent and thus, using Section 4, that as in the case when the three lines are parallel, $\mathscr{E}$ is a reflection, which is still impossible. Hence the only Euclidean transformation with a single fixed point is a rotation.

5. Suppose q is the fixed line of the Euclidean transformation, $\mathscr{E}$. The only rotation which has a fixed line is one through 180° for which every line through the center of rotation is fixed. A reflection has infinitely many fixed lines. Then using the argument of the answer to Exercise 3 with the notation, we have $q\mathscr{R} = q\mathscr{N} = q'$. This shows that the intersection of q and q' is on n. Let P be a point on q and P' its image on q' and continue the argument in answer to Exercise 3. Thus there is no Euclidean transformation with exactly one fixed line.

Section 8, p. 275

1. Using the theorem of Section 7 we see that if a Euclidean transformation preserves orientation it must be the identity or a succession of reflections in two lines l and m. If the lines are parallel, the transformation is a translation; if the lines intersect, it is a rotation. A rotation not the identity has exactly one fixed point and a translation none.

3. If two transformations separately preserve orientation, then it is preserved also when first one and then the other is applied. The identity preserves orientation and also the inverse of an orientation-preserving transformation. The associative property holds. Hence a group is formed.

5. Yes, a sequence of two rotations can be equivalent to a translation, for let l and n be parallel lines and m a line which intersects both. Then $\mathscr{L}\mathscr{M}$ and $\mathscr{M}\mathscr{N}$ are rotations and $(\mathscr{L}\mathscr{M})(\mathscr{M}\mathscr{N}) = \mathscr{L}\mathscr{N}$ is a translation. (Note that the angle from l to m is the supplement of that from m to n.)

7. A succession of three reflections changes orientation since each reflection does. By Exercise 2 of Section 7 this succession has at most one fixed point and, by Exercise 3, if it has exactly one fixed point it is a rotation. Hence if a Euclidean transformation is equivalent to a succession of three reflections but is not a reflection, it has no fixed points.

9. The set of translations form a group by Exercise 7 of Section 6. Translations and rotations are the orientation-preserving transformations—see Exercise 3 of this section.

Section 9, p. 279

1. Let $\mathscr{E}$ be expressed as a succession of e reflections and $\mathscr{E}'$ of e' reflections. Then $\mathscr{E}\mathscr{E}'$ can be expressed as a succession of $e + e'$ reflections. To complete the proof note that $e + e'$ is even if and only if e and e' are both odd or both even.

3.

Transformation	Fixed lines
Identity	All lines
Reflection in a line l	Line l (each point of which is fixed) and all lines perpendicular to l
Rotation	No fixed lines except for rotation through $180°$ in which case every line through the fixed point is fixed.
Translation	All lines in the direction of the translation
Rotation followed by reflection.	No fixed lines

All these results follow from previous theory. In particular, for the last case, see Exercise 5 of Section 7.

CHAPTER 10

Section 2, p. 284

(The graphs in Exercises 1 to 12 are left to the students.)

1. The images are $(a/3, 3)$ for $a = 1, 2, 3, 4, 5, 6$. All images are on the line $y' = 3$.

3. The points given are all on the line $y = \frac{1}{2}(x - 2)$ (see Section 8 of Chapter 7). Thus the images are all on the line $\frac{1}{2}y' = \frac{1}{2}(2x' - 2)$; that is, $y' = 2x' - 2$.

5. The images are all on the line $x' = 1$.

7. The given points all satisfy the equation $(x - 5)^2 + (y - 5)^2 = 25$. (Although the student might guess that the points lie on a circle, he is not likely to guess the equation.) The first four images are: $(\frac{5}{3}, 0)$, $(\frac{8}{3}, 3)$, $(\frac{10}{3}, 15)$, $(3, 24)$.

9. The given points satisfy $xy = 6$ and their images $x'y' = 6$.

11. The images of the first four points are $(7, 14)$, $(4, 16)$, $(3, 18)$, and $(\frac{5}{2}, 20)$. They are not collinear.

13. If $(x, y) = (2, 3)$, then $(x', y') = (2/k, 3k)$ for $k = 2, 3, 4, 5$.

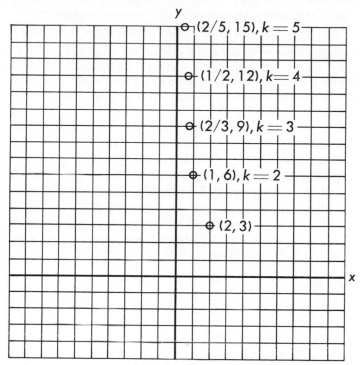

Figure 10:2:13

15. The equations $x = kx'$ and $y = y'/k$ imply that x is obtained from x' by multiplying by k and y is obtained from y' by dividing by k. That is, x' is obtained from x by dividing by k and y' from y by multiplying by k. Hence the inverse lorotation is $x = x'/k$ and $y = ky'$.

17. The equation of any line (see Chapter 7, Section 4) is $ax + by = c$. Using the transformation of Exercise 14, this becomes $akx' + (b/k)y' = c$, which is the equation of another line.

19. Suppose (x_1, y_1) and (x_2, y_2) are two points on the line $ax + by = c$. Then if $x_1' = x_1/k$ and $y_1' = ky_1$ and similarly for (x_2, y_2), we have $akx_1' + (b/k)y_1' = ax_1 + by_1 = c$ and similarly for (x_2, y_2).

Section 5, p. 290

1. Let $x' = kx$ and $y' = y/k$ be the lorotation and see that the image of $(4, 5)$ is $(4k, 5/k)$. The area of the rectangle determined as indicated by P is $4 \cdot 5$ and that for P' is $(4k)(5/k)$, both of which are equal to 20. The same result would hold for any point P.

3. For the lorotation of Exercise 1, the image of (a, b) is $(ka, b/k)$. Then a and ka are the x coordinates of the feet of the perpendiculars from the two points to the x-axis. As in Exercise 1, the areas are the same.

5. Replacing x and y by a and b, respectively, in the given equation we have $b =$

$ma - am + b$, which is true. Hence (a, b) lies on the line. To find the equation of the tangent line, note that the line $y = mx - am + b$ cuts the curve $xy = ab$ in the points whose x coordinates satisfy the equation $x(mx - am + b) = ab$; that is,

$$mx^2 - (am - b)x - ab = 0.$$

Since we know that one solution is $x = a$ [the point (a, b) lies on the curve], we see that the left side of the equation is equal to $(x - a)(mx + b)$. Hence the line will cut the curve in only one point if and only if the two factors are the same except for a numerical factor, $a = -b/m$; hence $m = -b/a$. Thus the equation of the tangent line is $y = (-b/a)(x - a) + b$.

7. The "perpendicular" in our new sense to the tangent line at (a, b) is the line through the origin and this point. Its equation is $ay = bx$.

9. If we replace x and y in the given equation by kx' and y'/k, respectively, we get $y'/k = -2kx'/3 + 4$, that is, $y' = (-2k^2/3)x' + 4k$. To find the x coordinates of the intersection of this line with $x'y' = 6$ we need to solve $x'(-2k^2x'/3 + 4k) = 6$. This reduces to

$$-(kx' - 3)^2 = 0,$$

which has just one solution. Hence the image line is tangent. Thus the image of the tangent line to $xy = 6$ is tangent to $x'y' = 6$.

Section 6, p. 293

1. $x'y' = (x/\sqrt{2} - y/\sqrt{2})(x/\sqrt{2} + y/\sqrt{2}) = \frac{1}{2}(x^2 - y^2)$. Thus if $x^2 - y^2 = 2a$, then $x'y' = a$.

3. We know that for every point (x', y') on the lorcle $x'y' = 6$, 6 is the area of the rectangle two of whose sides lie on the lines $x' = 0$ and $y' = 0$, and whose fourth vertex is (x', y'). For the transformation, $x' = 0$ corresponds to $x - y = 0$; $y' = 0$ corresponds to $x + y = 0$. Hence the area, unchanged for all points of the lorcle $x^2 - y^2 = 12$, is the area of the rectangle two of whose sides lie on the lines $x = y$ and $x = -y$ and whose fourth vertex is (x, y). In fact, just as $|y'|$ is the distance from $x' = 0$ to the point (x', y'), so $|x - y|/\sqrt{2}$ is the distance from the line $x - y = 0$ to the point (x, y); similarly, for the other perpendicular distance.

CHAPTER 11

Section 1, p. 298

1. In the first network of Fig. 11:3 all vertices are odd. For Fig. 11:4, network (a) has four odd vertices and one even; network (c) has no odd vertices; network (f) has exactly two odd vertices.

Section 2, p. 301

1. Conclusion (2) applies to parts (c) and (d) of Fig. 11:4; conclusion (3) to parts (b) and (f) of Fig. 11:4 as well as the second part of Fig. 11:3; conclusion (4) to parts (a) and (e) of Fig. 11:4 as well as the first part of Fig. 11:3.

3. To prove statement (5), first connect all but one pair of odd vertices by an extra arc. Then, by conclusion (3), we can start at one of the remaining odd vertices, A, and traverse the entire network including the extra arcs, in one path ending at the other odd vertex *B*. Now, if, instead of doing this, whenever we come to one of the extra arcs we "jump" to the other end, we will have $n - 1$ jumps and so will have traversed the given network, removing the pencil from paper $n - 1$ times. The number of paths necessary to traverse the network cannot be less than n because each path has exactly two odd vertices and there are $2n$ odd vertices to be accounted for.

5. Since each arc has exactly two ends whether or not a network is connected, conclusion (1) holds without restriction. Also (4) holds for disconnected networks. However (2), (3), and (5) are false. One may see this, for instance, for (2) by considering a square and a triangle without common points. All vertices are even, but one cannot trace the complete network without removing pencil from paper.

7. Consider the network consisting of a square and a triangle with no points in common. There will then be three regions: *A*, the interior of the square; *B*, the interior of the triangle; and *C*, the rest of the plane. The dual will be connected since vertex *C* will be connected with *A* by four arcs and with *B* by three.

Section 4, p. 310

1. **a.** Twelve states, including Maine and California.

    b. Nine states, including Florida and Washington.

    c. For instance, Kentucky is surrounded by seven states and hence four colors are needed for this part.

    d. New Hampshire is one state which, when deleted, "separates" the continental United States. There is one more which has this property.

    e. Missouri has common boundaries with eight states. No other state has more.

3.

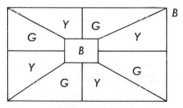

Figure 11:4:3

5. For the map $F = 7$, $V = 10$, $E = 15$, and for the dual, $F = 10$, $V = 7$, $E = 15$.

7. Denote the regions of the dual by lower case letters and have the three networks shown. Notice that the dual of the dual is the original network altered to make it connected by giving the triangles a common vertex. Intuitively there is a kind of "stability" about a connected network. The interested reader might like to explore this for a network disconnected into three pieces.

Given network   Dual network   Dual of the dual

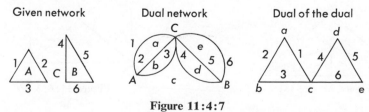

Figure 11:4:7

Section 5, p. 313

1. Five of the pairings are as follows: *a* and *r*, *c* and *f*, *e* and *g*, *h* and *m*, and *o* and *q*.

Section 6, p. 316

1. The network for part a is as shown.

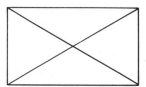

Figure 11:6:1

3. **a.** The network is planar, as is shown.

    b. Let *A*, *B*, *C*, and *D* be the vertices of the base of the cube and *I* the midpoint of its top face as in Fig. 11:23. Then part of the network will be as in the accompanying figure, where the curved lines correspond to the diagonals of the base of the cube. Here it is seen that each of the *A*, *B*, *C*, *D*, and *I* is joined to each of the others by lines which do not cross. By the statement in Section 6, the network is not planar.

    c. The network is planar; it is as shown.

    d. If the vertices of the tetrahedron are denoted by *A*, *B*, *C*, and *D* and *P* is the point within, each of the five will be connected by nonintersecting arcs with each of the other four, which is nonplanar.

    e. Let the vertices of the triangle be *A*, *B*, and *C* and the interior vertex *D*. In this case the line *DE* can, intuitively, be "shrunk" and "lifted over" one of the sides, without cutting or detaching, to put *E* inside the triangle *ABC*. Hence the given network is topologically equivalent to a planar network. Here one really needs a more precise definition of topological equivalence. On the way to more precision would be: Two figures are said to be topologically equivalent if every two points "close to" each other in one figure have corresponding points in the other figure which are also close to each other. Here the lack of precision is in the term "close to." But it is best for us to leave it at that.

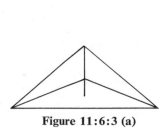

Figure 11:6:3 (a)

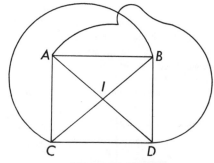

Figure 11:6:3 (b)

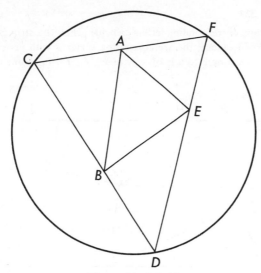

Figure 11:6:3 (c)

Section 7, p. 321

1. For the case $r = 3$, $s = 5$, we have $E(\frac{2}{3} + \frac{2}{5} - 1) = 2$, which implies $E = 30$. Then $V = 2E/r = 20$ and $F = 12$.

3. The planar networks are duals. In fact, we may say that the polyhedra are themselves duals.

5. The resulting polyhedron will have six square faces and eight triangular ones (one at each corner of the cube). It does not satisfy property (1) of the theorem of this section.

7. By the conditions of the exercise, properties (1) and (2) of the theorem are satisfied. Property (3) holds since the sides of each face of the polyhedron are straight lines, which implies that each face must have at least three sides. Hence formula (C) holds with $s = r$. Then $E(2/r - \frac{1}{2}) = 1$, or $E(4 - r)/2r = 1$. Since E is positive, $r \le 3$. Thus $r = 3$ and the polyhedron is a tetrahedron.

Section 8, p. 325

1. Suppose we consider one side of a strip of paper to be colored black and the other white. If it is twisted an odd number of times (that is, n is odd) the black side will become attached to the white side and the surface will be one-sided. If it is twisted an even number of times, the surface will be two-sided.

3. If a 1-strip is cut down the middle, we have a single piece which is a 4-strip and therefore two-sided. If we cut it again, we have two interlocking 4-strips. The easiest way to see what happens is to experiment.

5.

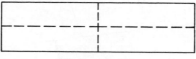

Figure 11:8:5

Section 9, p. 327

1. In the diagram, corresponding letters coincide when the strip is joined together to make a torus. For instance, the arc connecting vertex 3 with vertex 5 is by way of point E. The arc connecting 3 to 6 is by way of F_1, F_2, F_3, F_4. None of the arcs cross, even on an edge.

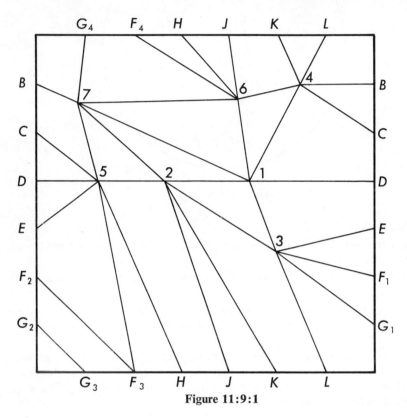

Figure 11:9:1

3. In the diagram, corresponding points (designated by letters) coincide when the strip is joined to make a Möbius strip. For instance, CB is the arc common to regions 3 and 5.

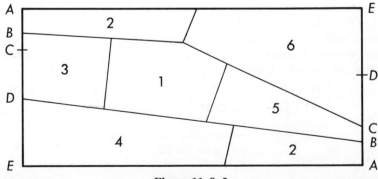

Figure 11:9:3

Index